Pests and Parasites as Migrants

Pests and Parasites as Migrants

Edited by
A.J. Gibbs and
H.R.C. Meischke

CAMBRIDGE UNIVERSITY PRESS

Cambridge
London New York New Rochelle
Melbourne Sydney

Published by the Press Syndicate of the University of Cambridge.
The Pitt Building, Trumpington Street, Cambridge CB2 1RP
32 East 57th Street, New York, NY 10022, USA

© Australian Academy of Science 1985

First published 1985

Cover design: Ross Buxton. Image supplied by the
　　　　　　　　Australian Landsat Station,
　　　　　　　　Department of Resources & Energy,
　　　　　　　　Division of National Mapping.
Typeset by The Typesetting Centre, Canberra, ACT.
Printed in Australia by Ruskin Press, North Melbourne, Victoria.

Pests and parasites as migrants.
　1. Micro-organisms, Pathogenic
　I. Gibbs, Adrian　II. Meischke, Roger
　576'.165　QR175

ISBN 0-521-30163-7

QR
175
.P47
1986

PREFACE

Australia was colonized 200 years ago by exotic human beings from Britain. These were British free-men and British convicts and with them came the diseases of freemen and convicts. But it was not just exotic human diseases that came to Australia with the first fleet; stowed away on the ships were organisms that were capable of causing diseases in plants and animals that had never occurred in the evolutionary history of the new colony.

For a variety of reasons the worst of the old world animal plagues and plant diseases did not become established in the new colony. It was difficult enough for human beings to survive the gut-rending six-month voyage around the Cape and through the roaring forties; any animals or plants suffering from a severe infectious malady inevitably found the voyage too difficult and were committed to the quarantine of the deep before a landfall was reached in the colony. This propitious circumstance did not occur in every case and over the years a variety of diseases, once exotic to the country, became endemic. But Australia still remains one of the most disease-free countries in the world and it is this particular situation that has given our agricultural and livestock industries the special advantages of lower production cost and preferred access in international trade.

The idea of having a symposium on exotic diseases of human beings, animals and plants arose out of dissatisfactions experienced by some people who took part in a meeting sponsored by the National Farmers Federation and run by CSIRO at Geelong in November 1982. The symposium was held so that primary producers might have the opportunity to hear arguments for and against the proposal by CSIRO to import exotic viruses into Australia, in particular foot-and-mouth disease virus (FMDV), in order to set up a diagnostic, training, research and vaccine producing facility in the Australian National Animal Health Laboratory. In the event the symposium turned out to be an occasion at which the case for importing exotic viruses was put on CSIRO's behalf by a group of exotic scientists imported to Australia for the occasion without any real discussion of the implications of exotic diseases for the Australian community.

The scientific bases of the arguments that live exotic viruses were essential for diagnosis and that FMD vaccines should be produced in Australia were shown at the Geelong meeting to be without merit and these proposals were rejected by Industry and by Government. Apart from this however the deficiencies of the Geelong meeting pointed out the need to have a properly organised scientific symposium on the general topic of exotic diseases, not just in regard to the economically important livestock diseases such as FMD, but also in regard to human and plant diseases from both an historical and contemporary viewpoint. From this idea, a proposal for a symposium on exotic diseases was developed in which the topic would be presented and discussed in a scientific, interesting and understandable way. The Academy of Science joined with the Australian and New Zealand Association for the Advancement of Science to organise a major symposium on exotic diseases at the ANZAAS meeting held in Canberra in May 1984. The organisation of the meeting was not without its difficulties as the sensitivities of the CSIRO had been disturbed by the outcome of the Geelong meeting and it required the good offices of the Chairman of the organising committee to bring together a spectrum of people, including some members of CSIRO, to plan the symposium. Its successful outcome was due in large measure to this Committee. The cost of bringing overseas and local participants to Canberra was met by a variety of fiscal strategies. It says something for the organising committee that, 8 overseas speakers were brought from the U.K., Europe and the U.S.A. and 20 speakers from around Australia, and the papers presented in the symposium were published as this book within a budget of $18,000.

The organising committee wishes to thank the following donors whose financial support made the symposium and the publication of this book possible.

ANZAAS
Australian Bureau of Animal Health
The Commonwealth Department of Primary Industry
Rural Credit Development Fund of the Reserve Bank
Australian Wool Industry
Biotechnology Australia Pty Ltd
Monoclonal Antibodies Pty Ltd
Australian Pig Industry Research Committee
Poultry Research Advisory Council
Australian Chicken Meat Research Committee
Australian National University
Walter and Eliza Hall Institute
Commonwealth Serum Laboratories

Bede Morris
Canberra, May 1985

THE ORGANISING COMMITTEE

Dr Max Day A.O., B.Sc., Ph.D., F.A.A.
formerly Chief, CSIRO Division of Forest Research
and member CSIRO Executive (Chairman)

Professor Bede Morris, B.V.Sc., D.Phil., F.A.A.
Head, Department of Immunology,
John Curtin School of Medical Research,
Australian National University (Treasurer)

Mr Michael Tracey, A.O., M.A., F.T.S.
Director, CSIRO Institute of Biological Resources,
(representing ANZAAS organising committee)

Dr Graeme Laver, M.Sc., Ph.D.
Senior Fellow, Department of Microbiology,
John Curtin School of Medical Research,
Australian National University (Program convenor)

Dr Bob Dun B.V.Sc., Ph.D.
Director,
Australian Development Assistance Bureau,
Department of Foreign Affairs

Dr Roger Meischke M.V.Sc., Ph.D., MASM, MRCVS.
Principal Veterinary Officer,
Australian Agricultural Health and Quarantine Service
Department of Primary Industry (Secretary)

Professor Douglas J. Whalan, LL.B., LL.M., Ph.D., M.I.Env.Sci., F.A.I.M.
Faculty of Law and Chairman, Board of the Faculties,
Australian National University

Dr Adrian Gibbs, B.Sc., Ph.D., A.R.C.S.
Head, Virus Ecology Research Group,
Research School of Biological Sciences,
Australian National University (Publication Convenor)

Dr Tony Robinson, B.V.Sc., Ph.D.
Director, The Medical Research Council of New Zealand,
Virus Research Unit,
University of Otago (New Zealand liaison)

Dr Ian Marshall, B.Ag.Sc., Ph.D.
Senior Fellow, Department of Microbiology,
John Curtin School of Medical Research,
Australian National University
(Liaison with Geelong Symposium Committee)

Dr Peter Bridgewater, B.Sc., Ph.D.
Director, Bureau of Flora and Fauna,
Department of Home Affairs and Environment

Professor Frank Fenner, C.M.G., M.B.E., M.D. (Adel.), D.T.M. (Syd.),
Hon.M.D. (Monash), F.R.A.C.P., F.R.C.P., F.A.A., F.R.S.,
formerly John Curtin School of Medical Research,
Australian National University (Academy nominee)

Dr Frank E. Peters, B.Sc., M.Sc., Ph.D., F.R.A.C.T., F.A.I.F.S.T.
formerly Australian Government Analyst,
Australian Government Analytical Laboratories

Mr Laurie Erwin, B.Ag.Sc., Dip.Ed., F.A.I.F.S.T.
Codex Contact Point for Australia,
Department of Primary Industry

Dr Jim Shelton, B.V.Sc., Ph.D.
Senior Fellow, Department of Immunology,
John Curtin School of Medical Research,
Australian National University
(Australian Veterinary Association Nominee)

Dr Ken Old, B.Sc., Ph.D.
Principal Research Scientist,
CSIRO Division of Forest Research

Dr A.J. Della-Porta, B.Sc. (Hons.), Ph.D.
Senior Research Scientist,
CSIRO Australian National Animal Health Laboratory,
(Representing CSIRO)

Dr P.J. Price, B.Sc., Ph.D.
Principal Project Officer,
Australian Science and Technology Council
Department of Prime Minister and Cabinet

CONTENTS

Preface   v

The Organising Committee   vi

Opening address   xiii
John Kerin, Minister for Primary Industry

SECTION A   The cast

Chapter 1
Australia — the isolated continent   3
Nigel Wace

Chapter 2
Exotic human diseases   23
H. Newton-John

Chapter 3
Exotic animal diseases   28
H.R.C. Meischke and W.A. Geering

Chapter 4
Exotic plant pathogens   40
J.W. Randles

SECTION B   Planning for an exotic disease outbreak

Chapter 5
Quarantine — a primary defence   45
K.A. Doyle

Chapter 6
Exotic disease emergency plans   50
R.W. Campbell

Chapter 7
Exotic diseases and the law   53
Douglas J. Whalan

Chapter 8
Economic effects of an exotic disease outbreak in Australia   58
Joe Johnston

Chapter 9
The Australian response to the threat of an exotic disease incursion: the establishment of ANAIIL   63
Ron Johnston and Pam Scott

Chapter 10
Viral pathogen importation — the risks and benefits   75
A.J. Della-Porta and R.I.B. Francki

SECTION C   The study of pests, parasites and their hosts

Chapter 11
Agent detection and identification: new and traditional techniques   83
G.W. Burgess

Chapter 12
New developments in the use of DNA probes for the rapid detection of viral pathogens   85
R.H. Symons

Chapter 13
Agent detection and identification — new immunological techniques   91
M.R. Brandon and H. Bults

Chapter 14
The study of genetic variation: a basic tool for pest and parasite studies   96
Adrian Gibbs and Rodney Mahon

Chapter 15
Host-pathogen relationships: struggle of the genes   104
J.J. Burdon and D.R. Marshall

Chapter 16
The bovine major histocompatibility system and disease   108
M.J. Stear, S. Bath, J. Mackie, C. Dimmock, S.C. Brown, F.W. Nicholas, and B. Morris

Chapter 17
Novel vaccines   112
F. Brown

Chapter 18
Pathogen importation for biological control — risks and benefits   115
R.J. Milner

SECTION D   Cautionary tales

Chapter 19
Acquired Immunodeficiency Syndrome (AIDS)   125
Walter R. Dowdle

Chapter 20
Bluetongue   128
B.M. Gorman

Chapter 21
Foot and Mouth Disease   135
Ulrich Kihm

Chapter 22
Lethal avian influenza (H5N2) in the USA:
is there potential for a similar outbreak in Australia or New Zealand?   140
*Robert G. Webster, Yoshihiro Kawaokal, William J. Beane,
Clayton W. Naeve, John M. Wood and William G. Laver*

Chapter 23
Myxomatosis   147
Frank Fenner

Chapter 24
Echium; Curse or Salvation?   152
Linton Briggs

Chapter 25
Western Gall Rust of Pines   160
K.M. Old

Chapter 26
The persistence of infectious diseases within wildlife populations: rabies and bovine tuberculosis   164
Roy M. Anderson

Chapter 27
From exotic disease to occupational hazard: the recombinant DNA safety debate   168
Ditta Bartels

Chapter 28
Beneficial use of an exotic phytopathogen, *Puccinia chondrillina*, as a biological control agent   171
for skeleton weed, *Chondrilla juncea*, in Australia
E.S. Delfosse, S. Hasan, J.M. Cullen, and A.J. Wapshere

Chapter 29
Potato Spindle Tuber Viroid and Avocado Sunblotch Viroid   178
J.L. Dale

SUMMARY

Reflections on the symposium; an agricultural producer's view 184
R.G. Warren

List of Contributors and Addresses 186

Index 188

OPENING ADDRESS

15th May 1984, Melville Hall
Australian National University

John Kerin, *Minister for Primary Industry*

Thank you for the opportunity to speak at the opening of this Symposium on exotic diseases of plants, animals and man. I notice that you have arranged an impressive interdisciplinary program with some forty speakers, many of whom have come from overseas — North America, Europe and the United Kingdom.

The possible introduction of exotic diseases has always been regarded with fear in Australia. Let me assure you that the Australian Government is giving high priority to the control and exclusion of foreign diseases requiring dedicated and sometimes laborious work by trained scientists. They are, to borrow a phrase, the true "quiet achievers" whose contributions are generally not appreciated by the community except when the introduction of a disease causes hardship, economic loss, stern Government controls and, in certain cases, social disruption.

It is difficult to overstate the potential damage which exotic diseases can cause. In some cases, complete collapse of the domestic market could follow the closure of our major export markets for livestock and their products. In other cases, wildlife populations could be decimated due to the outbreaks of new diseases.

One of the difficulties associated with the work of specialists is that they frequently become isolated from others with similar responsiblities but different field of action. To the best of my knowledge, this meeting is the first where scientists with an interest in the control of exotic diseases of man, plants and animals have joined together to share ideas and experiences. I congratulate the organisers in enabling this useful concept to develop.

In declaring this meeting open, I hope that you will all benefit from constructive discussion of an interdisciplinary nature — there can be no doubt that Australia's national interest will be served best through such collaboration.

I have much pleasure in declaring the meeting open.

Section A

The Cast

This book has a cast of millions — the inhabitants of Australia, past, present and future. The first chapter outlines what we know of the succession of inhabitants of the past. It is followed by three chapters describing the most worrying and unwanted potential inhabitants of the future — the known exotic diseases of plants, man and other animals.

The adjective *exotic* is used here in the formal sense defined in *The Shorter Oxford English Dictionary* as "Alien; introduced from abroad, not indigenous", rather than the colloquial modern sense of "outlandish and therefore somewhat special, even desirable". Thus exotic diseases, to an Australian, are those diseases not yet found in Australia and which, in most instances, are actively discouraged from entering the continent.

Chapter One

AUSTRALIA — THE ISOLATED CONTINENT

Nigel Wace

"What an occasion of felicity on the part of the inhabitants of New South Wales is the introduction of the pumpkin! — yet I could not tolerate those blue-bottle flies which blow the meat even while it is trundling on the spit. But New South Wales is the place to go to and live at for ever, without disease."

Jeremy Bentham, 1827 (Bowring 1843)

Except for Antarctica, Australia is the most isolated of all the continents. It was also the last inhabited continent to be discovered and settled by Europeans, and it was the only continent in which the indigenous people did not practise agriculture. But Australian agriculture is now based upon the use of plant and animal species which have been introduced from other parts of the world, only during the last two hundred years. This combination of natural circumstances and human history has lent a peculiar interest to the arrival and establishment of new species in Australia: whether these species are plants, animals or pathogens; whether they have arrived naturally, or were imported intentionally or by accident; and whether they were initially regarded as useful, harmless or noxious. The geological history of Australia has determined the origins and evolution of its native biota. The course of its evolution has in turn been greatly influenced by human activities. Following the separation of Australia from Antarctica in the final breakup of Gondwanaland, the southern 'super-continent', four periods are characterized by increasing interchange between Australian biota and that in other parts of the world.

THE FORMATIVE PERIOD: PALEOCENE TO LATE PLEISTOCENE
ABOUT 50 MILLION YEARS TO 50,000 YEARS AGO

Since Australia separated from Antarctica in Paleocene times (50–60 million years ago), it has been isolated from other continental lands to the west, east and south by stretches of ocean (Crook, 1981). Northward movement of the Australian plate (i.e. Australia and its continental shelf) fused it with Papua New Guinea and with arcs of islands at the edge of the Melanesian sea in late Miocene times, about 10 million years ago (Powell *et al*, 1981). Through all this span of 40–50 million years during which the flowering plants and the mammals were diversifying, and asserting their dominance of the land world wide, Australia was isolated and biogeographically on its own (Figure 1). During this period, only the vagaries of long-distance dispersal across the oceans can be invoked to explain the arrival of terrestrial plants and animals in Australia from other parts of the world. This long period of isolation has left its stamp upon the fundamental characteristics of the Australian native biota, and therefore of the biotic environment in which any newly arrived species had to compete to establish themselves.

Our existing native plants still surviving today from this long formative period came from four sources:

a. *Gondwanic plants*
   These came from the mesic forests that covered much of Australia and Antarctica when they separated. These forests probably persisted at least until 25 to 30 million years ago, until the Antarctic coasts became cold and the central areas of Australia became arid (Kemp, 1981; Lange, 1982).

b. *Autochthonous or Endemic plants*
   Species which evolved from the Gondwanic plants and adapted to the more arid and probably more seasonal climates in Australia that developed over the last 25 to 30 million years (Galloway and Kemp, 1981).

c. *Indo–Malayan plants*

Species that invaded the north of Australia from the tropics, since Miocene times when Australia approached Sundaland. Despite the narrowing sea gaps between the Australian land mass and the Indo-Malayan lands, the exchange of species between Australia–New Guinea and lands to their north has always been across sea gaps.

d. *Cosmopolitan plants*

Species that have entered Australia by long distance dispersal from other parts of the world. Many of these plants inhabit environments easily accessible to the vectors of long range dispersal (e.g. wide-ranging birds or the sea), such as freshwater habitats and coasts.

The native flora can therefore be considered as an ancient and land-derived "Gondwanic" group (a + b) and a more recent "Intrusive" group (c + d) derived from overseas (Nelson, 1981). Comparable groupings are recognized amongst the native fauna (Keast, 1981), and, in addition, there may be some animals that are pre-Gondwanic and come from the ancient world continent Pangaea.

Biogeography and evolution

It is important to realize that throughout this long formative period of isolation, Australia was separated from other regions with thermally seasonal climates, in the northern hemisphere, by the tropics, which have lesser season differences in temperature. Animals and plants preadapted to Australian climates had to traverse great distances of inhospitable land and ocean, well beyond the range of most groups of land animals. None but the most vagile of the placental mammals, such as bats, rats and seals, entered Australia. There were probably no opportunities for easy exchanges of land plants and animals between Australia and other lands in temperate latitudes during this formative period. Thus, Australia has several native plant genera which are also found naturally in other southern continents (Wace, 1965; Specht, 1981a), but fewer so-called "bipolar" genera that are shared with northern temperate lands. If we compare only native vascular plant genera, we find that those Australia shares with either northern or southern temperate lands, mostly inhabit coastal, freshwater or alpine environments. By contrast, all the temperate and cold areas of the northern hemisphere share many genera of vascular plants (including dominant forest trees), terrestrial mammals, birds, reptiles and freshwater fish (Darlington, 1957). Some widespread and vicarious species are found throughout the latitudinal zones of the northern hemisphere, whereas similarly widespread species are almost unknown in the souther continents, where the dominant terrestrial species are mostly endemic.

The origins and genetic relationships of the native biota of Australia are important in any study of introduced species, (including pests and parasites) because they set the terms of competition and the opportunities of infection for newly arriving species including those species later introduced by man. Australian ecosystems contain different proportions of species of the four elements outlines above, and so show differing degrees of ecological integration, and hence resistance to infiltration by exotic organisms brought in by man. The extent to which native vegetation is invaded by exotic plants is generally interpreted in terms of the degree to which it has

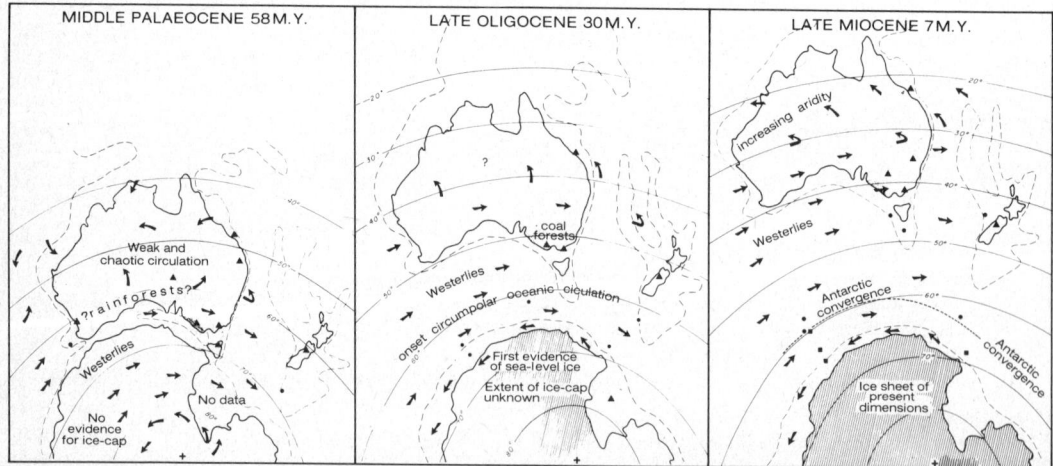

Fig. 1. The formative period in the evolution of Australia's biota. Paleogeographic reconstructions of the southeast Indian Ocean during the Cainozoic Period, showing the northward movement of the Australian continental plate, tentative atmospheric circulations and environmental trends, with ice formation in Antarctica by Late Oligocene and cooling and drying climates in Australia from Late Miocene. Symbols show sites for which palynological or sedimentological data available. Simplified from Kemp, 1978.

been disturbed by human activities. It is striking that even when disturbed by man, the 'Gondwanic' temperate rain forests of Tasmania and the 'autochthonous' kwonkan heaths of southwest Australia have few exotic invading plants, compared to the open eucalypt woodlands and humid grasslands of the central and eastern Australia which contain a larger proportions of cosmopolitan plant genera. The 'Gondwanic' and 'autochthonous' elements may have a higher proportion of k-selected species (i.e. ones that maintain large stable populations) than the others, and communities dominated by them may thus be able to exclude the mainly r-selected exotic invading species (i.e. ones that reproduce quickly when space permits) introduced by man.

When two regions, which have closely related animals and plants, have been only partially or recently isolated from one another, the native species of one region may be extremely vulnerable to the pathogens which infect their relatives in the other region. This appears to be true for the native biotas of the northern temperate regions, and those of temperate North and South America, which were linked by a land bridge in Pliocene times, some 10 million years ago. Example of plant and animal pathogens exchanged recently between the native temperate biotas of different regions which have had spectacularly destructive effects on vicarious hosts are:

The Asian chestnut blight fungus (*Endothia parasitica*) which has little effect on its longterm host the Chinese chesnut (*Castanea mollissima*), but severely affects the American chestnut (*Castanea dentata*).

The Dutch elm disease fungus (*Ceratocystis ulmi*) probably from East Asia, which destroys American elm (*Ulmus americana*) and the English elm (*U. campestris*), spread by barkbeetles (*Scolytus* spp.).

The myxoma poxvirus of South American leporids (*Sylvilagus* spp.) used to control European rabbits (*Oryctolagus* spp.).

The closeness of the relationship of the biotas of any two regions may therefore have a profound influence on the susceptibility of the native biota of one region to the pathogens of another, if they should gain entry. If the native biotas of two regions, which have been isolated from one another for a long time, are genetically distinct, most of their native species will lie outside the potential host range of pathogens that may be moved by man from one region to the other. This may be the reason why many of the species of temperate Australia are immune to pathogens introduced from other temperate biotas. This constraint on infection applies only to *native* Australian plants and animals: naturalized or feral species may prove very susceptible to pathogens introduced from regions where they are native (e.g. poplar rusts, myxoma poxvirus, blackberry rust).

Vavilov (1951) recognized the correlations between host/pathogen interactions, biogeographic origins, genetic relatedness and selection, in formulating his famous rule (see White, 1981, p. 91):

"Where a pathogen is indigenous in a geographic region, varieties of the host plant and strains of the pathogen have reached a state of equilibrium through the gradual process of natural selection, during which the more susceptible hosts are eliminated, leaving mostly tolerant and resistant hosts that have 'learnt to live' with the pathogen over millenia".

In Australia we know little about the exotic diseases affecting our native fauna or flora. Fenner (1979) put the situation succinctly:

"There is virtually no information available about free-living and saprophytic micro-organisms that are native to Australia, against which to determine the fate of introduced species. In studying exotic micro-organisms, attention has to be focussed on pathogens of animals and plants. Within this restricted field there is a further limitation: the only exotic pathogens that have been recognized are species that cause disease in exotic animals and plants".

Despite some advances since Fenner made this observation about our ignorance of *native* pathogens, there is the important question of how to recognize the source of any pathogen of the *native* biota which we had not brought in to Australia intentionally. Given the spectacular success of some exotic diseases, pests and weeds in Australia — and our resulting xenophobia concerning them — we are probably inclined to regard any aggressive pathogen as *ipso facto* exotic. The debate of the origins of the widespread root-rot fungus *Phytophthora cinnamomi* is a case in point (Shepherd, 1975).

Some residual effects of the Cainozoic period on the opportunities for exotic species to establish

Another important feature of Cainozoic Australia results from the climates it experienced when isolated from other lands, and rafting slowly northwards while its plants and animals diversified. There is general agreement that for a large part of this time, soils became deeply leached over most of the continent (Specht, 1981b; Nix, 1981), and that the broken up, stripped, redistributed and re-weathered remains of those deep-weathered soils formed the parent material for many of todays soils (Figure 2). Having thus undergone several cycles of leaching and erosion, most Australian soils are very deficient in plant mineral nutrients — especially in phosphates and nitrates, and often in trace elements. These soil deficiencies are now a major impediment to the establishment of exotic plants, not only weeds or harmless volunteers, but also useful pasture and crop plants. Few exotic plant species invade unfertilized native plant communities where soil nutrient levels are low (Amor and Piggin, 1979). When nutrients are added to soils in supporting native vegetation in Australia, (whether by fertiliser application, or as pollution), the growth of exotic plants is favoured. This widespread constraining effect of impoverished soils also affects organisms that depend on plants, notably the pathogens that infect exotic plants, and the animals which depend on them for food.

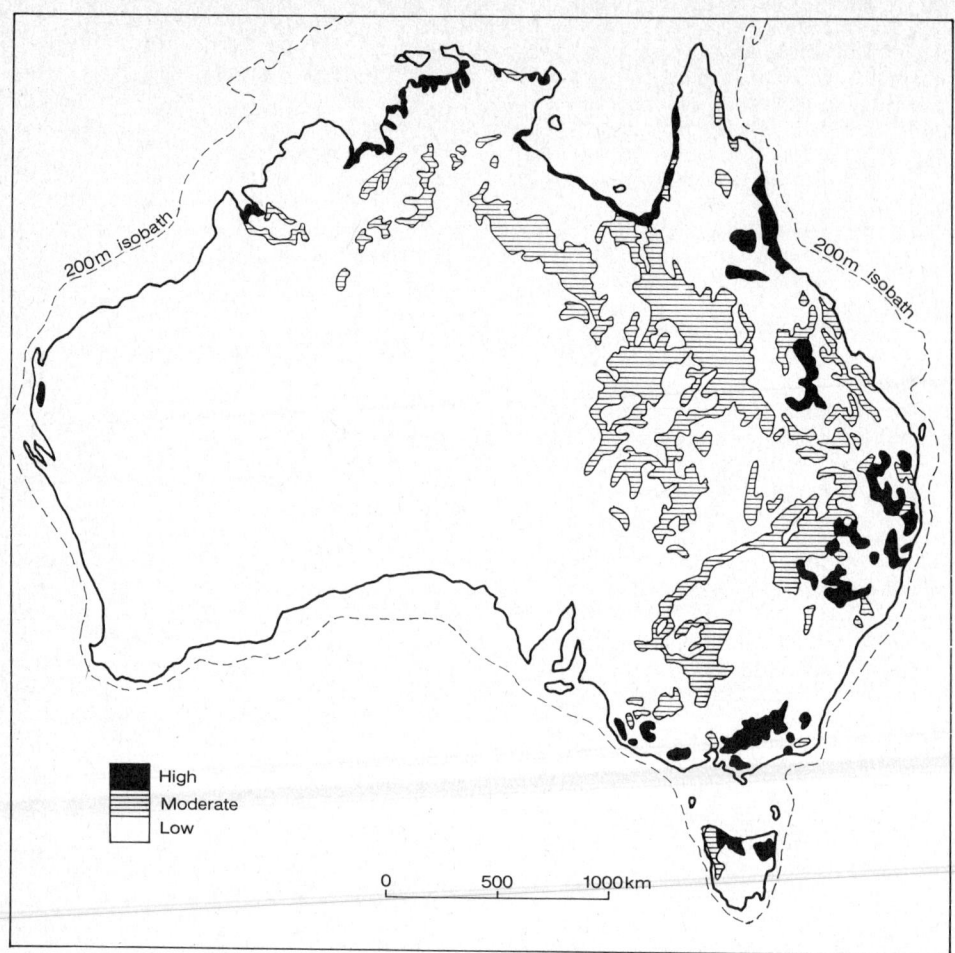

Fig. 2. Australia: soil nutrient status. Most of Australia has soils with low or very low nutrient status. Melbourne, Brisbane, Townsville and Launceston are the only large settlements near extensive areas with naturally fertile soils. The edge of the continental shelf at 200m below present sea level also indicates the extent of land for most of the Late Tertiary and Pleistocene, when New Guinea and Tasmania were joined to Australia. Redrawn from Nix, 1981, p. 114.

Another important feature of the Cainozoic environments of Australia was increasing aridity. Galloway and Kemp (1981) concluded that the transition from wet to locally drier climates took place in the Pliocene, when the continent was most isolated, and therefore least able to acquire from elsewhere biota adapted to desert conditions. As aridity increased so too did the incidence of fire. The influence of fire and of 'firestick farming' (Jones, 1969) employed by the Aboriginal people is discussed in the next section.

It was previously assumed that the Australian land mass traversed latitudinal climatic belts similar to those present today, as it drifted north through Cainozoic times, and thus each belt experienced winter rain/erratic rain/summer rain in sequence as the land moved north (Axelrod, 1976; Beard, 1977). Bowler (1982) argued that there is no evidence of past aridity in those parts of northern Australia now subject to summer rainfall, and that the winter rainfall ("Mediterranean") climates may be of rather recent occurrence anywhere in Australia, the ecological success of numerous species of exotic plants introduced during the last 500 years from winter rainfall regions in California, the Mediterranean, South Africa and Chile is more easily explained. Native grasses and sclerophyll shrubs in the winter rainfall zones of Australia seem less well-attuned to rapid early seasonal growth than many exotic species, (Specht, 1981b); the exotics may have had a longer competitive experience of such "Mediterranean" climates elsewhere, and thus be better adapted than the native species in those parts of Australia now most intensively used by man.

THE FIRST HUMAN INVASIONS: LATE PLEISTOCENE (50,000 YEARS AGO) TO 'HISTORIC TIMES'

The time of arrival of the first Aboriginal people in Australia is unknown. It is not known whether there was one or more migrations, but they certainly occurred before human beings spread to the Americas, and must have taken place during the late Pleistocene, when (at present rates of movement) the continents were, at most, within a few kilometres of their present positions (Figure 3). Sea levels were lower than today, and northern Australia was connected to New Guinea during most the period after 120,000 years B.P. (*Before Present*) (Chappell, 1976, 1983). This period includes the earliest dated Aboriginal remains in Australia at Lake Mungo, 32,000 years B.P. Since this date lies near the extreme age of reliable radiocarbon dating, and is at a site far removed from the likely points of entry, there has been much discussion (briefly reviewed by Thorne, 1981) on when man first entered Australia.

Aboriginal people must have increased the opportunities for dispersal and establishment of plants and animals from Indonesia to Australia. Golson (1971) noted that a number of genera of plants used for food are shared between Arnhem Land and Cape York on the one hand, and Malaysia on the other. He suggested that the use of plants by Australia's earliest people may have been influenced by traditions established in Malaysia, although the species involved had not necessarily been deliberately introduced from Malaysia to Australia, or vice versa. Few coastal archaeological deposits in the Australian tropics have been analysed, and little is known of the economy of the people who first reached Australia. We can only guess which species of plants first introduced by Aborigines are now part of the flora naturalized in Australia (Wace, 1978). Some plants may have arrived with Aboriginal people in late Pleistocene times, but their speed of transit from Indo-Malaysia would have been little if any greater than by natural means of dispersal.

The dingo, which is the only domesticated species that was undoubtedly introduced by Aborigines, is thought to have been in Australia only about 5,000 years. It ran wild over mainland Australia, but never reached Tasmania or other offshore islands, as these were isolated from the mainland by the postglacial rise in sea levels some 6,000 years ago. The dingo was the first large terrestrial placental carnivore to reach Australia, and had a profound effect on the native marsupials on the mainland. It also affected the native placental carnivores, notably seals on their breeding beaches and other marine mammals which strand on

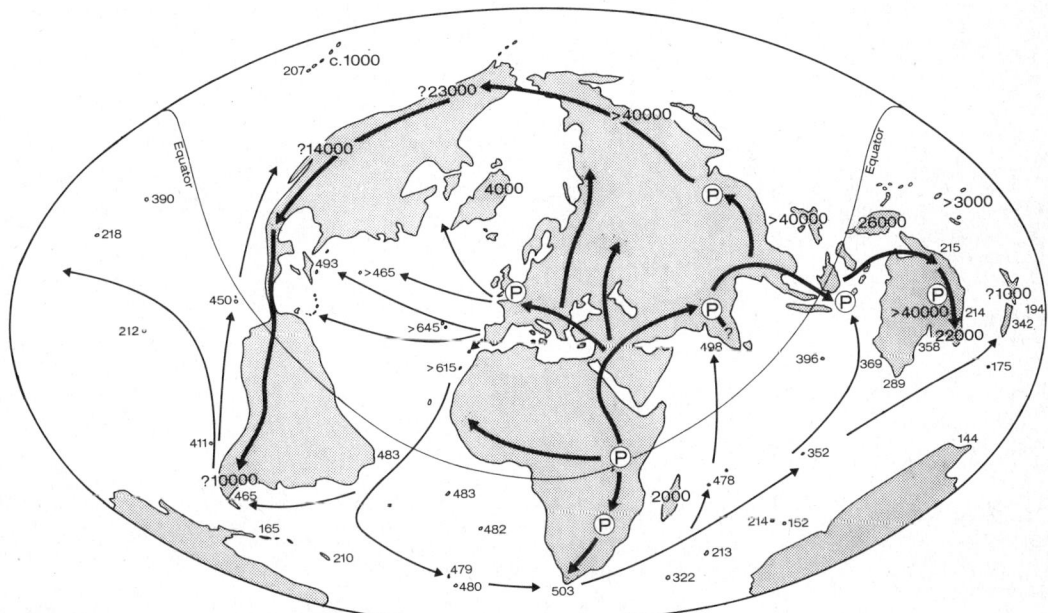

Fig. 3. The global spread of homo sapiens. The symbol ⓟ indicates the evidence of human presence on the continents of Africa, Eurasia, Indo-Malaysia and Australia in Pleistocene times, but beyond the range of reliable radiocarbon dating. Large figures show evidence of prehistoric human presence in years before present, before the post-Columbian spread of the Europeans out of the Atlantic during the last 500 years. Smaller figures indicate the number of years before 1985 in which various continental coasts and oceanic islands were first visited by voyagers from Europe during the last 500 years, or shortly before. Arrows diagrammatic only. Adapted from Wace, 1978b.

beaches. Shaughnessy (1982) has described scavenging and predation by jackals, infected with rabies, on furseals breeding on Namibian beaches. Vedros (1984) has mentioned the possible importance of marine mammals as carriers of *Salmonella*, leptospirs, and other pathogens of terrestrial animals or man, and has suggested that the beaching of cetaceans may be associated with such diseases. The marine mammals could have become vectors for exotic diseases of terrestrial mammals, particularly when beach-scavenging carnivores, such as the dingo, became established in Australia.

Even though Aborigines did not practise systematic tillage, or domesticate plants (the hallmarks of 'agriculture', they undoubtedly provided new opportunities for immigrant plants over a period of at least 40,000 years in three ways:—

a. by repeatedly disturbing vegetation and soil near their campsites;

b. by local eutrophication of soils, and accumulation of shells and other debris in their middens;

c. by greatly increasing the frequency of bush fires thus changing vegetation soils, hydrology, and fauna.

The Aboriginal occupation of Australia must have been of importance to plant and animal immigration mainly by altering environments and disturbing native vegetation, especially by the use of fire. Singh (1982) made detailed analyses of the charcoal fragments in sediments at Lake George (in the south-east of the continent) and noted a sudden increase in the incidence of fire about 128,000 years B.P., and has suggested that human activity is the most probable explanation of that increase.

SETTLEMENT AND AGRICULTURE; SEA LINKS TO THE TEMPERATE ZONES — 1788 to 1960

Australia was the only one of the six habitable continents, whose aboriginal inhabitants did not practise agriculture or livestock herding. Once migrants and traders from Europe, and from the coasts of the Americas, had established colonies in different places around the southern and eastern coasts, then the conditions were created for a massive invasion of exotic plants, animals and micro organisms, including pathogens of the native and exotic biota. The Settlement at Sydney Cove in 1788 thus started a new era in the biogeographical history of Australia. For the first time, plants and aimals arrived in Australia direct from the temperate zones of the northern hemisphere without having had to adapt to the aseasonal climates of the equatorial tropics on the way. Suddenly, organisms began to arrive from overseas regions with climates similar to those in southern Australia where most of the colonies were being established. Native ecosystems near the new settlements suffered the greatest disruptions as a result of human activities. Thus European settlement not only breached the tropical barriers to the dispersal of temperate zone plants and animals — it also created the disturbed environments of tillage, livestock rearing, urban settlement, mining, highway construction and industrialization which had not previously extended to the continent of Australia, and to which many of the newly introduced species were already adapted.

Some idea of the opportunities immediately presented for both the intentional and involuntary transport of new organisms to Australia are given by the Judge-Advocate of the new colony in his account of its foundation (Collins, 1798). Collins mentions the acqusition of living plants and seed by the First Fleet at Rio de Janeiro and at Cape Town, to add to those embarked in England. Grapes, figs, oranges, pears and apples were growing within two months on the east side of Farm Cove in Sydney, where cereal crops had also been planted. Horses, cattle, sheep, pigs, goats, rabbits and poultry (432 head of stock in all) were ashore, and some pigs and cattle had run wild only a few months after the landing. From the very start of the first settlement, there was plenty of scope for the establishment of all sorts of plants and animals from Northwest Europe, South America and South Africa, together with their pathogens.

Transit Times and Routes from Europe: Numbers of Immigrants

The human dispersal of new organisms to Australia across the tropics by sea from the northern hemisphere was slow, (at least by today's standards): but it was much faster and more reliable than natural dispersal by wind, ocean drift or oversea carriage by birds.

Sailing from England to Australia by way of Brazil and Cape Town, the First Fleet took about 35 weeks (of which 26 were spent at sea) to reach Sydney (Bateson, 1969). Forty years later, this time had been halved. At the start of free settler emigration from Europe to Australia, the naval surgeon, Peter Cunningham (1827) wrote:

"About eighteen weeks is the average passage from England to Sydney, if the ship proceeds direct, the distance covered by ship's course being about sixteen thousand miles. Many vessels touch, however, at the Canaries, the Cape Verds, Cape of Good Hope, or Brazils, to replenish their stock and their water, which both a numerous body of passengers and the consumption of water by live stock tend soon to exhaust".

The average length of voyage from Europe probably remained at more than 100 days (14 weeks) until the Gold Rushes (Bach, 1976). At this time, Matthew Fontaine Maury (1855) was claiming that increased knowledge of the wind systems in the Southern Ocean had reduced the average time for a voyage from England to Australia to 97 days. The larger and faster American-built clipper ships could sail further south along the shorter but stormy 'great circle' routes in the westerlies, and the clippers began to replace the smaller English-built vessels for much passenger transport (Wace, 1980; Charlwood, 1981). Passage times of little more than two months were soon achieved by the clippers on the outward run from England to Australia. The famous *'Marco Polo'* sailed from Liverpool to Port Philip Heads in 68 days in 1852, and the *'Thermopylae'* from Plymouth to Port Philip in 62 days in 1869 (Blainey, 1966). These were amongst the fastest passages made in their times; average times for voyages from England to Australia in sailing ships via the Southern Ocean probably fell from over 100 to about 80 days in the period 1850 to 1880. Passage times for 300 vessels sailing from Europe to Melbourne between 1858 and 1863 probably averaged 95 days or more, with a mere 20% taking less than 80 days (Charlwood, 1981, p. 137). After 1869, when the Suez Canal was opened, and when steamships became more reliable on long voyages, and slowly began to replace sail, passage times from Europe to Australia via Suez fell to about 50 days, and finally to about a month. This was still the length of voyage from Europe to the east of Australia by the 1950's when aircraft were replacing ships as the principal passenger carriers (Figure 4). These times relate only to voyages on the principal immigration routes from Britain, whence most immigrants arrived; transit periods from other places were considerably less.

Transit times are of importance in plant and animal immigration to Australia as they affect the possible survival of 'hitch-hiking' species (including weeds and pests) or their pathogens, and those of man and his domesticated animals. The absence in Australia of some widespread diseases of livestock before the imposition of effective animal quarantine regulations (after Federation), must be partly because the transit periods for people and imports exceeded the survival periods of the pathogens in question. Thus, foot-and-mouth disease may have been effectively excluded from Australia during the era of seaborne trade and immigration because the incubation period is 14 days. Rabies may also have been excluded because any rabid animals aboard ship would have died or been destroyed at sea before arrival in Australia.

Australian immigration and trading policies of the time must also have minimized the invasion of exotic organisms from the tropics, and from Asian and African countries. Many sailing vessels carrying emigrants and their livestock and other chattels, did not call at any ports on the voyage from Britain to Australia (Wace, 1980). Exotic organisms and disease most likely to invade Australia over this period of seaborne immigration were from Britain — a region which is climatically quite unlike most of Australia. Nevertheless, given the contacts with India, Malaysia and Africa, it is remarkable that Australia apparently escaped the establishment of many disease of livestock which were endemic there, and from whose absence we still benefit. The sea voyage to Australia was itself a period of enforced quarantine against many human and animal pathogens.

The number of people entering the country each year is one measure of the degree to which isolation is broken down. It might be expected to bear some relationship to the rate of introduction of exotic species (whether or not these succeed in establishing themselves). During the period of long sea voyages, most people entering Australia were immigrants, and most of these were from Britain (including Ireland), whence the total numbers between 1850 and 1900 fluctuated from 86,183 in the peak Gold Rush year of 1852 to 7,277 in 1872 (Crowley, 1951). The numbers of people (excluding convicts) leaving Britain and Ireland with the stated purpose of going to Australasia (including New Zealand) each year from 1821 and 1950, varied from 320 in 1821 to 87,811 in 1852 (Carrier and Jeffrey, 1953). Mean annual figures for each decade during this period are shown in Table 1.1.

Immigrants from California, China, Melanesia, and elsewhere, added somewhat to these numbers at times, but were no more than a few thousands each in any one year. Above the beach at Robe in South Australia is a remarkable monument with the bland inscription:

> "During the years 1856 and 1858, 16,500 Chinese landed near this spot and walked 200 miles to Ballarat and Bendigo in search of gold".

Incursions of this sort, as well as the continuing flow of immigrants from Europe, must have provided plenty of opportunities for the import of exotic plants and animals into Australia, especially from lands bordering the Pacific. Such incursions of people, related to activities such as gold mining and sugar growing, must have had a relatively small potential for importing exotic species when compared to the continuing influx of settlers from Europe.

Exotic species whose dispersal was favoured by ship-borne immigration
Many exotic organisms were intentionally imported aboard ship, and some of these became naturalized or feral in Australia (Rolls, 1969), but are now regarded as more-or-less of a nuisance (e.g. cats, rabbits, sparrows, many ornamental plants, such as soursob, Paterson's Curse and lantana). Three groups of organisms are especially likely to have been introduced accidentally by long sea voyages:—

a. Those having a resistant or quiescent phase of a month or more in their life cycles. This group includes most vascular plants and many cryptogams, with spores, seeds, or vegetative dispersal units; and the arthropods with resistant eggs, pupae or larval stages.
b. Human inquilines (i.e. 'nest-sharers') such as rats and mice, and especially insect pests of stored food such as weevils and flour beetles, which may also have quiescent dispersal phases.

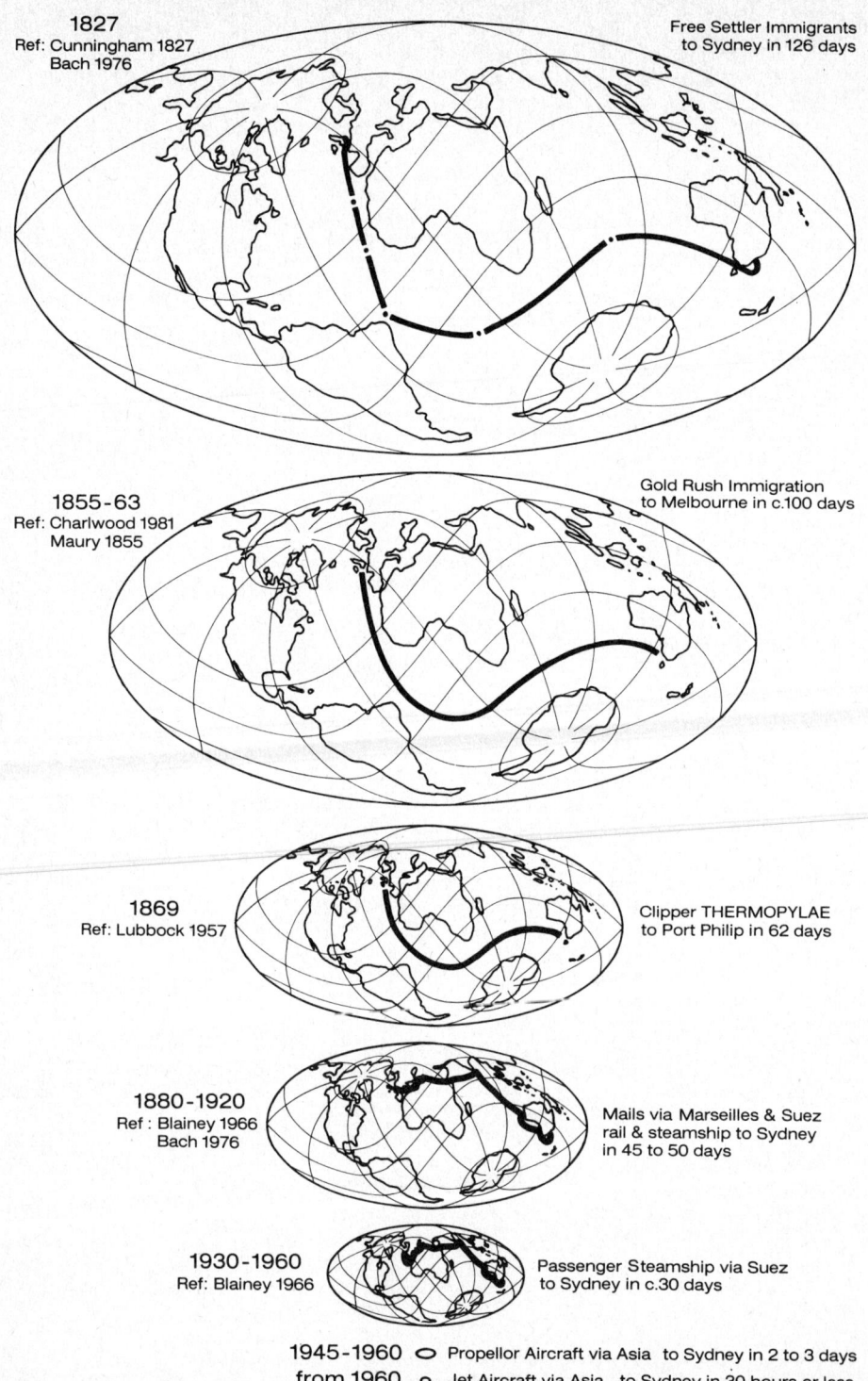

Fig. 4. The size of the world in travel time from England to Australia showing routes taken from 1827 to present day. Size of the globe is proportional to the time of passage from England to Australia. (Oblique Mollweide projection, canted 45° and centred on intersection of Equator and prime Meridian. After Staley and Benveniste, 1959, p. 8).

TABLE 1.1. *Ten year averages of the annual number of migrants from the United Kingdom to Australasia (1821 to 1938), and year-by-year (1946–50) also to Australia alone (1850–1900)*

Years	Australasia[1]	Australia alone[2]
1821–1829 (9 years)	854	
1830–1839	5,327	
1840–1849	12,633	
1850–1859	46,932	46,512
1860–1869	27,267	22,171
1870–1879	29,571	17,063
1880–1889	37,575	34,058
1890–1899	13,394	10,105
1900–1909	20,070	
1910–1919	36,814	
1920–1929	45,503	
1930–1938 (9 years)	7,111	
Total to World War II	2,822,159	
1946	15,020	
1947	18,930	
1948	41,372	
1949	62,320	
1950	64,746	

Sources: [1]Carrier and Jeffrey (1953); [2]Crowley (1951).

c. Pathogens of man or 'domesticates' (i.e. domesticated organisms) with incubation or quiescent periods exceeding the length of the voyage at the time, and thus escaping detection aboard ship.

Two types of cargo would be particularly likely to harbour many exotic plants and other organisms. Fodder for livestock, as any surplus was likely to be unloaded with the stock at the end of a voyage; and solid ballast, which was frequently carried by sailing ships on their outward voyages from Europe or America and dumped at ports where ships took on cargo. Dumping of solid ballast was characteristic of tramp steamers on speculative voyages (Bach, 1976, p. 155; Blainey, 1966, p. 285). Schelpe (1980) considered the possibility that some Australian coastal dune plants may have been taken sand ballast from Cape Town to Australia, or cast ashore with ballast in shipwrecks. The use of solid ballast by ships is now obsolete, and 'ballast floras' in other parts of the world are said to be declining (Ouren, 1968; 1978).

The creation of new environments: opportunities for the establishment of exotic species during the era of ship-borne immigration

European settlement created many new opportunities for the establishment and spread of exotic species within Australia. Some of these were summarized by Adamson and Fox (1982). During the era of ship-borne human immigration, these major disturbances were:—

a. *Agriculture and horticulture.* All the crops and almost all the weeds of cultivation in Australia are exotic species, many of which arrived with the earliest European migrants (Kloot, 1984). In South Australia, most of these early arrivals which became naturalized were intentional imports which ran wild (Kloot, 1979). Similarly,

> "Most of the diseases of cultivated plants in Australia are caused by introduced pathogens ... Almost from the start, attempts at growing food crops in Australia were plagued with diseases referred to as 'blights' in the early records". (White, 1981).

b. *Grazing of livestock and introduced feral animals.* Enormous ecological changes were initiated in Australian vegetation with the release of exotic animals. The disturbed and dunged environments created by domesticated and feral animals are preferentially invaded by exotic plants. All 25 species of plants dubbed 'thistles' in Australia are exotic plants, and must have been imported by ship, probably with stockfeed or seed mixtures for 'improved' pastures. Thistles were the first plants to be outlawed in Australia from the 1850's under the various "Thistle Acts" of the colonial governments (Amor and Twentyman, 1973).

c. *Bushfires.* Displacement of Aboriginal people by white settlers generally altered the fire regime. Frequent light burns were in general replaced by occasional but severe wildfires. Much of the native biota was killed in these severe fires, and opportunities created for the spread of exotic species; in the winter rainfall zones, these commonly included winter and spring-growing annual grasses which displaced the native species and dried off by early summer, producing tinder for summer fires.

d. *Highway and Railway Construction.* This created opportunities for the spread of exotic plants and animals, along avenues of intermittently disturbed environments unlike anything existing before European settlement. The mobility of the new settlers and their stock along these highways provided continuous opportunities for dispersal of the new arrivals. The comparatively rapid movements of the human immigrants to distant parts of Australia, must have increased the likelihood of establishment of the exotic species whose disseminules they carried with them. Such local transport within Australia of newly arrived flora and fauna greatly exceeded the speed with which Aboriginal people moved over the country.

e. *Mining and Quarrying.* Open cut, alluvial and deep mining produced areas of overburden, outwash, and mullock respectively. Destruction of native vegetation and the creation of such new surfaces locally increased the opportunities for the spread of colonizing plants. Most colonists of mullock at Bendigo and Ballarat are exotic plants, and this is probably true of other mining areas.

f. *Urbanisation and Industrial Development.* The construction of cities and the establishment of industries (especially the disposal of human and industrial wastes) provided environments unlike anything hitherto existing in Australia, and their creation near the ports greatly increased the opportunities for establishment of exotic species which had become adapted overseas to exploit such man-made situations.

THE MOBILE SOCIETY: AIRBORNE MIGRATIONS TO A CAR-BORNE COMMUNITY; 1960 TO THE PRESENT DAY

The latest phase in the breakdown of Australia's biogeographical isolation, has brought even greater opportunities for the rapid importation of plants and animals from other parts of the world. This modern phase started with the increase of intercontinental air travel after the Second World War. During this period, Australians have also become more mobile within their own country, thus increasing the opportunities for the rapid transport of organisms from place to place. The establishment of newly-arrived species in suitable habitats within Australia therefore became more likely.

The numbers and provenance of people entering Australia and their transit times from overseas

Throughout the 1950's up to 250,000 people entered Australia each year. For the 20 years from 1960, the numbers of these arrivals rose by about 100,000 each year, to 2.28 million entrants in 1980 (Figure 5). These figures include all the people entering Australia whose passports were examined; immigrants, businessmen, foreign tourists, or Australians returning home. They give a measure of the potential number of vectors for human pathogens, "hitch-hiking" organisms, and for other organisms which are intentionally but informally imported. The ten-fold increase in the number of people entering Australia since 1959 is because of the growth in commercial air transport. In 1965 about 30% of the 520,000 people who then entered the country came by sea, but by 1982 this seaborne proportion had fallen to a mere 0.04% of the 2.4 million entrants.

The parts of the world from which all these people started their journeys, and from which new exotic organisms might therefore be introduced, appear to fall into three very differently sized groups. In 1982 the largest group, comprising nearly half of all entrants, came from South East Asia (25%), and New Zealand plus Norfolk Island (23%). The second major group comprising 42% of all entrants, came from the northern temperate lands of North America (12%), Britain and Ireland (10%), mainland Europe (8%) and also from Hong Kong/China (6%) and Oceania (mostly Fiji and South West Polynesia; 8%). Other parts of the world, from each of which fewer than 100,000 people arrived in 1982, were Japan (3%), Papua New Guinea (2%), Africa and India (1% each), and Latin America (0.03%). These numbers are based on the responses to questionnaires which ask travellers, where they had spent most time overseas, and where they boarded the vessel (ship or aircraft) which brought them to Australia. Parts of the world from which there are no direct flights to Australia (e.g. Latin America) will thus be under-represented. Correspondingly, places such as Hong Kong and Singapore where many passengers change aircraft on long journeys from elsewhere, will tend to be over-represented.

Nevertheless, the data show with which parts of the world Australia has the most direct and rapid biological contacts, and how these contacts have been changing from 1965 to 1982 (Figure 6). Contacts with South East Asia, (including Indonesia and Philippines) have increased much more rapidly than those with distant lands which have been the traditional sources of migrants for most of the last 200 years. Biogeographically, the changing influx of people into Australia seems to be anticipating the geographical future, and magnifying the long term trend over the last 20 million years since the Australian continent approached the Asian plate and

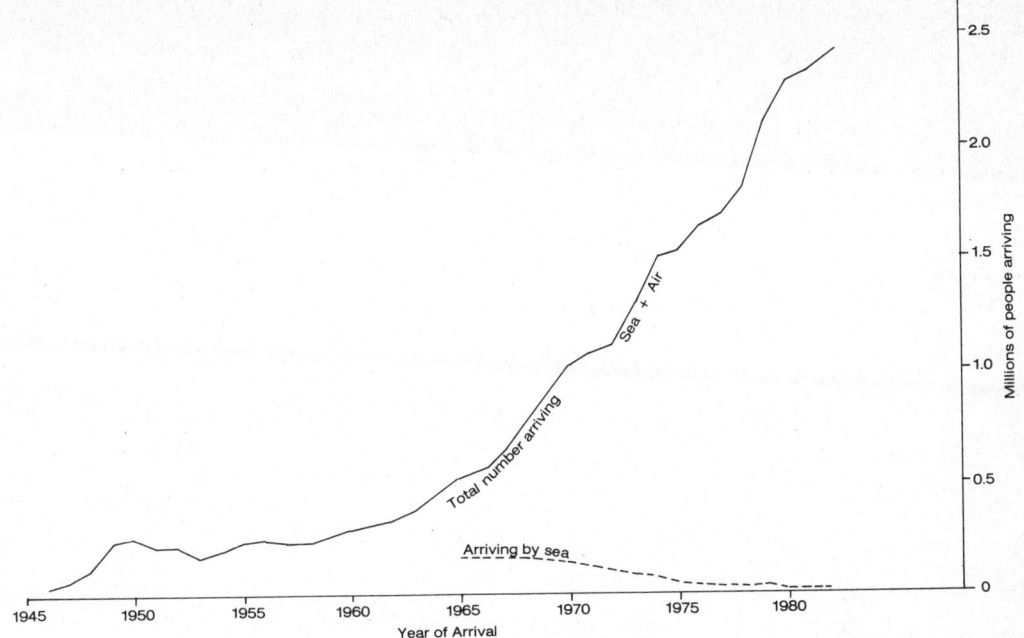

Source: *Australian Bureau of Statistics, Quarterly & Annual Reports, Overseas Arrivals and Departures, 1965-82*

Fig. 5. Numbers of people entering Australia annually, 1946–1982. Includes all people entering Australia legally; immigrants, tourists, long-and-short term visitors, and Australian residents returning home, but does not include transit passengers who do not leave airports. (One person making several overseas trips may count more than once in a year).

merged with Sundaland. With the exception of New Zealand, Australia's biogeographic links with the other remnants of Gondwanaland are small, and static or declining. Links with the temperate lands of the Northern Hemisphere declined from 42% (in 1965) to 30% (in 1982) of the total number of people entering Australia.

Inter-continental air travel on this scale presents different opportunities for the entry of new organisms into Australia, when compared to the older and slower modes of surface travel. It is now possible to reach Australia from any other place on each served by a major airline, in less than 36 hours. Because of this speed, and because the environment in which the passengers travel is conditioned to minimise physical stress, "hitch-hiking" organisms can travel in an aircraft cabin in a physiologically active phase, and not only in the inactive or dormant stages favoured by sea travel. Even diseases spread by short-lived respiratory aerosols and carried by humans such as foot-and-mouth disease virus (French and Geering, 1978), could be readily transmitted by passengers entering Australia by air. For free-living species, air travel perhaps favours the transport of generalized r-pests that reproduce fast and in a range of habitats, rather than the specialized k-pests of stored foodstuffs and other bulky products carried aboard ship (Conway, 1976).

Laird (1951) found that 36% of some 250 propellor-driven aircraft arriving in New Zealand in 1946–48 had insect "stowaways" aboard, some of which had travelled on the outside of the machine. External carriage is now more difficult because of extreme cold at the higher altitudes flown by passenger jets, and standardized methods have been evolved to treat aircraft cabins to prevent the entry of exotic organisms. Wace (1968) found venomous spiders being dispersed over very long distances in an aircraft on an unscheduled flight, and no doubt other organisms travel in this way.

Along with the increase in air transport of people to and from Australia, the inflow of ships and their cargoes has also been increasing. Figure 7 shows the annual numbers and tonnages of ships arriving in Australian ports from overseas since 1917. A similar plot for shipborne cargo imported from overseas (Figure 8) shows an overall but fluctuating rise in volume but a decline in cargo tonnage since 1970: perhaps these data imply that packaging has become more important in relation to contents, or that Australia has been importing bulky but lighter cargoes since 'containerization' set in, in the early 70's? The implications of these data for the possible import of new species by sea are not obvious, though if opportunities for imports increase in direct proportion to volume of imported cargo, then they have doubled in the 20 years 1960 to 1980. Similarly, if new species have been arriving by sea in proportion to the numbers of ships visiting Australia, then the opportunities have trebled since the inter-war years. But technical developments in ship design, unloading facilities and containerization,

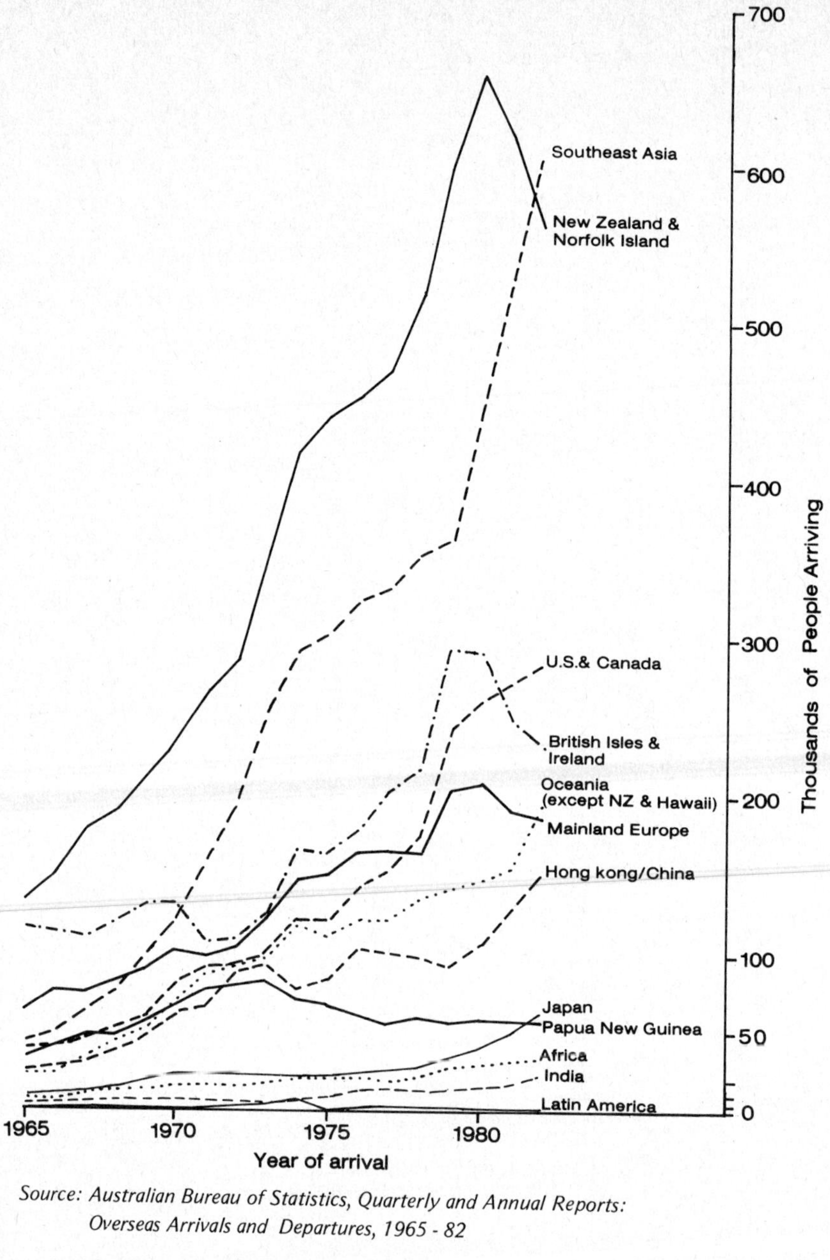

*Source: Australian Bureau of Statistics, Quarterly and Annual Reports:
Overseas Arrivals and Departures, 1965 - 82*

Fig. 6. Provenance of people entering Australia, 1965–1982. All people entering Australia legally, grouped by the region or country of embarkation aboard the ship or aircraft in which they arrived.

and the responses of quarantine authorities, are probably important in diminishing the likelihood of new species getting ashore, despite increases in the numbers, size, volume or weight of ships or cargo arriving. These latter figures, (like the figures for passenger arrivals) do indicate crudely some of the problems faced now by the quarantine authorities in excluding unwanted organisms.

Travel by modern jet aircraft represents an enforced containerization of people. At least the most undesirable "hitch-hiking" pests can be treated in ways comparable to those used on containerized seaborne cargo, by enforced fumigation of the containers themselves (if not of their contents). The numbers of aircraft landing Australia from 1961 to 1982 from overseas, and granted pratique, is shown in Figure 7. It is now approximately twice the number of ship arrivals from overseas.

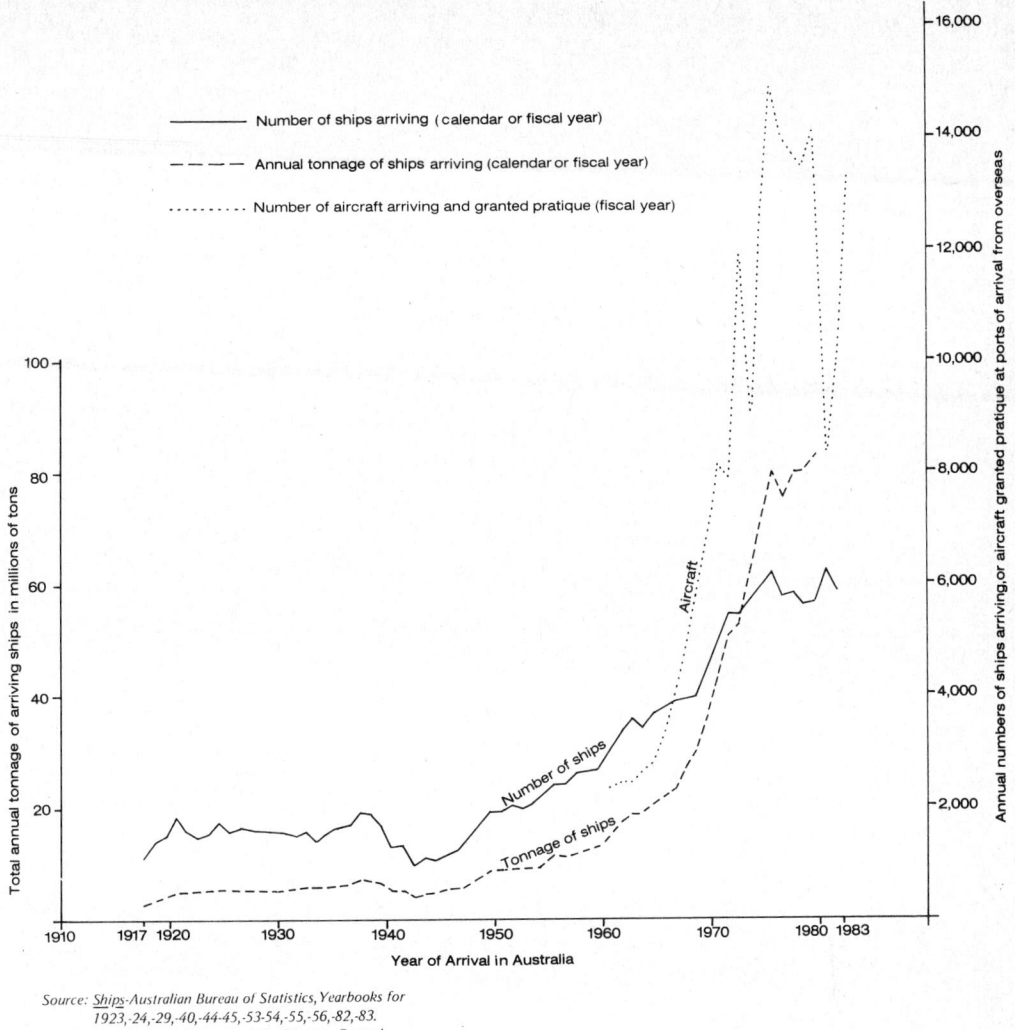

Source: Ships-Australian Bureau of Statistics, Yearbooks for 1923,-24,-29,-40,-44-45,-53-54,-55,-56,-82,-83.
Aircraft-Annual Reports of the Director General of Health, 1960-61 to 1982-83

Fig. 7. Ships arriving in Australia from overseas, 1917–1982 and aircraft granted pratique, 1961–1983.

There has been concern that these large increases in the numbers of people and volume of imports entering Australia, threaten a corresponding influx of exotic species, including the pathogens of economically important livestock and crop plants. This concern was mostly clearly expressed by the Senate Standing Committee on National Resources report on *The Adequacy of Quarantine* (Thomas et al, 1978), which stated:

"In more recent times these natural and historic barriers [to the import of exotic organisms] have been eroded by ever increasing international movement of people and goods, movement which has become speedier and more flexible with the growth of cheap mass air transportation and containerization. The former advantages of being an island continent no longer exist". (page 3).

It also states more specifically:

"The quarantine importance of arriving passengers is now largely as vectors of animal and plant disease. Any food they carry, souvenirs, clothes &c may be the vehicle for introduction of exotic diseases". (page 62).

Such a threat was used as an argument to support the construction of the Australian National Animal Health Laboratory at Geelong (Johnston and Scott, this volume). However there is little quantitative evidence to support these alleged threats.

Some pest or pathogen incursions appear to have taken place recently; poplar rusts in 1972, lucerne aphids in 1977 and a cereal aphid in 1983.

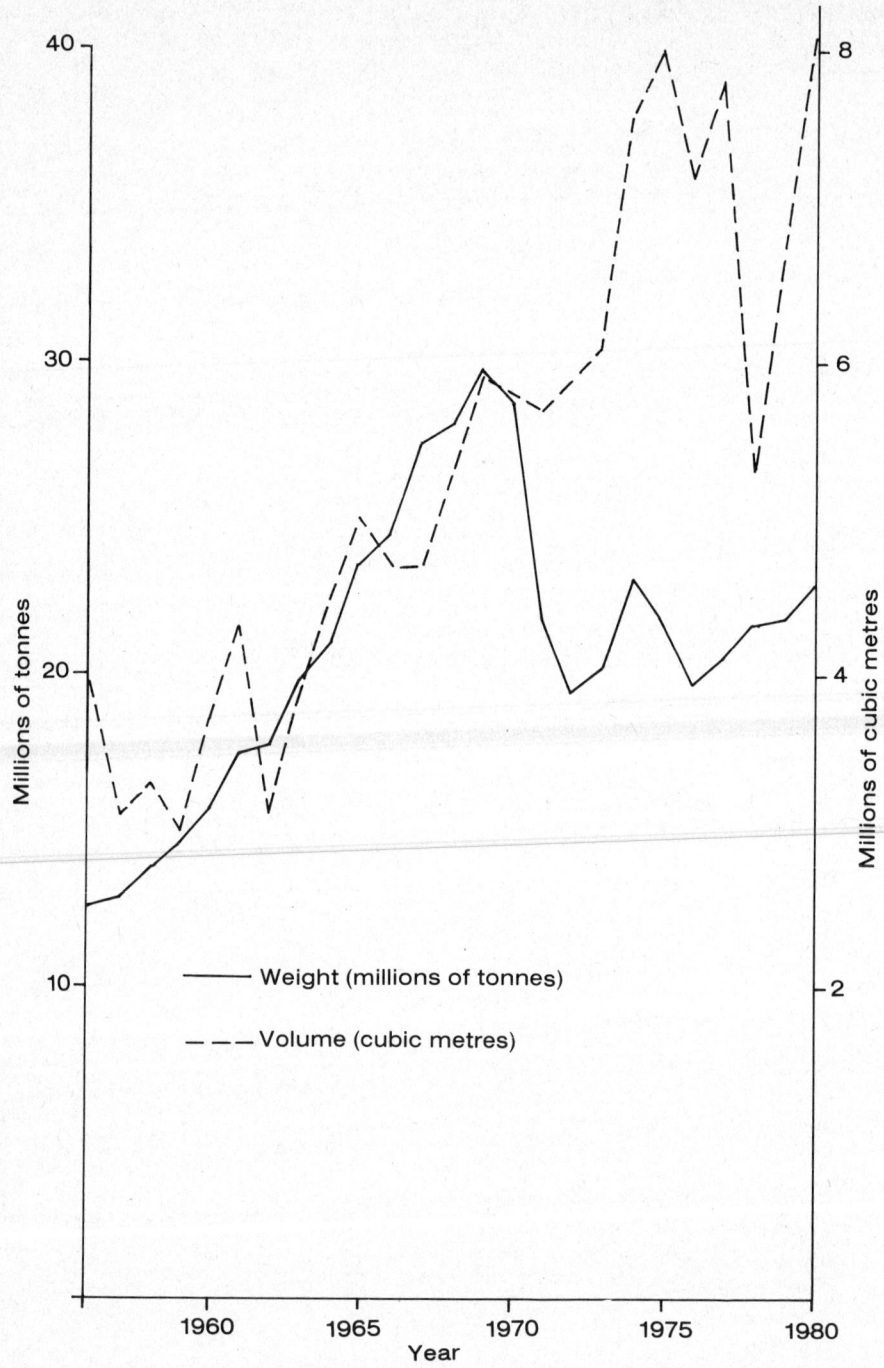

Source: Australian Bureau of Statistics, Yearbook 1982, p. 744].

Fig. 8. Seaborne cargo from overseas, discharged in Australia, 1956–1980.

Introduced vascular plants are one group of exotic organisms for which estimates of the rates of naturalization have been made over a number of years in different States (Figure 9). But even from these meagre data, there is no sign yet of a relation between the rate of establishment of exotic plants as naturalized species (however recognized), and the numbers of people or quantity of goods arriving from overseas. Data on passenger numbers and cargo landings thus remain merely suggestive, so far as the immigration of pests and parasites are concerned. Research on the numbers and types of organisms being inadvertently carried by travellers and in cargo from overseas despite quarantine procedures would be useful. We should monitor the 'uninvited' if we are to maximize the efficiency of quarantine. One category of such unwanted organisms includes plants which may become troublesome weeds. Seeds germinated from the socks, shoes and trouser cuffs of co-operative travellers, who were kind enough to give me the brushings from these garments shortly after they entered Australia (or other Pacific countries), are ecologically an ill-assorted collection of species, most of which are already established in Australia (Table 1.2).

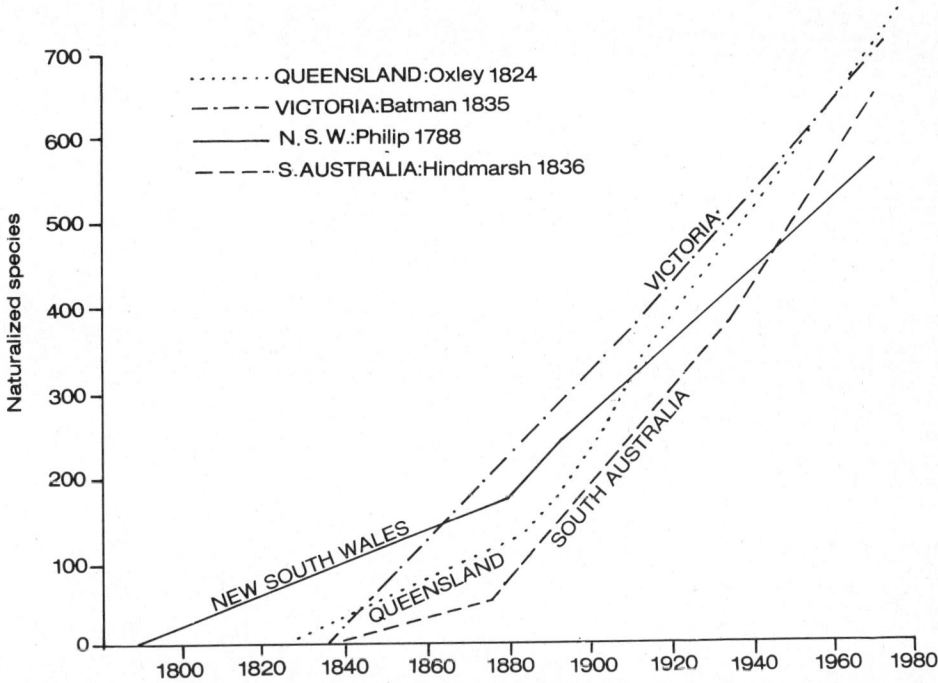

Source: Redrawn from Specht 1981a

Fig. 9. Numbers of species of vascular plants not native to Australia recorded as naturalized in the four eastern mainland states. Since there is no concensus on the criteria for accepting species as naturalized, these figures (which are derived from State floras and other lists) can give only an approximate measure of the rates of establishment of exotic plants. Dates of the settlements indicated are not necessarily the dates of first introduction of exotic plants, but represent the first major disturbances of the environments by European settlers.

TABLE 1.2. *Plants germinated from the clothing of tourists recently arrived in Australasia or other parts of the Pacific, 1981–82*

Species	Count	Origin	Date	Code
Aira caryophyllea	2	Kashmir	4/81	NMW
Agrostis ?tenuis	4	Argentina	2/81	NMW
Cenchrus echinatus	75	New Guinea	11/81	MN
Cenchrus echinatus	27	New Guinea	11/81	MN
Cenchrus echinatus	6	New Guinea	11/81	JL
Cyperus sp.	1	New Guinea	5/82	MN
Danthonia sp.	1	Lord Howe Is.	5/82	AG
Digitaria sanguinalis	1	Kashmir	3/81	NMW
Echinochloa crus-galli	1	New Guinea	11/81	JLW
Holcus lanatus	12	NZ/Antarctic Cruise	5/82	NMW
Juncus sp.	43+	NZ/Antarctic Cruise	5/82	NMW
Lolium perenne	3	NZ/Antarctic Cruise	5/82	NMW
Lolium perenne	2	New Guinea	11/81	MN
Poa annua	1	Kashmir	5/81	HTB
Poa pratensis	5	New Zealand Islands	9/82	NMW
Poa pratensis	1	New Guinea	11/81	MN
Setaria verticillata	1	Kashmir	3/81	NMW
Vulpia sp.	1	New Guinea	5/82	MN
Vulpia sp.	3	NZ Islands	9/82	NMW
Vulpia sp.	5	NZ/Antarctic Cruise	5/82	NMW
Stenotaphrum secundatum	1	Lord Howe Is.	5/82	NMW
Unknown grass 1	2	Kashmir	4/81	NMW
Unknown grass 2	21	Argentina	3/81	NMW
Iris Hookeriana	2	Kashmir	11/81	NMW
Acaena sp.	1	NZ/Antarctic Cruise	5.82	LOU
Bidens pilosa	6	Lord Howe Is.	5/82	AGW
Bidens pilosa	10	New Guinea	6/81	SHW
Centarium minus	1	Polynesia Cruise	7/81	HC
Conyza sp.	1	Polynesia Cruise	7/81	HC
Conyza sp.	2	NZ/Antarctic Cruise	5/82	NMW
Hypochoeris radicata	2	NZ/Antarctic Cruise	3/82	NMW
Oreomyrrhis sp.	1	New Guinea	12/81	MN
Paronychia brasiliana	1	New Zealand Islands	8/82	NMW
Plantago lanceolata	7	Polynesia Cruise	3/81	HC
Rumex sp.	1	Kashmir	4/81	NMW
Salix sp.	1	Nepal	3/83	ORM
Senecio hispidus	1	New Guinea	12/81	MN
Sonchus oleraceus	6	Lord Howe Is.	5/82	AG
Stellaria angustifolia	12	New Zealand	8/82	NMW
Trifolium repens	1	New Zealand	8/82	NMW
Trifolium sp.	1	Kashmir	4/81	NMW
Verbascum thapsus	2	Kashmir	4/81	NMW
Verbena bonariensis	20	Polynesia Cruise	6/81	HC
Veronica persica	1	Kashmir	5/81	HTB
Unknown 1	2	Kashmir	7/81	NMW
Unknown 2	4	New Guinea	5/82	JL

Opportunities for dispersal and establishment of exotic species within Australia in a mobile, urban, industrialised society

Personal mobility is not confined to international travellers. Along with the steady increase in the number of people entering the country each year since the late 1950's, there has been a similar increase in human mobility within Australia. Passenger air transport activity within Australia grew at an average rate of 7.8% per year from 1972–1981 (Anonymous, 1981). This growth in internal mobility is also illustrated by the numbers of motor vehicles on register, which increased slowly from about 0.25 million in 1924 to one million in 1948, and then rapidly to 7.5 million in 1980 (Figure 10). In June 1981 there were 2.5 people to every car in Australia (Anonymous, 1981).

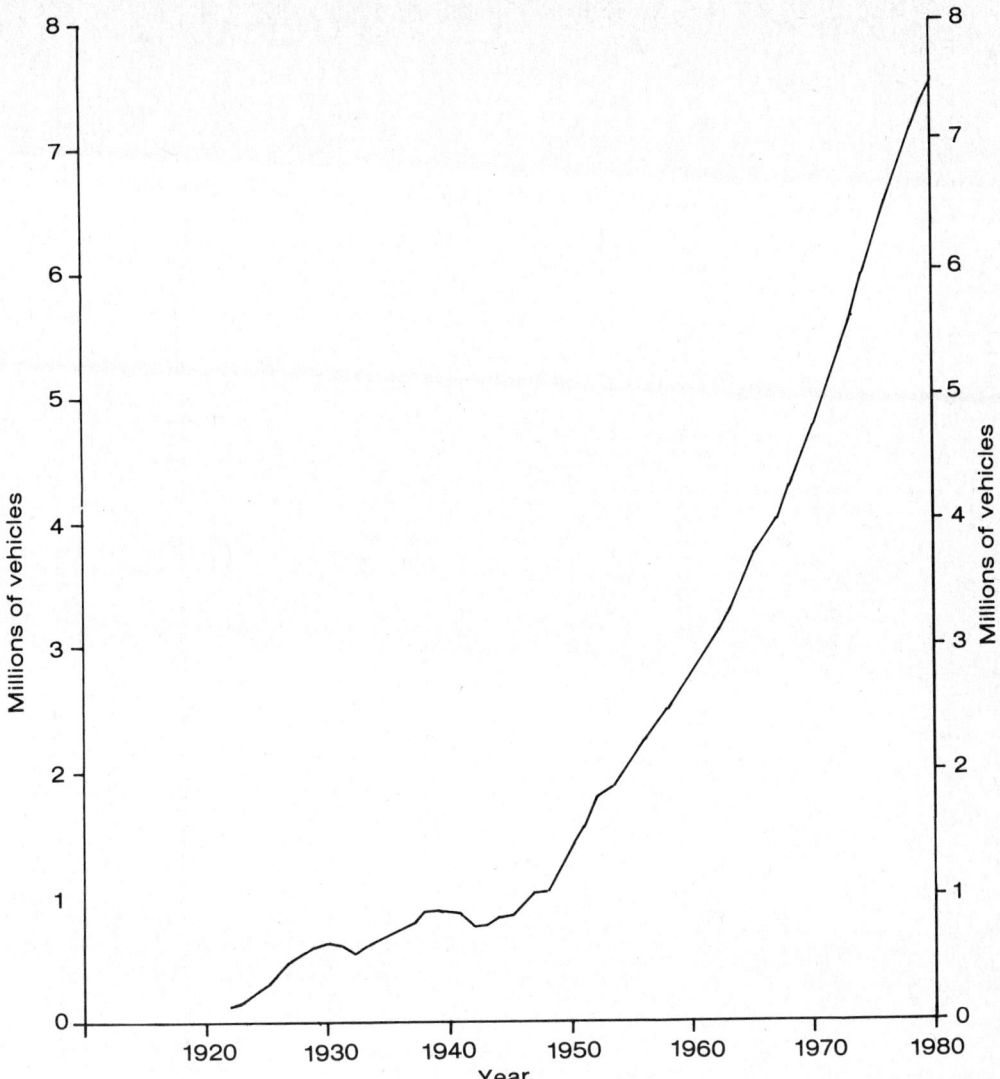

Source: Australian Bureau of Statistics, Yearbooks, 1922-1982

Fig. 10. Motor vehicles on register in Australia, 1924–1981.

If these motor car numbers are indeed a measure of increased personal mobility within the country, they may affect the ease with which exotic organisms can be spread from their points of entry or first establishment. An internally mobile society makes the establishment of unwelcome organisms easy because cars (and other human tools) easily and rapidly disperse them to favourable sites, and therefore make containment difficult. The claimed spread of the root-rot fungus, *Phytophythora cinnamomi*, from eastern to western Australia has been attributed to the dispersal of resistant chlamydospores by man. Its establishment in south-western eucalypt forests is associated with road construction works (Weste and Taylor, 1971; Shepherd, 1975). Studies of the botanical content of 'carwash sludge' in Canberra (Wace, 1979, 1983) has shown that motor cars have a very considerable ability to move viable plant propagules around. The same probably applies to many soil organisms, some insect eggs and larvae, and other weed and pest species. The study suggested that of the many plant species dispersed by cars, few grow in places most heavily infested by motor traffic. To arrive is one thing: to establish is another. Perhaps we should be paying more attention to the conditions for *establishment* of our least wanted exotic organisms, rather than merely trying to keep them out of the country?

Smith (1982a, b and *in press*) has studied the fruit trees spread by people in cars to roadsides, and found that these untended trees are often hosts to pathogens. Just as feral animals in Australia may pose problems in the control of exotic diseases of livestock by acting as reservoirs of infection, so the naturalized exotic plants play a similar role for the pathogens of domesticated plants.

All the opportunities for establishment of exotic species noted during the era of ship-borne migration have themselves increased during this latest era of increased human mobility. Six types of ecological disturbances were mentioned:

a. agriculture and horticulture.

b. grazing of livestock and feral animals.

c. bushfires.

d. highway and railway construction.

e. mining and quarrying.

f. urbanization and industrial development.

These have all continued, or increased, in extent and intensity. In the latest phase of the mobile human society, these should now be added:

g. *The widespread use of man-made molecules (herbicides, insecticides &c) to control vegetation and pests.* These greatly alter the competitive environment between native and introduced species. Since most herbicides and insecticides have been developed and first used overseas, exotic species are most likely to have achieved a measure of adaptation to them, and to have evolved a selective advantage over native organisms which have not been previously exposed to them.

h. *The increasing popularity of wild lands for recreation.* The penetration of wilderness by tourists and bushwalkers increases the invasion pressure of exotic species on previously undisturbed regions. It leads to the management of paths, campsites etc., to the construction of roads and tourist accommodation and hence to ecological disturbance and increased opportunities for exotic species to establish and spread.

CONCLUSIONS

The following conclusions can be drawn from this historical view of Australia, the island continent, whose biogeographic isolation has been broken down by human activities.

1. There are still some very considerable advantages to be derived from isolation. Biologically, too, Australia has had its share of luck. We should not squander the good fortune that our former isolation has bestowed on us in a headlong rush to bring in more and more visitors without some consideration of the likely side effects of such an activity. There is an inherent contradiction between the insistent demands of the tourist industry to bring in more people, and the requirements of quarantine to keep out those organisms that would be harmful to human beings and to established primary industries, and which may also be damaging to our native flora and fauna.

2. There is a need to be far more active in studying organisms coming to Australia, but not becoming established here. Although we should not lower our guard against the few pathogens of man and other animals and plants whose introduction would be socially or economically catastrophic, we should be doing research to establish which other organisms are evading our existing controls. We may well find that this includes some of those that we are trying hardest to exclude, but that the barriers to their establishment in Australia are ecological rather than geographical.

3. The total exclusion of uninvited and unwanted new exotic plants, animals and pathogens is impossible. Present and likely future human activities ensure this. We must be careful not to develop a Maginot Line mentality towards such uninvited newcomers, and to rely only on stopping them before they get here. We should be far more active in developing strategies and tactics to recognize and to eliminate infestations and outbreaks of pests and pathogens after they have penetrated our first line of defences. We need defence in depth, not just a *cordon sanitaire*.

The new Australian National Animal Health Laboratory is probably one of the most sophisticated animal health laboratories in the world, and could be used to build up overall defences in diagnosis, pathogen strain identification, and counter-attack by vaccine production in Australia against some of the exotic diseases we are most anxious to keep at bay. There is a need for more plant disease studies, and studies on the exotic biota that are penetrating quarantine, but not becoming established. Sooner or later these and other pathogens may establish themselves. We should understand how to act when they do.

Historically, it is appropriate that the new Australian National Animal Health Laboratory should be situated at Geelong. It was near Geelong, at Barwon Park in 1859, that Thomas Austin released the rabbits that were to have such an enormous effect on Australian native plants and animals, and on our primary industries.

ACKNOWLEDGEMENTS

It am most grateful to Dr Philip Hallett, Ms Jill Paterson, Ms Mary Pescott and Mr Richard Warren for the provision of data, and to Mrs Jenny Tode for typing the manuscript with despatch and accuracy.

REFERENCES

Adamson, D.A. and Fox, M.D. (1982). In: *A History of Australasian Vegetation.* (J.M.B. Smith, ed.) Chapter 5, p. 109, Sydney, McGraw Hill.
Amor, R.L. and Piggin, C.H. (1977). *Proc. Ecol. Soc. Australia 10,* 15.
Amor, R.L. and Twentyman, J.D. (1973). *Jour. Aust. Inst. Agric. Sci. 40,* 194.
Anonymous (1981). *Transport Indicators, (June Quarter).* Australian Govt. Publishing Service, Canberra.
Axelrod, D.I. (1976). In: *Historical Biogeography, Plate Tectonics and the Changing Environment.* Oregon State University, p. 435.
Bach, J. (1976). *A Maritime History of Australia.* Nelson, Sydney.
Bateson, C. (1969). *The Convict Ships, 1788–1868.* Glasgow.
Beard, J.S. (1977). *J. Biogeog 4,* 111.
Blainey, G. (1966). *The Tyranny of Distance.* Sun Books, Melbourne.
Bowler, J.M. (1982). In: *Evolution of the Flora and Fauna of Arid Australia.* (W.R. Barker and P.J.M. Greenslade, eds) Chapter 4, p. 35, Peacock Publications, Adelaide.
Bowring, J. (1843). *The Works of Jeremy Bentham — Volume X Memoirs, including autobiographical conversations and correspondence.* William Tate, Edinburgh.
Carrier, L.K. and Jeffery, J.R. (1953). *External Migration: A Study of the Available Statistics.* H.M.S.O., London.
Chappell, J.M.A. (1976). *The Origin of the Australians.* (R.L. Kirk and A.G. Thorne, eds) p. 11, Australian Institute of Aboriginal Studies, Canberra.
Chappell, J.M.A. (1983). *Proc. of the First CLIMANZ Conf., February 8–13 1981.* p. 121, Dept. of Biogeography & Geomorphology, A.N.U., Canberra.
Charlwood, D. (1981). *The Long Farewell — Settlers Under Sail.* Allen Lane, Melbourne.
Collins, D. (1798). *An Account of the English Colony in New South Wales.* Cadell & Davies, London (Australian reprint 1975; Reed, Sydney).
Conway, G. (1976). In: *Theoretical Ecology.* (R.M. May, ed.) Chapter 14, p. 257, Blackwell, Oxford.
Crook, K.A.W. (1981). *Ecological Biogeography of Australia.* (A. Keast, ed.) Chapter 1, p. 3. Dr W. Junk, The Hague.
Crowley, F.K. (1951). Unpublished D. Phil. thesis, Oxford University.
Cunningham, P. (1827). *Two Years in New South Wales.* Reprint 1966 (ed. D.S. MacMillan) Sydney, Angus & Robertson.
Darlington, P.J. (1957). *Zoogeography: The Geographical Distribution of Animals.* Wiley, New York.
Fenner, F. (1979). *Proc. Ecol. Soc. Australia 10,* 39.
French, E.L. and Geering, W.A. (1978). *Exotic Diseases of Animals.* 2nd Edition, Australian Govt. Publishing Service, Canberra.
Galloway, R.W. and Kemp, E.M. (1981). *Ecological Biogeography of Australia.* (A. Keast, ed.) Chapter 4, p. 53. Dr W. Junk, The Hague.
Golson, J. (1971). In: *Aboriginal Man and Environment in Australia.* (D.J. Mulvaney and J. Golson, eds) A.N.U. Press, Canberra.
Jones, R. (1969). *Austr. Nat. Hist. 16,* 224.
Keast, A. (1981). In: *Ecological Biogeography of Australia.* (A. Keast, ed.) Chapter 26, p. 814. Dr W. Junk, The Hague.
Kemp, E.M. (1981). In: *Ecological Biogeography of Australia.* (A. Keast, ed.) Chapter 3, p. 33. Dr W. Junk, The Hague.
Kloot, P.M. (1979). *Proc. 7th Asian–Pacific Weeds Sci. Conf.* p. 387.
Kloot, P.M. (1984). *Jour. Biogeogr. 11,* 63.
Laird, M. (1951). *Publications in Zoology II.* Victoria University College, New Zealand.
Lange, R.T. (1982). In: *A History of Australasian Vegetation.* (J.M.B. Smith, ed.) Chapter 3, p. 44. McGraw Hill, Sydney.
Maury, M.R. (1855). *The Physical Geography of the Sea and its Meteorology.* Harper, New York.
Nelson, E.C. (1981). In: *Ecological Biogeography of Australia.* (A. Keast, ed.) Chapter 25, p. 735. Dr W. Junk, The Hague.
Nix, H.A. (1981). In: *Ecological Biogeography of Australia.* (A. Keast, ed.) Chapter 6, p. 103. Dr W. Junk, The Hague.
Ouren, T. (1968). *Norsk Geogr. Tidsskrift 22,* 245.
Ouren, T. (1978). *Geojournal 22,* 123.
Powell, C.M.A., Johnson, B.D. and Veevers, J.J. (1981). In: *Ecological Biogeography of Australia.* (A. Keast, ed.) Chapter 2, p. 17. Dr W. Junk, The Hague.

Rolls, E.C. (1969). *They All Ran Wild*. Angus & Robertson, Sydney.
Schelpe, E.A. (1980). *Veld & Flora*. December 1980, p. 127.
Shaughnessy, P.D. (1982). In: *Mammals in the Seas — F.A.O. Fisheries Series 5*, 383.
Shepherd, C.J. (1975). *Search 6*, 484.
Singh, G. (1982). In: *A History of Australasian Vegetation*. (J.M.B. Smith, ed.) Chapter 4, p. 980. McGraw Hill, Sydney.
Smith, J.M.B. (1982). *Establishment of exotic woody plants along roadsides in New England (NSW)*. Dept. Geography, University of New England, Armidale.
Smith, J.M.B. (1982b). *Search 13*, 207.
Smith, J.M.B. (1985). *The Australian Geographer — in press*.
Specht, R.L. (1981a). In: *Ecological Biogeography of Australia*. (A. Keast, ed.) Chapter 8, p. 318. Dr W. Junk, The Hague.
Specht, R.L. (1981b). In: *Ecological Biogeography of Australia*. (A. Keast, ed.) Chapter 9, p. 357. Dr W. Junk, The Hague.
Staley, E. and Benveniste, G. (1959). *Possible non military scientific developments and their potential impact on foreign policy problems of the United States: a study prepared for the Committee on Foreign Relations*, U.S. Senate (36th Congress), Washington.
Thomas, A.M. et al (Committee Members) (1979). *The Adequacy of Quarantine and other control measures to protect Australia's pastoral industries from the introduction and spread of exotic livestock and plant diseases*. Report by Senate Standing Committee in National Resources. Govt. Printer, Canberra.
Thorne, A.G. (1981). In: *Ecological Biogeography of Australia*. (A. Keast, ed.) Chapter 62, p. 1749. Dr W. Junk, The Hague.
Vavilov, N.I. (1951). *Chronica Botanica 13*, 364.
Vedros, N. (1984). Broadcast on ABC Science Show. (R. Williams, ed.) 12 May 1984.
Wace, N.M. (1965). In: *Biogeography and Ecology in Antarctica*. (P. van Oye and J. van Mieghan, eds) Chapter 6, p. 201. Dr W. Junk, The Hague.
Wace, N.M. (1968). *Austr. J. Sci. 31*, 189.
Wace, N.M. (1978). In: *Biology and Quaternary Environments*. (D. Walker and J.C. Guppy, eds) Australian Academy of Science, Canberra.
Wace, N.M. (1979). *Proc. Ecol. Soc. Australia 10*, 167.
Wace, N.M. (1980). In: *Of Time and Place*. (J.N. Jennings and G.J.R. Linge, eds) Chapter 2, p. 28. A.N.U. Press, Canberra.
Wace, N.M. (1983). In: *The Best of the Science Show*. (R. Williams, ed.) Chapter 9, pp. 70-79. Thomas Nelson, Sydney.
Weste, G. and Taylor, P. (1971). *Austr. J. Bot. 19*, 281.
White, N.H. (1981). In: *Plants and Man in Australia*. (D.J. and S.G.M. Carr, eds) P. 42. Academic Press, Sydney.

Chapter Two
EXOTIC HUMAN DISEASES
H. Newton-John

Peter lived in Zambia and had worked on a tobacco farm for twenty years. He decided to visit relatives in Australia, and arrived in Melbourne on 3rd April 1984. He became ill *en route*, and was admitted to Fairfield Hospital six days after the onset of chills, headache, vomiting, muscle pains and cough.

Malaria parasites were found in his blood, and treatment for this disease was started immediately. However, because he had developed a febrile illness within three weeks of arrival in Australia from rural Africa, he belonged to a category of people classified as being at moderate to high risk of having viral haemorrhagic fever (VHF). It is well known that this infection can co-exist with malaria (Commonwealth Department of Health, 1980).

From that point until his final release from hospital, he was managed in strict isolation as a potential case of VHF. This entailed spending eight days inside a large air-conditioned plastic bag, known as a 'bed isolator'. This equipment is designed to prevent any virus escaping into the working environment of the nurses and doctors caring for the patient.

In fact, Peter did not have VHF. Tests for antibodies against Lassa fever, Marburg and Ebola virus disease, Rift-Valley fever and Crimean–Congo haemorrhagic fever were all negative on the fourteenth day of his illness. Nevertheless, the fact that such stringent isolation precautions were considered necessary does highlight both the vulnerability of Australia to the importation of exotic human pathogens and also the extreme precautions which are needed to ensure the safety of those caring for patients with such infections.

An exotic infection is one caused by a pathogen which does not occur in Australia unless imported, and to which the population possesses little or no natural immunity. To most people, the idea of acquiring an exotic infection is alarming; we have not developed the contempt borne of familiarity. In this paper, I will examine the following issues:

1. Which exotic infections pose a threat to this country?
2. What are the risks of a 'plague' occurring as a result of the importation of an exotic pathogen?

WHICH INFECTIONS POSE A THREAT?
Any infection which can be acquired by an Australian who is travelling overseas poses a threat to that individual. In nearly all cases, even though the individual concerned may have an unpleasant illness, he/she will recover without ill effects. In addition, in nearly every case, there will be no spread of the pathogen to other people, and the disease will not become established in Australia. We will not discuss the many exotic infections which affect individuals in this way.

Our concern is with diseases which have one or more of the following characteristics:

1. The pathogen can be transmitted directly from person to person.
2. The pathogen spreads via an insect or animal vector which is found in Australia.
3. The disease is fatal in a significant proportion of infected persons and/or no vaccine is available for the disease.

These are characteristics which make an exotic disease justifiably frightening, but fortunately are shared by a relatively small number of such infections.

An additional feature of many exotic diseases is that their incubation period is often longer than just a few days. The implication of this is that a well person harboring an exotic pathogen, and perhaps able to transmit that pathogen, can move freely within Australia and have contact with a large number of susceptible individuals before the diagnosis is first considered.

Quarantinable infections

These are internationally recognised diseases which have an established reputation for serious human illness and usually a potential for spread; international notification is required. They are:

Cholera
Ebola virus disease
Epidemic typhus
Korean haemorrhagic fever
Lassa fever
Leprosy
Marburg virus disease
Plague
Yellow fever

Smallpox was on the list until 1979 when the World Health Organization officially declared the world free of the disease; a landmark in international co-operation. Leprosy poses no threat of spread and perhaps could be left off the list.

Australia has been free of all these diseases, except cholera and leprosy, although plague did occur earlier this century as a result of poor rat control around our major ports. Small cholera outbreaks have occurred in the last 15 years among tourists returning by aeroplane, due to contaminated food provided unwittingly by the airlines. Fortunately, and predictably, there were no secondary cases at all, from a total of 36 primary cases.

Modes of Spread

This brings us to a discussion of the factors which determine whether a disease (such as cholera) introduced in this way is *likely* to spread, for an infectious agent (bacteria, virus or protozoon) can be transmitted from person to person in several ways:

1. *Direct* person to person spread.

   This is the most dangerous, and can occur by aerosol (airborne infective particles), sexual transmission or by inoculation of blood (by accidental needle prick or through a skin cut or abrasion).

   Of these, aerosol spread is by far the most likely to lead to a widespread epidemic. Sexual spread (such as with A.I.D.S.) can certainly cause an epidemic, but transmission is confined to a restricted age group and is not as indiscriminate as aerosol spread disease for obvious reasons. Blood which contains infectious agents, such as the viruses of VHF, may be extremely dangerous to health care workers, but does not pose any risk of epidemic disease.

2. *Indirect spread* by biting *vectors* such as insects.

   In Australia, both dengue fever and malaria, which are mosquito spread, have the potential for establishment in the north where the vectors can flourish. Indeed, dengue has recently become endemic on the Queensland coast once again, but the disease has fortunately remained fairly mild and no deaths have been reported. While such infections are of great significance, their epidemiology, diagnosis and treatment are well known, and this reduces the anxiety associated with their introduction. It is also true that because there is *no direct* man to man transmission of vector-spread diseases (yellow fever, dengue and malaria are good examples) there is no chance of an epidemic occurring throughout the country. These diseases can only occur where the insect vector occurs. Recent surveys have shown for example that *Aedes aegypti*, the specific vector for yellow fever and dengue viruses, is absent in N.S.W. (Cooper et al., 1982) (Environmental Health Branch Dept, 1984).

A *zoonosis* is a disease transmitted from vertebrates to man. Good examples are psittacosis (a respiratory disease of man acquired by aerosol from infected birds) and rabies (see Anderson, this volume). For a zoonosis to have any chance of spreading in Australia, there must exist a suitable vertebrate host (such as a rodent) which lives in proximity to humans. Epidemics of a zoonosis are likely only in countries where there is a large reservoir of infected animals, and where there are opportunities for the disease to spread from animals to man.

Rift Valley Fever (Burnet and White, 1972) is a good example. This is a viral infection which occurs in Africa, recently especially in Egypt. It is mainly spread by mosquitoes from its animal hosts (ruminants such as sheep, camels and others) to man; there is no evidence for man to man transmission (Anon, 1980). Though there have been large outbreaks of human disease in Africa associated with epizootics amongst the animal hosts, there is little chance of this ever happening in Australia. There is no doubt that the virus could be introduced here, yet control measures would make human outbreaks extremely unlikely.

For an exotic zoonosis to be able to cause significant human disease in this country, it must first become established. Could this happen?

The answer is a qualified 'yes'. Rabies is one of our biggest fears, because of its potential for serious human disease (Kaplan, 1977). Rabies virus causes a form of encephalitis, which has a case fatality rate of very nearly 100%. It is preventable in both animals and man by immunization. In Europe the animal reservoir is the red fox; in southern U.S.A. increasingly it is the raccoon. In Australia the red fox has become very common, so if rabies virus were introduced it could readily become established.

Korean haemorrhagic fever (KHF) is another zoonosis with a potential for establishment (Anon, 1982). This virus infection, also known as 'haemorrhagic fever with renal syndrome', occurs widely throughout

Europe and Asia. Human disease may be serious, although rarely fatal, and is characterised by acute kidney failure. The virus is spread by rats and other rodents. The rat has been shown to shed the virus in urine, which may accidentally infect man by aerosol. Two out of ten rats caught in the Melbourne dockyard area in recent months have been shown to have antibodies to the virus which causes KHF (Hantaan virus). While this virus will probably become established here eventually, it is unlikely to spread widely because of the restricted habitat of the main rodent vector.

The viral haemorrhagic fevers

Viral haemorrhagic fever is the name given to a group of viral diseases all of which may cause a fatal human illness characterised by bleeding. All have an animal reservoir and are therefore zoonoses, although the reservoir has not been determined for all.

VHFs have probably caused more panic in recent years than all the other exotic diseases put together. This can be illustrated by quoting the blurb from the condensed (*Readers Digest*) version of John Fuller's book *'Fever'*, written about the Lassa fever scare of 1969:

> "The threat of plague has always been a terrifying one, but in modern times the threat has seemed distant, even unreal. Yet in the hinterlands of South America and Africa, and in certain other remote regions, lurk lethal viruses, each fully capable of triggering a world-wide calamity".

What are the viral haemorrhagic fevers and how real is their threat?

There are only four VHFs which are known to be able to spread from person to person. They are:

1. *Crimean–Congo haemorrhagic fever*, which has so far not been found outside the USSR and Africa (Anon, 1982);

2. *Marburg virus disease.* Named for the town in West Germany where this disease was first recognised, this infection is caused by a new type of virus which has been classified provisionally as a filovirus. In 1967 a total of 31 people in Marburg as well as in Belgrade, Yugoslavia, became ill with a severe febrile illness. Twenty-five were laboratory workers who were in contact with the organs, tissues or blood from African green monkeys, *Cercopithecus aethiops*. These had been imported from Uganda. The other six people were close contacts of the primary cases. It is interesting in retrospect that this species of monkey does not appear to be the natural reservoir of infection, and may also have been accidentally infected from an unknown animal source. Twenty of the cases occurred amongst the personnel of one Marburg pharmaceutical company which was producing cell lines for vaccine production; five of them died.

    Since then there have been very few further cases. In 1975 an Australian man, aged 20, died in Johannesburg, having caught Marburg virus infection on a hike through Rhodesia. His girlfriend and a nurse looking after him also became ill but recovered. The last reported single case was in South Africa in 1982. A total of nine deaths have occurred in the 37 described cases (25% fatality).

3. *Lassa Fever.* Fear about VHF probably began in U.S.A. after Lassa fever developed unexpectedly in two laboratory workers. They were employed in a laboratory where viruses were being isolated from specimens from patients who had developed the disease in Africa in 1969. The story of Lassa fever has since become widely known and will only be summarised here. Three nurses in Nigeria became ill in 1969 with an unexplained serious illness, and two died. Since then there have been a total of 15 outbreaks of Lassa fever associated with secondary cases in hospital personnel, including doctors, nurses and laboratory staff. In these outbreaks, which have all occurred in sub-Saharan Africa, many infected individuals did not in fact become very ill. The case fatality rates of sick patients admitted to hospital (as high as 52%) gives a false idea of the severity of the illness; the actual case fatality rate in community-acquired cases has been calculated to be closer to 3%.

    The disease is a zoonosis, with the multimammate rat, *Mastomys natalensis,* the principal reservoir. However, other peridomestic rodents appear also to be able to act as hosts. Infected rodent urine is probably the source for human beings.

    The disease appears to be more severe if transmitted by blood (as in hospital-acquired cases, where accidental inoculation is a frequent event), because a very large dose of virus may be introduced by this route. This is important to realise in the Australian setting, as secondary cases in laboratory workers could prove to be more severe than any primary case. Aerosol spread appears to play little if any role in human to human transmission (see below).

4. *Ebola virus infection.* In 1976 Sudan and Zaire experienced devastating outbreaks of a new haemorrhagic disease caused by Ebola virus. Over 500 cases occurred, with case fatality rates of 53% and 88% respectively. Subsequently studies suggested that the virus was endemic in these two countries as well as Kenya and the Central African Republic.

    The disease is similar clinically to Marburg virus disease, although it may be more severe. There is evidence of person-to-person spread, but not by aerosol; again inoculation of infective blood seems a particularly important route of infection.

Is there any evidence for spread of VHF outside Africa?

The only VHFs which have so far caused disease in people outside Africa are Lassa fever, a single laboratory-acquired case of Ebola fever in Great Britain in 1976 (no spread occurred) and the original Marburg outbreak. There have, up to this time, been at least eight incidents involving exotic importation of Lassa fever around the world. In four of these, the possibility of the patient having VHF was not considered initially, and no barrier nursing techniques were used to prevent spread. the best documented example was in St. Thomas' Hospital, London, in 1981; there were 173 contacts of a patient who ultimately proved to have Lassa fever (M.M.W.R. Supp., 1983). Despite the lack of precautions, *not a single person* developed the disease from such contact.

The evidence therefore strongly suggests that the viruses of VHF are not readily transmitted by aerosol, and that the main risk is to health workers concerned with the direct care of the patient, or to laboratory staff handling blood or virological specimens.

Therefore we can summarise the risks to Australia of VHF as follows:

1. There is little risk that one of the viruses could be introduced into a local animal reservoir, so establishment is most unlikely.
2. Human epidemic disease would not occur.
3. There is a real risk to health care and laboratory staff involved with a patient who has VHF.

There is some recent evidence that, at least with Lassa fever, a new antiviral drug called ribavirin holds some promise in treatment (Anon, 1982). This knowledge should help reduce the level of anxiety amongst health workers caring for patients with VHF from Africa.

PERSPECTIVE ON 'PLAGUES' IN AUSTRALIA

A plague can be defined as a contagious disease with epidemic potential and a fearsome reputation. Inherent in the concept of a plague are problems with control and treatment of the disease. None of the exotic infections I have mentioned so far, with the exception of A.I.D.S. (see Chapter 19, this volume) really fits this definition as far as Australia is concerned.

There are, however, a number of infections which regularly occur in Australia which fit the definition of a plague rather well. The sexually transmitted disease, genital herpes is a good example. It is extremely prevalent, control is almost impossible, it certainly has an alarming reputation, and there is no safe curative treatment and no vaccine. However, the best example of all is influenza. This virus has probably killed more people than all the other exotic viruses I have mentioned put together. In the great pandemic of 1918–19, an estimated 20 million people died, many of them young. Influenza is so well known that we casually talk of having 'the flu' when all we have is a bad cold. However, influenza A, the most important member of the group, is in fact an exotic pathogen.

Influenza as an exotic pathogen

Influenza virus is the most important of all the exotic viruses for Australia and yet we fear it least because it has become familiar. Yet why do I consider it to be an exotic virus?

Influenza A is notorious for its ability to alter genetically, and so present an entirely 'new face' to the world. Genetic recombination between human and animal strains of the virus has been postulated to occur in nature (Schild, 1984). Animal and human strains of the virus are certainly very closely related; both avian and swine strains have a genetic structure very close to that of human viruses. 'New' strains of the virus often appear to originate in Asia, perhaps in China, and spread from there all around the world. Pandemics occur every 10–15 years and result in attack rates of over 25% of the population. Many old people, and some young ones, die of pneumonia as a result. The economic effects are enormous. A quick calculation shows that if one quarter of the Australian workforce were off sick with influenza for five working days, the cost in lost time could exceed $500,000,000.

We still have no effective way of combating this infection. The current vaccines are pitifully ineffective in the face of a pandemic, although they do save lives. Influenza remains our greatest exotic disease challenge.

CURRENT STRATEGY FOR CONTAINMENT OF EXOTIC PATHOGENS

Lassa, Marburg and Ebola virus infections are quarantinable diseases. Australia has a central quarantine unit in Melbourne, and patients suspected of having VHF are flown from the point of diagnosis to Melbourne using a specially trained team and 'transit isolator' to contain the patient. The patient is then transferred to a high security unit for medical and nursing care. One of the subsidiary aims of such a contingency plan are to allay anxiety and to protect staff. I believe that there is no real risk to the community from VHF.

It is ironic that the virus of A.I.D.S., also an exotic pathogen from Africa, should have neatly bypassed this net and become a true plague under our very noses. There is no effective containment strategy yet for this alarming disease.

HOW SECURE ARE WE?

The *status quo* is unlikely to be altered unless there is a major sociopolitical upheaval (such as war), the unexpected emergence of a completely new pathogen, or the breakdown of our vigilance and our surveillance systems. New pathogens can and do emerge, so we must be very cautious before we go along with Burnet and White, who stated in 1972:

"The most likely forecast about the future of infectious disease is that it will be very dull".

REFERENCES

Anonymous. Workshop proceedings (1980). *Rift Valley Fever*, Herzlia, Israel.
   Anonymous. Muroid virus Nephropathies (1982). *Lancet*, 2: 1375–1377.
Burnet, M. and White, D.O. (1972). *Natural History of Infectious Diseases*. 4th ed., Cambridge, p. 263.
Commonwealth Department of Health (1980). Handbook on measures to control Marburg Virus Disease, Ebola Virus Disease or Lassa Fever. Australian Govt. Publishing Service, Canberra.
Cooper, C.B., Gransden, W.R. and Webster, M. (1982). *A case of Lassa Fever. Brit. Med. J.*, pp. 1003–1005.
Environmental Health Branch Dept (1984). *Aedes aegypti surveillance,* N.S.W. Communicable Diseases Intelligence, Canberra, 5.
Kaplan, C. (ed.). *Rabies, the Facts,* Oxford, 1977.
M.M.W.R. Supp. (1983)., *Viral haemorrhagic fever*, 32: 25.
Schild, G.C. (1984). 'Influenza'. In: Brown, F., Wilson, G. (eds). Principles of bacteriology, virology and immunity, 7th ed., London. W. Arnold, Vol. 4: (96) pp. 315–322.

Chapter Three

EXOTIC DISEASES OF ANIMALS

Roger Meischke and Bill Geering

INTRODUCTION

Australia is free of more than 40 major infectious or parasitic diseases of animals. They are listed at Table 1. These diseases have the potential to cause serious production losses to livestock industries, decimate exports of livestock and livestock products and/or cause major public health problems should they penetrate our quarantine barriers.

A number of these diseases are discussed briefly in this paper. These particular diseases have been selected for two reasons. Firstly, they are probably some of the most serious diseases with which we would have to contend. Secondly, between them they cover a wide spectrum of types of disease, risks of entry to Australia, and challenges for eradication.

TABLE 1.

African horsesickness	Jembrana disease
African swine fever	Louping ill*
Aujeszky's disease	Lumpy skin disease
Avian influenza	Maedi-visna
Bluetongue (and related viruses)	Nairobi sheep disease
Borna disease	Newcastle disease
Brucella melitensis infection	Peste des petits ruminants
Canine brucellosis	Pulmonary adenomatosis (Jaagsiekte)
Chagas' disease*	Rabies
Swine fever	Rift Valley fever
Bovine pleuroneumonia	Rinderpest
Dourine	Scrapie
Duck viral hepatitis	Screw-worm fly
Duck viral enteritis	Sheep pox
East Coast fever	Sheep scab
Epizootic lymphangitis	Surra
Equine babesiosis	Swine influenza
Equine influenza	Transmissible gastro-enteritis
Equine viral arteritis	Trichinosis
Equine viral encephalomyelitis	Tropical canine pancytopaenia
Glanders	Vesicular diseases
Haemorrhagic septicemia	Warble fly
Heartwater*	Wesselsbron disease
Japanese encephalitis	

* These diseases may present in a very attenuated form as suitable vector species may not occur in Australia.

FOOT-AND-MOUTH DISEASE

Foot-and-mouth disease (FMD) is a highly contagious virus disease of cloven-hoofed animals. Although not often lethal it causes serious production losses and is a major constraint to international trade in livestock and livestock products.

Aetiology

FMD viruses constitute the Aphtovirus genus of the Picornaviridae. It is an RNA virus with a naked nucleocapsid, approximately 25 nm in diameter, with icosahedral symetry, and composed of four major structural polypeptides.

There are seven serotypes, O, A, C, SAT1, SAT2, SAT3 and Asia 1, differentiated by complement fixation and serum neutralisation tests. There is no cross-immunity between the serotypes. Each type is further divided into a number of subtypes on the basis of quantitative rather than qualitative differences in serology and cross-protection.

FMD virus is stable at low temperatures, but is very rapidly inactivated above 50°C (although residual fractions with greater heat resistance may survive). It is sensitive to direct sunlight and desiccation. It is also sensitive to acid and alkaline conditions, being rapidly and completely inactivated below pH4 or above pH11.

In infected carcases, the virus is rapidly inactivated in muscle tissue, through acidification processes, but may persist for long periods in blood, bone marrow, lymph glands and viscera. The virus can survive for up to several weeks in fomites if protected from sunlight and drying.

Hosts

Cattle, buffaloes, pigs, sheep, goats, antelopes, deer and several other wildlife species.

World Distribution

Africa, several countries in Europe, the Middle East, Asia and South America. The incidence and geographical distribution of the disease in Europe has declined very substantially in the last 20 years, through national vaccination campaigns. In the South-East Asian region, FMD is endemic in Burma, Thailand, Laos, Kampuchea, Vietnam and the Philippines. Malaysia has had long periods of freedom, but had several outbreaks in the period 1978–82. A national eradication campaign was carried out in Indonesia from 1974–82 with Australian assistance. There were no reported outbreaks from January 1979 to July 1983, but a serious outbreak of type O has since occurred in Java.

Occurrences in Australia

There is a report of a localised outbreak near Melbourne in 1872 however it is likely that this was not FMD.

Epidemiology

Infected animals may excrete virus up to several days before the appearance of vesicles. Virus is excreted in the exhaled air, all secretions and excretions and from ruptured lesions. Direct contact is a most important method of transmission. The movement of animals through saleyards may lead to rapid spread of infection over long distances.

Transmission by indirect contact is also important. This may be through such things as contaminated yards and buildings, trucks, and equipment. Bulk milk tankers, unless equipped with filters, have been incriminated in spread of infection amongst dairy herds. Infection may be carried from farm to farm by persons through contaminated clothing and equipment.

Pigs are very susceptible to infection by ingestion, and many outbreaks have been started by feeding pigs on swill containing contaminated food scraps. As pigs excrete large amounts of virus by the respiratory route, they are an important amplifying host.

Aerosol transmission occurs. Under some climatic conditions infected aerosols can be carried for long distances on the wind.

A proportion of clinically recovered cattle and sheep continue to carry the virus in their pharynx and oesophagus for long periods. The epidemiological significance of such carriers is uncertain.

Clinical Signs

In cattle, after an incubation period of 2 to 8 days, the first signs are a fever (40–42°C), severe depression, anorexia and a sharp drop in milk production. An acute, painful stomatitis is quickly followed by the appearance of vesicles on the tongue, dental pad and buccal mucosa. These vesicles, which contain a thin straw-coloured fluid, increase rapidly in size and rupture within 24 hours leaving a raw painful surface. This is accompanied by excess salivation and lip smacking. Vesicles also occasionally appear on the muzzle and inside the nostrils. Tongue lesions heal fairly rapidly and eating may be resumed in a few days.

Concurrently with the appearance of the oral lesions, vesicles appear on 1 or more feet. These occur particularly in the interdigital cleft, on the coronet and bulbs of the heel. They rupture within 24 hours and this is accompanied by acute lameness and perhaps recumbency. Secondary bacterial infection may lead to severe involvement of the deeper structures of the foot.

Vesicles may also appear on the udder and teats, often leading to severe mastitis and permanent damage of the teats. Pregnant cattle may abort in the acute stage.

The mortality rate in adult cattle is usually less than 2%. There is a prolonged convalescent period — up to 6 months.

Up to 20% of young calves may die suddenly, without premonitory signs, as a result of cardiac lesions.

Pathology

In addition to the external lesions already described, vesicular lesions may also be found on the ruminal pillars. In young calves there may be focal necrosis of cardiac and skeletal muscle tissues. Cardiac lesions may give rise to a 'tiger heart' appearance.

Differential Diagnosis
- Vesicular stomatitis
- Swine vesicular disease
- Vesicular exanthema

These vesicular diseases cannot be differentiated on clinical grounds, although the last two only infect pigs. Any vesicular disease should be treated as FMD until proven otherwise.

After FMD vesicles rupture and the epithelial covering is lost, the lesions appear as erosions or ulcers that are gradually covered by a fibrinous coating. At this stage the vesicular diseases may be confused with other diseases such as mucosal disease, rinderpest, bluetongue, and papular stomatitis. Ruptured foot lesions with secondary bacterial infection may be confused with footrot.

Diagnosis

If possible collection of diagnostic specimens should be left to an official diagnostic team.
The specimens required are:
- Vesicular fluid
- Epithelial covering of unruptured vesicles and epithelial tags from ruptured lesions
- Serum and whole blood samples
- Oesophageal-pharyngeal fluid obtained with a cup probang
- Specimens of lymph nodes, thyroid, adrenal, kidney and heart at autopsy

Fresh tissues should be despatched to the laboratory preserved in glycerol buffer or frozen.
Laboratory diagnosis is based on:
- Complement fixation tests to detect antigen directly from lesion material or on passaged virus
- Virus isolation in tissue culture or laboratory animals
- Serological tests

In some cases animal transmission tests may also be carried out on the suspect property.

Eradication

Contingency plans have been prepared for control of an outbreak of any of the vesicular diseases (including FMD) in Australia. The policy is for eradication as quickly as possible. This would be done by the conventional methods of quarantine, slaughter, disinfection and movement controls.

RINDERPEST

Rinderpest is an acute, highly contagious virus disease which is known to have caused devastating livestock losses for at least the last 1500 years. However, it has been brought under better control in many regions with the advent of efficient vaccines.

Aetiology

Rinderpest virus belongs to the Morbillivirus genus of the Paramyxoviridae. It is closely related serologically and in its physico-chemical properties to canine distemper and measles viruses. It is also closely related, although not identical, to the virus of peste-des-petits ruminants.

There is only one serotype of rinderpest virus. However, virus strains may vary considerably in their virulence.

Hosts

Rinderpest is primarily a disease of cattle and buffaloes, although many cloven hoofed animal species are susceptible to a degree. Pigs are an important host in south-east Asia. Rinderpest occurs on rare occasions in sheep and goats.

World Distribution

The remaining strongholds of rinderpest in Africa are in West Africa and in the Sudan, Ethiopia and Somalia region. The incidence of the disease has increased in West Africa in the last five years. The disease extends through several countries of the Middle East through to the Indian sub-continent. The only known infected countries in south-east Asia are Laos and Kampuchea but information from the Indo-China region is rather sketchy.

A limited outbreak occurred in 1923 in the Fremantle district of Western Australia, introduced by cattle offloaded from a ship that had also carried pigs from an Asian port. The disease was quickly eradicated by quarantine, slaughter and disinfection measures.

Epidemiology

Most strains of rinderpest virus are highly contagious. The disease is usually spread by direct contact between animals. The virus is excreted in the expired air, nasal discharges, saliva, faeces and urine. Pigs may become infected through eating contaminated offal.

Rinderpest virus survives poorly in the environment and indirect transmission such as through contaminated feedstuffs and water is unimportant.

Major epidemics occur when infected animals are introduced to susceptible populations in rinderpest free areas. In endemic areas the disease tends to spread slowly, affecting mainly younger animals, with mild epidemics occurring when nomadic herds mingle at watering places.

The incubation period usually ranges from 3–9 days, but may be longer in cattle with a high innate resistance. The clinical syndrome varies from peracute to subacute.

For peracute cases there is sudden fever, collapse and death within a few days. Rarely, there are nervous signs. In the acute disease, there is a sudden onset of fever that lasts 2 to 7 days. Clinical signs appear a day or so after the onset of fever and these at first include depression, restlessness, partial loss of appetite, nasal discharges which are initially serous but later muco-purulent, congested mucous membranes, shallow, rapid respiration and constipation. Three to four days after the onset of fever lesions appear on the mucosae of the mouth, nostrils and urogenital tract, firstly as raised necrotic pin-points which rapidly enlarge and coalesce. The necrotic tissue sloughs leaving irregular, well demarcated, shallow erosions. At this stage the animal is very obviously ill, its breath foetid, and respirations are laboured with a characteristic grunting expiration. Diarrhoea commences a day or two after the appearance of mucosal lesions. The fluid faeces are profuse, dark and foetid and may contain mucus, blood and fragments of necrotic mucosa. Straining is severe and frequent. There is a rapid dehydration, collapse and death. Some animals recover but there is a prolonged convalescence.

Subacute cases are generally seen in endemic areas, but could occur in susceptible populations with the introduction of lower virulence virus strains. Varying combinations of the above clinical signs may occur, but they are milder.

Pathology

The carcase is dehydrated, emaciated, soiled and foetid. Mucosal erosions are present in the mouth, pharynx and anterior oesophagus. Congestion, ulcerations and haemorrhages are prominent in the abomasum and small and large intestines. 'Tiger striping' of the colonic mucous membranes is characteristic. Peyers' patches are prominent, being swollen, haemorrhagic and necrotic.

Mucosal erosions may also be present in the vagina, vulva or prepuce. The turbinates are coated with a thick muco-purulent discharge and the trachea commonly contains longitudinal streaks of congestion.

Diagnosis
- Histopathogy
- Virus isolation in tissue culture from blood, lymph node or spleen specimens
- Identification of antigen in lymph nodes or necrotic mucosa by agar gel diffusion or complement fixation tests

Control

Contingency plans have been prepared for the eradication of an outbreak of rinderpest in Australia. The principles are similar to that in the vesicular diseases plan and are based on slaughter, quarantine, movement controls and disinfection. Vaccine buffer zones may be used in remote areas.

Tissue culture attenuated vaccines have proven remarkably effective. They are safe in all cattle breeds and provide a durable immunity.

SCREW-WORM FLY

Screw-worm flies (SWF) are obligate parasites of all warm-blooded animals, including man. They present the most direct threat of all the major exotic animal diseases because of their proximity to Australia.

Aetiology

There are two SWF — the old world SWF *Chrysomya bezziana* and the new world SWF *Cochliomyia hominivorax*. These species belong to different genera and have apparently evolved independently in different parts of the world, but are very similar in their biological characteristics.

Hosts

Freshly wounded animals of all species. Although strikes have been reported in all domestic animals, the greatest economic losses are in cattle and sheep.

World Distribution

Chrysomya bezziana — sub Saharan Africa, the Middle East, Indian sub-continent, south-east Asia, Papua New Guinea. The fly is distributed throughout PNG and recent surveys have demonstrated significant populations in the Papuan swamplands adjacent to Torres Strait.

Cochliomyia hominivorax — Mexico, Central and South America. The fly has virtually been eliminated from USA.

Epidemiology

The adult flies prefer well-wooded riverine areas and moist well shaded country. The optimum temperature range for the fly is 20–30°C. The flies will survive in hot and dry conditions providing shade is available.

The female fly mates only once, during the first four days of life. Following mating, it is attracted to fresh wounds on animals, which may have resulted from trauma, dehorning, castration, branding, and navels of new born animals.

Egg masses are laid in the edges of wounds. The larvae which hatch within 12–24 hours, enter the wound and feed on the underlying tissues. Tissue destruction results in deep penetration by the larvae. Larval development is completed within 5–7 days after which the larvae drop off the host. The larvae penetrate into the soil and there pupate. Adult flies emerge after about a week.

If SWF became established in Australia, it would have a year round range in the wetter areas of the Kimberleys, Northern Territory and Queensland but in favourable summers it could extend throughout the sheep raising areas of eastern and southern Australia. It has been estimated by the Bureau of Animal Health that SWF, if allowed to spread unchecked in Australia, could result in economic losses of $65–155 million annually to the sheep and cattle industries.

Clinical Signs

Strikes occur anywhere on the body. Shearing, marking, dipping and fighting wounds provide excellent sites. Sheep are also frequently struck in the infra-orbital fossa. SWF strike wounds are quite deep and are foul smelling. The larvae are not very obvious. The wounds generally heal rapidly when the larvae are removed. However, if left untreated the wounds may be struck repeatedly resulting in debility or even death of the animal. Deaths may also occur in new-born calves and lambs as a result of navel strike. Strike in the genital area may lead to sterility in both sexes through disfigurement.

Diagnosis

Larvae from wounds should be preserved in 70% alcohol or methylated spirits and sent to the State diagnostic laboratory or a specialised entomology laboratory for identification.

Traps, using specific SWF attractants, can be used for monitoring SWF activity in the region. A surveillance program, based on such traps, has been established to detect a possible entry of SWF into northern Australia.

Control

SWF strikes can be treated by local application of ointments containing insecticides, such as coumaphos ("SWF grease"). In endemic countries animals are prophylactically treated at times of wounding such as castration, dehorning, and navels of new-born animals.

The sterile insect release method (SIRM) has been very successfully used in eradication campaigns against *Cochliomyia hominivorax* in North America. Field trials carried out by the CSIRO Division of Entomology using the SIRM technique for *Chrysomya bezziana* have given promising results.

The Commonwealth Government is taking steps to maintain the capability to mount a SIRM campaign at short notice in the case of an outbreak in Australia.

BLUETONGUE

Although the presence of bluetongue virus infection of cattle and buffaloes was discovered in northern Australia in 1977, there has been no evidence of clinical disease in any livestock species in the field in Australia associated with this virus. Virulent bluetongue disease should therefore be still regarded as exotic to Australia and precautions taken accordingly.

Aetiology

Bluetongue virus is a member of the orbivirus genus of the Reoviridae. In common with other orbiviruses it is an arthropod-borne virus. There are at least 20 serotypes of blutongue virus. There are very considerable differences in virulence between strains of bluetongue virus, and this is not necessarily related to serotype.

Hosts

Sheep, goats, cattle, buffaloes, deer and antelopes.

World Distribution
Africa, Middle East, Indian sub-continent, USA and Mexico. Epidemics have also occurred in Spain and Portugal but the disease has disappeared from these countries. There is also serological evidence of infection in south-east Asia (including Papua–New Guinea) and Brazil.

Occurrences in Australia
Five serotypes of bluetongue virus have thus far been isolated in Australia. The first, CSIRO 19 virus, was isolated from Culicoides insects collected near Darwin in 1975 and identified in 1977 as bluetongue virus of a new serotype (serotype 20 or BTV 20). The other two viruses, CSIRO 154 and 156, were isolated from healthy sentinel cattle in the Northern Territory in 1979. CSIRO 156 virus has been found to belong to bluetongue serotype 1 (BTV 1) but the other virus has yet to be fully typed. In summary 5 serotypes have been identified in Australia — BTV 1, 15, 20, 21, 23. BTV 15, 20, 23 occur in far north of WA, NT and Qld only BTV 1 and 2 extend as far south as NE corner of NSW.

Experimentally, each of the Australian bluetongue viruses has been shown to be only mildly to moderately pathogenic for sheep. In cattle they produce a febrile reaction and viraemia, but no clinical signs.

Serological surveys have indicated that BTV 20 has only been active in the top end of the Northern Territory over the last two years, but that it has also been active in the far northern areas of Western Australia and Queensland at some time past. CSIRO 154 and 156 viruses have a wider distribution that BTV 20, but both are restricted to the northern half of the continent.

The potential for further spread of bluetongue viruses in Australia is probably limited and this is determined by the distribution of the insect vectors. An as yet unnamed species of Culicoides (described as C. avaritia species no. 5) has been found to be the most competent vector of BTV 20. This species only occurs in the coastal flood plains of northern Australia.

As well as bluetongue virus, a number of other orbiviruses occur in Australia, and these cause inapparent infections of livestock and/or marsupials. The serological and biochemical relationships of these viruses to bluetongue and to each other are complex.

Epidemiology
Bluetongue virus is transmitted biologically by Culicoides insects. It is not transmitted by direct contact. In temperature areas (e.g. South Africa, USA) bluetongue is of seasonal occurrence, with outbreaks occurring in late summer and autumn but ceasing with the onset of winter frosts. The level of bluetongue activity is related to the population density of the insect vector and this in turn is influenced by environmental temperature and rainfall (the insects requiring warm, moist conditions). In tropical areas the virus is active throughout the year but is most active after seasonal rainy periods.

Infection is generally sub-clinical in cattle, but some cattle may remain virus carriers for long periods. Cattle are the main maintenance and amplifying hosts for the virus.

Clinical Signs
In sheep, after an incubation period of 4–7 days there is a variable and fluctuating febrile reaction which may last about a week. Within 24–26 days of the onset of fever, hyperaemia of the buccal and nasal mucosae become apparent, and this is accompanied by excess salivation and a clear nasal discharge. Over the next few days this discharge becomes muco-purulent and may be blood-stained.

In acute cases, over the next few days the lips and tongue become very swollen. This oedema may extend over the face to include the ears and inter-mandibular space. The hyperaemia becomes more intense and petechial haemorrhages appear on the oral, nasal and conjunctival mucosae. The tongue is discoloured to purplish-blue in only a small percentage of cases.

In 5–8 days after the onset of fever, necrotic ulcers develop in the gums, cheeks and tongue. These heal slowly under a diptheritic membrane. Breathing becomes difficult. A profuse haemorrhagic diarrhoea occurs in some cases.

Foot lesions (on 1–4 feet) appear towards the end of the febrile reaction. There is acute reddening and petechial haemorrhages on the coronary band. Because of pain the affected animal stands with an arched back and is very reluctant to move.

There is rapid emaciation, due to loss of appetite and specific muscular necrosis. Torticollis sometimes appears as a late sign.

The mortality rate is very variable but generally is of the order of 5–20%, although some virulent virus strains cause up to 70% mortality. Death may occur at any stage up to a month. There is a prolonged convalescence in surviving sheep.

It must be emphasised that the severity of the clinical disease is very variable.

Pathology
The external lesions have already been described. Internally, there is catarrhal inflammation of the intestinal tract with petechial haemorrhages in the mucosa. There is widespread hyperaemia, oedema and haemorrhages in other tissues. Haemorrhages in the tunica media at the base of the pulmonary artery are very characteristic.

The lymph nodes and spleen are moderately enlarged and haemorrhagic. Petechiae and pale areas of necrosis are scattered irregularly through the skeletal muscles. There is a catarrhal inflammation of the upper respiratory tract and odema of the lungs.

Diagnosis
Bluetongue virus may be isolated from whole blood collected from affected sheep during the febrile stage — this is inoculated into experimental sheep, tissue culture and eggs. Serological diagnosis can be made by agar gel diffusion, complement fixation (group tests) and serum neutralisation tests (serotype specific).

Control
Attenuated polyvalent vaccines are used in Africa and USA. A degree of prevention can also be achieved by management practices such as separation of sheep from cattle, stabling sheep at night or pasturing sheep on high ground.

BRUCELLA MELITENSIS INFECTION
Brucella melitensis causes abortions and mastitis in sheep and goats and is also pathogenic for man.

Aetiology
Brucella melitensis — of which there are three biotypes.

Hosts
Sheep, goats, man.

World Distribution
France, Italy, Malta, Spain, Portugal, Albania, Greece, Cyprus, Turkey, USSR, throughout Middle East to India, southern USA, Mexico, Latin America.

Occurrences in Australia
Never in livestock, but there are occasional human cases in migrants from endemic regions.

Epidemiology
Large numbers of organisms are excreted from the uro-genital tract for from 1–4 months after parturition or abortion. The method of infection is similar to that of *Brucella abortus* in cattle. Man may become infected by handling infected livestock or by consuming raw milk and cheese or inadequately cooked meat.

Clinical Signs
Primary infection of a flock may result in a trickling of abortions at first but this usually progresses to a severe abortion storm. Abortions occur from the third month of pregnancy to term. Mastitis commonly occurs. Less common signs are unthriftiness, bronchitis, lameness, hygroma, and orchitis in rams.

Pathology
Significant lesions are usually absent but abscesses in the spleen and costochondral cartilage may occur.

Diagnosis
Isolation and identification of organisms from milk, vaginal discharge, placental and foetal tissues, and blood. A range of serological tests similar to that for *Brucella abortus* is available.

Control
The princples of control are similar to that for *Brucella abortus*. Both killed and live vaccines are used, and these give good immunity.

SCRAPIE
Scrapie is a progressive, and invariably fatal, degenerative disease of the central nervous system of sheep and goats. It is one of a group of classical slow virus diseases (which also includes Kuru and Creutzefeld-Jakob disease of man and transmissible mink encephalopathy) for which the nature of the aetiological agent has yet to be determined.

Aetiology
Scrapie is infectious in origin. It can be experimentally transmitted to sheep, goats, monkeys, mice and rats by inoculation of brain and other tissue suspensions from affected animals. The agent will also replicate in some cell culture systems.
The scrapie agent is not a typical virus. Infectivity is strongly associated with host cell membranes, but the agent has neither been purified nor visualised by electron microscopy. The agent does not stimulate may immune responses in the host.
The scrapie agent is very resistant to inactivation by physico-chemical treatments. These include wet or dry heating, irradiation, exposure to alkylating agents, organic solvents and many detergents.

The available evidence on the nature of the scrapie agent is consistent with there being a small scrapie specific nucleic acid (perhaps of the same size as viroid RNA), functionally associated with hydrophobic proteins to given an infectious complex which *in vivo* is located in cell membranes.

There is a school of thought which considers that scrapie is primarily a genetic rather than an infectious disease. There is certainly a strong familial pattern to scrapie in sheep. However, this is due to a combination of maternal transmission of the agent and genetic determinants from the parents influencing host resistance and the length of the incubation period.

Transmissible mink encephalopathy (TME) was recognised in USA after scrapie had been introduced to that country. It has been speculated that TME is scrapie that has been adapted to mink as a result of feeding infected sheep tissues to the latter.

Hosts
Sheep and goats are the only natural hosts.

World Distribution
Scrapie has been known in Europe for over 200 years. There is little accurate data on its prevalence because it is not a reportable disease in most countries. However, it is economically important in parts of the United Kingdom and continental Europe.

Scrapie was introduced to Canada, USA and India in the 1930's and 1940's and is still endemic in these countries. There have also been introductions to other countries such as South Africa, Kenya and New Zealand but it has been eradicated on each occasion. Spread of the disease to new countries has been associated with importation of sheep and goats for Europe, particularly the United Kingdom.

Occurrences in Australia
Scrapie occurred in a small group of Suffolk sheep imported from the United Kingdom and released on to a property near Melbourne in 1951. The disease was eradicated by slaughter of the imported and in-contact sheep, movement restrictions on sheep from the affected property (to slaughter only) and resting of affected paddocks for one year.

Epidemiology
Maternal transmission is the main method of spread. There is reasonable evidence that this can occur pre-natally but it is not known whether it is at the semen and ovum stage or later in embryonic life. Transmission from ewe to lamb during the early post-natal period certainly occurs. Foetal membranes from scrapie affected ewes contain appreciable amounts of the agent. Because of the long incubation period, ewes may have two or more lamb-crops before being recognised as infected. Contact transmission probably occurs. There is also strong evidence that sheep may become infected by grazing contaminated pastures. Because of its resistance to inactivation it is likely that the agent would remain viable in the environment for several years.

All breeds of sheep are susceptible to scrapie, but the prevalence varies in different lines and breeds of sheep and in different areas. In the United Kingdom, 10–15% of sheep may die before natural culling age in severely affected flocks.

The devising of safe methods for the importation of sheep from endemic countries is complicated by the fact that the disease may not become overt until the second or later generations.

Epidemiological evidence suggests there is no direct causal link between scrapie and human encephalopathies.

Clinical Signs
Scrapie has a very long incubation period — usually 1–3 years or longer. The clinical disease is most often seen in sheep 2–5 years of age, although occasionally cases occur in younger or older sheep.

The onset of clinical signs is insidious. There is no febrile reaction. Initially, affected sheep either lead or trail when the flock is driven. As the disease progresses the sheep becomes hyperexcitable, and tends to carry its head high, has a fixed stare, and runs with a high-stepping gait. There is inco-ordination and ataxia of the hind legs. A fine tremor or convulsions may occur when the animal is handled and when the sheep is rubbed over the loins it often makes vigorous chewing movements. In later stages sheep rub vigorously against fences and other fixtures, and nibble at the affected areas of skin, leading to wool loss and denudation of skin. Appetite is not greatly impaired and there is little loss of condition until late in the course. In the terminal stages there are pronounced nervous signs and debility.

Death may occur as early as 2 weeks or as long as several months after the onset of signs.

Pathology
There are no gross lesions. Scrapie is a non-inflammatory vacuolating encephalopathy. The vacuolation of neurones is bilaterally symetrical and is prominent in the medulla, pons and mid-brain. Single or multiple vacuoles may occur, often causing distention of the nerve cell. Although vacuolation may be found in cells which in other respects look normal, the surrounding cytoplasm usually shows signs of degeneration. Interstitial spongy degeneration is often found in the same areas as neuronal vacuolation.

Diagnosis

This can only be made on the basis of clinical signs and histopathology. There are no serological tests. Sheep incubating scrapie cannot be identified.

Control

No country in which scrapie has become well established has succeeded in eradicating the disease, despite intensive efforts by some countries. Breeding for genetic resistance against scrapie has generally been disappointing as genetic resistance appears to be specific for each strain of scrapie agent. The only practicable method of controlling scrapie in endemic areas at present is by culling programs. This may range from complete slaughter of affected flocks to slaughter of all bloodline relatives of scrapie cases to limit the effects of maternal transmission. It is also recommended that different lambing areas should be used each year, and that afterbirth should be removed promptly.

SHEEP POX

Sheep pox is a highly contagious and often fatal viral disease in which there are generalised skin lesions. It is the most severe of the pox diseases of domestic animals.

Aetiology

A poxvirus. Sheep pox virus is closely related to goat pox virus and there is substantial cross-protection between the two viruses. As with other poxviruses, sheep pox virus is resistant to inactivation. It can survive in dried scabs for many months, in unused sheep pens for six months and in pastures for two months.

Hosts

Sheep are the only natural hosts. There are considerable breed differences in susceptibility. Merinos are the most susceptible; coarse-wooled breeds are less susceptible and some Middle East breeds are very resistant. Goats and cattle can be experimentally infected.

World Distribution

Middle East, USSR, Indian sub-continent, north Africa, east Africa as far south as Kenya.

Occurrences in Australia

Never.

Epidemiology

Sheep pox is particularly prevalent in the nomadic sheep herding areas of the Middle East and north Africa where infection is transmitted by direct contact between sheep and indirectly by sheep entering contaminated sheep camps. Infection could also be spread by contaminated vehicles, utensils and other fomites. Biting insects are also thought to have a role in mechanical transmission of the disease.

Clinical Signs

After an incubation period of 4–8 days there is fever, lachrymation, swollen eyelids, and a mucous nasal discharge. A day or two later there is an intense inflammatory reaction in the dermis and eruption of lesions, particularly in the sparsely woolled areas of the skin — e.g. groin, scrotum, underneath the tail, udder, eyelids, lips, cheeks and nostrils. These lesions follow the typical progression of pocks — papules, micro-vesicles, pustules and finally scabs. They cause considerable irritation, scratching, loss of wool, inappetance and loss of condition. The course of the disease in uncomplicated cases is about 3–4 weeks and the mortality rate 5–10%.

There is a malignant form of sheep pox which is more prone to occur in lambs. In this there are severe and more generalised skin lesions which coalesce, as well as internal lesions in the gastro-intestinal and respiratory tracts. Secondary functions are common. The mortality rate in this form may be as high as 80%.

Pathology

The external lesions have been described. Internally, there may be haemorhagic inflammation and mucous membrane ulceration in the respiratory and intestinal tracts. Additionally there may be caseous nodules in the lungs and kidneys, lymphadenopathy, and petechial haemorrhages in most organs.

Diagnosis

Visualisation of pox virus in scab material by electron microscopy; isolation of virus in tissue culture or eggs; sheep transmission tests.

Attenuated live vaccines are used to control the sheep pox in endemic countries.

TRICHINOSIS

Trichinosis, caused by the nematode *Trichinella spiralis*, occurs in many parts of the world, but has not been reported in Australia. A large survey of pig carcases at Australian export abattoirs in 1975–76 did not reveal any evidence of the parasite.

The pig is the main host, but trichinosis also occurs in many other mammalian species, particularly omnivores and carnivores. Pigs tolerate infestation well and rarely exhibit clinical signs.

Man acquires trichinosis by eating raw or improperly cooked infested meat. In man, penetration of the larvae through the intestinal wall and their subsequent migration through tissues causes variable, and sometimes serious, clinical disease.

AUJESZKY'S DISEASE
(Synonyms: Pseudorabies, infectious bulbar paralysis)

Aetiology: Herpesvirus.

Natural hosts: Pigs, cattle, sheep, dogs, cats, mink, foxes, raccoons, rats.

Distribution: Europe, North and South America, Malaysia, Singapore, New Zealand, Samoa.

Epidemiology: Aujeszky's disease is highly contagious between pigs — outbreaks on farms usually originate from introduction of carrier pigs — rodents and wildlife may also be important in spreading infection — cases in cattle, sheep, cats usually sporadic and these species are terminal hosts although some evidence is accumulating that they may occasionally excrete virus — outbreaks have occurred in dog packs through eating virus contaminated pig meat — virulence of Aujeszky's disease virus has been increasing in USA and disease is estimated to cost the USA pig industry $US22m annually.

Clinical Signs
Pigs: In piglets there is fever, inco-ordination, muscle twitchings or trembling, convulsions, occasional respiratory signs, prostration and death. Severity is age related: mortality rate in piglets 10 days of age to 100% — older piglets 10–25% — adult pigs usually subclinical. Pregnant sows often abort or give birth to mummified foetuses, stillborn or weakened piglets.

Other species: Intense localised pruritis, self-mutilation, nervous signs, death.

Pathology: No gross lesions, diffuse lymphocytic, non-suppurative meningo-encephalitis and ganglio-neuritis — Cowdry type A nuclear inclusions.

Diagnosis: Histopathology — virus isolation in tissue culture — transmission tests in rabbits and piglets — serological identification by serum neutralisation tests.

Control: Vaccination (inactivated and live vaccines available). Quarantine and testing of new introductions to herd. Pilot eradication trial based on test and slaughter is under way in USA. National eradication policy is being debated in UK where eradication is estimated to cost 3.7 to 4.5 million pounds.

THE FERAL PIG PROBLEM AND EXOTIC DISEASES

Feral animals constitute one of the more serious problems in Australia for veterinarians, farmers and conservationists alike. As a conservative estimate feral pigs outnumber owned domestic pigs in Australia by a ratio of 4:1.

The highest concentration of feral pigs is to be found in the swamps, marshes, and drier pastoral areas of inland NSW and Queensland. However, they have a wide-ranging habitat from northern rain forests to southern mountainous areas. Whilst populations fluctuate, (and have probably declined over the last year or two with droughts in inland areas) there is estimated to be up to 3–5 million feral pigs in NSW, 8 million in Queensland, up to 200,000 in each of the Northern Territory and Western Australia, and small numbers in South Australia, Victoria, Tasmania and the ACT. Particularly disturbing has been the introduction of feral pigs to new areas by "sportsmen" for recreational shooting.

Feral pigs are frequently in close contact with domestic livestock and other feral animals. They could therefore acquire any of the many exotic diseases to which they are susceptible from domestic livestock (or *vice versa*) and act as an important reservoir of infection. The vesicular diseases and swine fevers are, of course, of greatest concern.

There are many situations in Australia where the primary foci of an exotic disease outbreak may be in feral pigs. This could happen through feral pigs gaining access to town rubbish dumps. A high risk area is northern Cape York where feral pigs are known to browse on rubbish washed ashore from ships — in an area which is a major shipping lane from Asian countries where many exotic diseases occur. If an exotic disease did occur in such pigs it could smoulder for a considerable time and become widespread before being recognised.

If an outbreak of foot-and-mouth disease occurred in an area of Australia where there are feral pigs it is unlikely that major trading nations would accept that Australia had eradicated the disease until it was proven conclusively that there were no remaining pockets of infection in feral animals.

Responsibility for control of feral pigs in Australia is fragmented between various local, State and Commonwealth authorities. It is fair to say that no significant long-term reductions are being made in feral pig populations anywhere in Australia at present. The Federal government recently undertook to develop a national feral pig control policy.

In an exotic disease emergency, such as an outbreak of foot-and-mouth disease, the requirement would be to quickly eradicate, or at least reduced to very low numbers, feral pig populations in the infected area. This may be an area of 100 square miles or larger. With current technology, the best way of achieving this may be by a mixture of helicopter shooting, poisoning and perimeter fencing and/or patrols. This would be expensive, but costs would pale into insignificance compared with lost export revenues should the country remain infected with foot-and-mouth disease.

POTENTIAL THREATS AND AUSTRALIAN POLICY

The huge value and importance of the Australian livestock industry leads to a high national awareness of exotic disease threats and high priority in keeping them exotic.

There are many ways that diseases could enter — legally, illegally, accidentally or even deliberately. Increasing interest in new or different genetic strains of sheep and goats increases the exotic disease entry risk. Methods continually need to be found to accommodate community desires and needs with minimal risk. If exotic disease gains entry it is less likely to be through a deliberate and well controlled legal quarantine introduction.

There are five major categories of avenues for potential introduction:

Live animals

Chances of illegal introductions of farm animals are low but not nil. One of the authors presided over the slaughter of 5 Indonesian goats on the beach at Melville Island, NT, in 1969. These goats were on a pirhau blown off course by a storm — but are an example of accidental illegal import risk.

All the diseases discussed in this chapter can, of course, be transmitted by farm animals so rigid precautions will need to be exercised with legitimate introductions. Problems can be seen with, for example, the slow viruses, where no satisfactory diagnostic tests have been developed.

Cattle, deer, pigs and rodents can be carriers of various diseases (bluetongue, rinderpest, FMD, Rift Valley fever, SWF) affecting sheep and goats. It could therefore happen that the first appearance of any of these diseases in sheep and goats might be due to imports of other species.

Man

Humans move more casually, in greater numbers and with much less restriction than the lesser animals. Being capable of carrying Rift Valley fever, Wesselsbron Disease and *Brucella melitensis* systemically and FMD in the nasopharynx (for a few days), man is a potential introducer of some diseases.

Rift Valley fever should be regarded as a candidate of relatively high risk of importation by this means. A likely scenario is an African safari park visitor returning by air to Australia while incubating RVF, being bitten by mosquitoes whilst viraemic and close to sheep or goats; alternatively, virus-infected insects escaping disinsection at an international airport.

Animal Products

Meat, hides and skins, meat products, wool and hair are potential long term vehicles for many of the diseases discussed above.

Semen, ova and embryos present new opportunities — and new risks. they can be smuggled in ampoules or thermos flasks. They may be coveted by importers with the resources to outwit the quarantine system. Legal procedures facilitate imports from "safe" countries but access to most countries is difficult because of a multiplicity of exotic diseases, unsatisfactory tests or unsatisfactory veterinary resources.

At least with semen an aliquot can be sacrificed for tissue culture, animal inoculation or other tests. But there are many problems to be solved with ova or embryos; we must assume that they can carry at least viruses, but we can scarcely test them without destroying them. Implant into recipients under quarantine supervision seems a safe way.

Ectoparasites

Ticks, flies, mosquitoes, culicoides insects and other arbovirus vectors could introduce Nairobi sheep disease, bluetongue, Rift Valley fever and Wesselsbron Disease.

Intercontinental wind drifts cannot be controlled but we can reduce the risks of aircraft introduction. Department of Health investigations has shown that disinsection is effective and that aircraft can carry surprising numbers and varieties of insects. It is justifiable to continue these procedures in spite of their unpopularity with passengers.

CONCLUSION

This chapter has gone to some length describing some varied threats to Australian livestock from exotic diseases. The precise risks and exact threat have never been determined, however, there can be no doubt that they are real. In certain cases, Australia has gone from being a country where a disease such as bovine pleuropneumonia has been endemic to a position where it is now exotic. Bovine brucellosis and tuberculosis will almost certainly be regarded as exotic diseases by the turn of the century.

Whether the exotic diseases be viral, bacterial or parasitic — they all represent either a threat to our trade in livestock and animal products or a direct threat to our animal production.

The prevention of entry of these diseases is one of the central roles of the national government through its quarantine and animal health services.

REFERENCES

Alton, G.G., Jones, L.M. and Pietz, D.E. (1975). 'Laboratory techniques in Brucellosis', 2nd edition, published by WHO, Geneva.
Anon (1978). 'Foot-and-Mouth Disease' — A series of 6 papers reviewing history, clinical aspects, control and vaccines. *Vet. Rec. 102,* 184–193.
Anon (1979). 'Screw-worm fly — possible prevention and eradication policies for Australia'. Australian Bureau of Animal Health.
Howell, P.G. (1963). 'Bluetongue'. In: *Emerging Diseases of Animals,* published by FAO, Rome, pp. 109–153.
Hunter, G.D. (1974). 'Scrapie', *Prog. Med. Virol.* 18: 289–306.
Kimberlin, R.H. (1976). 'Slow virus diseases of animals and man'. North Holland Research Monographs — Frontiers of Biology, 44: 209–378.
Martin, W.B. and Stamp, J.T. (1980). 'Slow virus infections of sheep'. *Br. Vet. J.,* 136: 290–295.
Periera, H.G. (1981). 'Foot-and-Mouth Disease'. In: *Virus Diseases of Food Animals* — Vol. 2, Edited Gibbs, EPJ, Published by Academic Press, 333–364.
Plowright, W., MacLeod, W.G. and Ferris (1959). 'The pathogenesis of sheep pox in the skin of sheep'. *J. Comp. Path,* 69: 400–413.
Plowright, W. (1968). 'Rinderpest Virus', *Virology Monographs, No. 3,* Ed Gard, S., Hallauer, C. and Meyer, K.F., published by Springer-Verlag, Vienna, 27–110.
Sabban, M.S. (1955). 'Sheep pox and its control in Egypt'. *Am. J. Vet. Res.* 16: 209–213.
Scott, G.R. (1981). 'Rinderpest and peste-des-petits ruminants'. In: *Virus Diseases of Food Animals, Vol. 2,* Ed Gibbs, E.P.J., published by Academic Press, London, 401–432.
Various authors (1975)., Symposium on bluetongue. *Aust. Vet. J. 51,* 165–231.

Chapter Four
EXOTIC PLANT PATHOGENS
J.W. Randles

The world distribution of plant pathogens, whether they are nematodes, fungi, bacteria, mycoplasmas, viruses, or viroids, is rarely the same as that of their hosts. Catastrophic epidemics often seem to occur when there is a 'new encounter' or 're-encounter' between a pathogen and a susceptible host species. The lack of recent selection for resistance in the plant host species, or for avirulence of the pathogen, is thought to be the reason why the resulting epidemics are so severe. In Australia, for example, great care must be exercised because many of the most important agricultural plants were originally imported from the northern temperate zone. They have been separated from their original pests and pathogens for some time and, in becoming adapted to their new environment, may have lost resistance to these parasites.

In this chapter I will outline the risks, methods of combat, successes and failures of attempts to exclude exotic plant pathogens from Australia.

PAST CATASTROPHES

Potato became a staple food of Europe during the 18th century, substituting for diseased and unproductive cereals. In 1845, a potato disease, now known as 'late blight', struck Europe (Lange, 1950) and caused collapse of the crop (Carefoot and Sprott, 1967). The pathogen, a fungus (*Phytophthora infestans*), sporulates rapidly and profusely and is spread by wind and raindrop splash. Epidemics of varying severity have occurred in most years since then. In Ireland, which was poor and dependent on the potato, starvation was rife. In 10 years, the human population decreased by 3 million; 1 million died and 2 million emigrated. The social effects of this emigration were far reaching, and it has been said that but for *P. infestans*, the USA would not have had J.F. Kennedy as President. The origin of *P. infestans* is uncertain but, like the potato, it is probably a native of the high Andes, and it continues to be a serious pathogen in the cool temperate potato growing areas of the world.

There are many other examples of catastrophic plant diseases caused by fungi. Chestnut blight (*Endothia parasitica*) is native to Japan, China, and Korea, and was first reported in New York in 1904. It killed chestnut forests in the east and west coasts of the USA and Canada. It also spread widely in Europe after its discovery in Italy in 1938 (Anon, 1982). Dutch elm disease (*Ceratocystis ulmi*) is of more recent interest as, since virulent strains were introduced from North America to England in the 1960's, it has killed 17 million elm trees, three-quarters of the U.K. population. Coffee rust (*Hemileia vastatrix*) was introduced into Ceylon in 1875 and, as a result, that country now exports tea not coffee. Needle-rust of pine (*Dothiostroma pini*) appeared in New Zealand in the 1960's where it is causing much reduced forest growth rates. Tobacco blue mould (*Peronospora tabacina*) was introduced to Europe from Australia around 1960 (see Figure 1) with consequent severe losses to the tobacco industry.

Fungi are not the only pathogens which spread and cause catastrophes. The bacterial disease, fireblight of apples and pears (*Erwinia amylovora*) has recently entered England (1958). Potato cyst nematode (*Globodera rostochiensis*) has been introduced to the USA, Canada and NZ (Anon, 1982).

Major plant disease epidemics also occur when plants are moved to new environments. For example, when cocoa (*Theobroma cacao*) was taken from the West Indies and grown as a crop in West Africa it was found to be susceptible to cacao swollen shoot, a mealy-bug transmitted virus of nearby native trees.

EXCLUSION AND QUARANTINE

In the 1870's the principle of plant quarantine was introduced to many parts of the world to stop the movement of exotic pathogens. The practice of plant quarantine has since been modified in order to cope with:

a. increased volume and speed of world trade;
b. changing needs of primary industry such as the introduction of new cultivars and breeding material, 'genetic engineering', production of seed for growers in the northern hemisphere;
c. increasing demands of city dwellers for exotic plants.

To combat these pressures, the following principles have been developed to minimize the risk of introducing dangerous pathogens:

a. Plant material is imported only from areas free from known pathogens;
b. Material is inspected and certified, and may be given additional tests, before dispatch;
c. Minimum quantities required for the establishment of cultivars are imported.
d. Plant parts least likely to carry pathogens are introduced, such as aerial parts, seeds, or cultured tissues;
e. After entry the plants are inspected and tested, sometimes for specified periods of time.

The intensity with which these principles are applied depends on the crop, and the potential danger of its pathogens. It is difficult to estimate the danger of particular pathogens because they may be more or less serious in a new environment, especially when infecting cultivars which they have not previously encountered. Nevertheless, because of previous problems, the quarantine services in most countries are conservative, and the Australian plant quarantine service is well known for its strict policies. However, loopholes do exist. Ornamentals generally enter with much less stringent control than cultivars of commercially more important crops, even though they carry pathogens which might spread to crops. Furthermore, 'commercial interests', for various reasons, frequently apply pressure for large imports of seed or propagules.

A BLACK LIST

The known exotic pathogens that are a threat to Australian crops include the following:

Fire blight bacterium of pome fruit (*Erwinia amylovora*; Anon, 1982);
Potato cyst nematode, potato black wart (Anon, 1982);
Plum pox virus of stone fruit (Anon, 1982);
Karnal bunt of wheat (*Tilletia indica*; Waller and Mordue, 1983);
Citrus canker; citrus dieback and greening (Anon, 1982);
Sugar cane smut (Anon, 1982);
Dutch elm disease (Anon, 1982);
Black sigatoka of banana (Anon, 1982);
Pierce's disease of grapevine (Anon, 1982).

Fig. 1. World distribution of tobacco blue mould (*Peronospora tabacina*) in 1962. Before 1942, the disease occurred in Australia, North and South America (Site 1), it appeared in Europe in 1961 (Site 2) and the Philippines in 1962 (Site 3). Adapted from Anon (1962).

Most epidemics are of course unpredictable, the result of fortuitous infections that are not ameliorated by resistance of the host or avirulence of the pathogen. Perhaps Jarrah dieback falls into this category, as the pathogen (*Phytophthora cinnamomi*) is thought to have a tropical origin. Its devastation of the jarrah trees and many other native species in West Australian forests suggest that there have been no previous encounters that have selected tolerance or resistance of the native species.

BREACHING THE BARRICADE

Pathogens with airborne spores, such as rusts, are very difficult to exclude from a country for ever. The recent introduction to Australia of poplar rust, stripe rust of wheat (Brown and Holmes, 1983) and rust of blackberry (Marks et al., 1984) illustrates not only the difficulty of excluding such pathogens but also the speed with which they become distributed throughout the country.

Pathogens borne in soil seed propagules or by vectors can be excluded by appropriate action provided that knowledge and facilities are available. Pathogens such as fire blight, plum pox, potato wart, and Karnal bunt can be denied entry to Australia provided that quarantine procedures are effective, and none of the basic principles of plant quarantine are broken. Two recent examples are worth recording as they illustrate the results of introducing more than the minimum amounts of planting material to establish cultivars. The first arose through the enterprise of people who perceived that since field bean (*Vicia faba*) was imported for high quality horse feed, there might be profit in first selling the seed to Australian contract growers for multiplication before selling it as feed. Three seedborne diseases were introduced in this way, and they may become established widely in crop, pasture, and weed species (Randles and Dube, 1977). The other example has more complex origins. Species of lucerne aphid previously unrecorded in Australia were found in 1977, and their rapid devastation of existing stands of lucerne led to the introduction of large quantities of seed of aphid-tolerant cultivars from the U.S.A. Approximately 2% of this seed carried alfalfa mosaic virus (AMW) (Garran and Gibbs, 1982), which is transmitted by all species of lucerne aphid. Subsequently there has been a widespread increase in the incidence of alfalfa mosaic virus, not only in lucerne but also in other crops (e.g. peas). Thus hasty measures to combat lucerne aphids have resulted in an irreversible decrease in lucerne productivity; the virus does not produce spectacular symptoms, but the losses it causes, though insidious, are nonetheless large in terms of absolute loss in carrying capacity.

CONCLUSIONS

Despite the geographical isolation of Australia, increasing and changing demands in the community will probably result in the introduction of more exotic plant pathogens.

Attempts to prevent introductions should not be relaxed. Information on the biology and world distribution of pathogens must be current, and trained plant pathologists need to be vigilant so as to detect new outbreaks and immediately isolate pathogens if possible. It is far cheaper to exclude pathogens than to try to control them after entry.

REFERENCES

Anonymous (1962). Distribution maps of plant diseases No. 23. Commonwealth Mycological Institute, Kew.
Anonymous (1982). *Plant quarantine leaflets*. Australian Government Publishing Service, Canberra.
Brown, J.S. and Holmes, R.J. (1983). *Plant Disease 67*, 485.
Carefoot, G.L. and Sprott, E.R. (1967). *Famine on the wind*. Longmans, Canada.
Garran, J. and Gibbs, A. (1982). *Aust. J. Agric. Res. 33*, 657.
Lange, E.C. (1950). *The advance of the fungi*. Jonathan Cape, London.
Randles, J.W. and Dube, A.J. (1977). *A.P.P.S. Newsletter 6*, 37.
Marks, G.C., Pascoe, I.G. and Bruzzere, E. (1984). *Australasian Plant Pathology 13*, 12.
Waller, J.M. and Mordue, J.E.M. (1983). *C.M.I. Descr. of Pathogenic Fungi and Bacteria* No. 748.

Section B

Planning for an exotic disease outbreak

The first five chapters of this section examine various aspects of the way in which Australian society has responded to the possible entry of unwanted pests and parasites. Quarantine is the primary defence (Chapter 5), but unwanted pests and parasites may evade this barrier, and so plans of various sorts have been made in case the community ever has to be mobilized to control an invasion. This involves not only planning to eradicate the unwanted migrant, as expeditiously as possible (Chapter 6), but also assessing the legal (Chapter 7) and economic (Chapter 8) consequences of such a campaign. Planning of this sort may result in recommendations for specific preemptive action. The building of the Australian National Animal Health Laboratory (ANAHL) is an action of this sort, but a controversial one which has provided an interesting 'case history' (Chapter 9) on how some important decisions are made.

The final Chapter of this Section advocates the value of deliberately importing particular exotic viruses to the laboratories of those responsible for diagnosis and control, so that diagnostic reagents may be prepared, and staff trained in their use. This is a thorny topic, which provokes arguments about relative risks and advantages, and has been at its most bitter over the proposal to import foot-and-mouth disease virus.

Chapter Five

QUARANTINE — A PRIMARY DEFENCE

Kevin Doyle

The term 'quarantine' is derived from the Latin word 'quarantum' meaning 'forty'. For centuries a forty day period of detention was applied to ships arriving in certain countries from areas in which diseases such as bubonic plague, cholera and yellow fever occurred. It is generally accepted that the first international quarantine measures were imposed in Venice in 1374 when the entry of travellers suspected of being infected with bubonic plague was banned. However quarantine as a concept is age old. It is said that the Book of Leviticus in describing the laws of Moses refers to laws which required lepers to be kept isolated and to notify their presence by the ringing of a bell; thus introducing the concepts of isolation and warning. In this paper I discuss the international quarantine measures used to prevent the entry of diseases of man, animals and plants into Australia.

The concept was introduced into Australia with the passing of the (N.S.W.) Quarantine Act in 1925 'to prevent the introduction of the disease called the Malignant Cholera or any other infectious diseases highly dangerous to the health of His Majesty's subjects into the Colony of New South Wales'.

Outbreaks of smallpox in the 1850s and 1870s, that culminated in a major outbreak in 1881, together with the discovery at Geelong in 1875 of *Phylloxera*, the vine destroying aphid, ensured that the emerging nation gave full attention to quarantine.

On 14 August 1932 Spring Cove in North Harbour, Sydney was proclaimed a quarantine station for ships and on 21 February 1933 an area of land was proclaimed as a quarantine station for the crews, passengers, cargo and other goods on board.

Internationally, a series of conferences in Paris, beginning in 1851, aimed towards standardising methods for isolating and checking travellers, for disinfection and for controlling rodents.

Port inspection of cargoes began in 1889 when the Export and Import Branch of the NSW Department of Agriculture was established and at that time the States aimed to introduce uniform quarantine legislation. Conferences at which the six colonies were represented were held in 1884 and 1896 and emphasized the need for uniformity. The issue was active in the period preceding federation. So that when the Commonwealth of Australia was formed, the new Constitution included quarantine as the only specific power of the Commonwealth Parliament. The Commonwealth Quarantine Act was passed in March 1908 and came into effect on 1 July 1909.

THE QUARANTINE SERVICE

The Australian Quarantine Service was created in 1909 within the Department of Trade and Customs. A loose network with State services supervised by a federally appointed Director of Quarantine, was found to be inadequate in handling the pandemic of influenza which followed World War I. Co-ordination proved almost impossible and Dr J.H.L. Cumpston was appointed head of a new Federal Department of Health as well as Director of Quarantine in 1921. W.A.N. Robertson of Victoria was appointed Director of Veterinary Hygiene within the new Department to co-ordinate, control and eradicate an outbreak of rinderpest which occurred in Western Australia in 1923. And so the Service as we know it gradually took shape.

A wise decision, probably arising from a shortage of graduates, left the operational elements of animal and plant quarantine with the States. This has avoided the costly and divisive duplication which occurs in many other federations. As a result of this decision a unique agency arrangement was developed under which State officers:

- are gazetted as Chief Quarantine Officers (Animals)
- are gazetted as Chief Quarantine Officers (Plants)

- are gazetted as Quarantine Officers
- work under Commonwealth funding
- work under Commonwealth policy
- work under Commonwealth directions

All operate under the Commonwealth Quarantine Act under direction of the Director of Quarantine.

States are reimbursed for the proportion of their time that their officers spend on Commonwealth work. There are more than 1,700 quarantine officers; some work full time, some part, some are State, some Local Government employees, and some are private practitioners. The General Quarantine Officers are officers of the Commonwealth Department of Health. Yet an integrated, uniformed, disciplined Service, co-ordinated by the Department of Health has functioned effectively as 'possibly the most advanced and efficient in the world' for more than seventy years at 'bargain basement prices' (Senate Committee on National Resources; 1979). The animal and plant quarantine services were transferred to the Department of Primary Industry in March 1985, and form a part of the Australian Agricultural, Health and Quarantine Service (AAHQS).

RESOURCES AT RISK

Australia has a population of some 15.5 million people with huge agricultural production, therefore a large proportion of the produce must be exported, and it provides almost 50% of our export earnings. It is well known that access to overseas markets depends upon the absence of certain disease of animals and plants. Examples of such diseases are foot and mouth disease and Newcastle disease for animals and fire blight, a bacterial disease of pome fruit, for plants. Other less well defined resources are also at risk. Freedom of the country from rabies is taken for granted, a change would affect our enjoyment of open spaces and the handling of animals. Other serious human diseases either occur in our region or, because of the speed of motor transport, present risks of being introduced into this country (e.g. dengue fever, malaria).

INTERNATIONAL OBLIGATIONS

As a major agricultural trading nation which seeks access to markets around the world, Australia has obligations regarding the scientific justification of quarantine measures.

Arrangements like the General Agreement of Tariffs and Trade (GATT) aim to avoid the use of health and quarantine conditions as non-tariff barriers to trade. It is in Australia's interest to support such principles with vigour as a means of ensuring access to markets, and at the same time justify all quarantine bans or limitations on sound scientific grounds, so that they may be defended in other countries. It is not possible to prevent the entry of animals, plants or their products on the basis of unspecified risks, though genuine risk of poorly defined diseases must be treated with caution. Naturally value judgements made by different people on the same scientific evidence may vary. However countries which are not as well protected as Australia is by isolation, and which do not have the same needs for access to international markets, may place less importance on some of the lesser quarantine risks.

A prime example is Australia's requirement that cheese from countries affected by foot-and-mouth disease must be manufactured by a process which produces conditions of pH, temperature or maturation that inactivate foot-and-mouth disease virus (FMDV), or otherwise that cheeses must be held in quarantine on arrival until the natural changes in the cheese would inactivate the virus. Other countries, especially those which vaccinate stock against FMDV, regard the risk from cheese to be negligible. Nevertheless by reviewing the process of manufacture of individual cheeses, risks can be assessed and restrictions applied intelligently; 97% of cheeses examined are acceptable and enter without quarantine storage. This example illustrates how disease may be controlled by selective rather than total bans.

PHILOSOPHY AND PRACTICE OF QUARANTINE

The philosophy and the practice of quarantine continually evolves. During and immediately after World War II many quarantine controls failed and, with improved transportation and increased demands for trade, there was an increase in the spread of diseases around the world. This led to a total ban by Australia on imports of many species and many products.

More recently control over illegal importation has been made more difficult by the sophistication of the methods available to those who wish to evade quarantine controls. For example semen and ova may be imported instead of adult animals, and fast light aircraft are easily obtained. Furthermore there is a constant demand for new products and improved species. Thus it has been realised that total bans may be unproductive, even counterproductive, as a means of preventing the entry of disease.

Since 1949 Australia has totally banned imports of birds from all countries except New Zealand, and New Zealand was added to that ban in 1972, largely because poultry was imported from there under conditions which were not acceptable to Australia. Nonetheless ornithologists report that during the past 10 years, 48 new types or species of birds have been introduced into Australia, and some new diseases of birds have been detected in the country in this period. These facts indicate that, for many species, it is best to control imports and use the best available techniques to detect diseases, as this provides far less risk than a total ban, which induces people to smuggle untested stock.

The International Court of Justice and the European Court have accepted this type of control, whereas total bans are only justified by the presence of known diseases.

POLICY FORMULATION

I described above the structure of the Australian Quarantine Service and the relationships between the Department of Primary Industry and its State counterparts and also the State Departments of Agriculture or Primary Industry. These relationships are the basis of the consulting machinery of the service and give it access to a wide range of opinion and expertise.

Annual conferences of all Chief Quarantine Officers are held. There, policies are developed, funds and staff allocation discussed and, if required, recommendations sent for approval by the Minister of Primary Industry. However such discussions continue throughout the year.

The various 'arms' of the quarantine service are also represented in national and international fora. Animal and Plant Quarantine are represented on the Standing Committee on Agriculture and participate in the affairs of the Australian Agricultural Council. General Quarantine, or at least its human health aspects, are discussed at the Health Ministers' Conferences.

The Australia/New Zealand Technical Committee on Quarantine was formed in 1964 and has met each year since. It has had increasing attention as a result of the 'Closer Economic Relations Agreement' between the two countries. The Committee's principal term of reference is to 'Consider and report on Plant and Animal Quarantine questions as they effect trade between Australia and New Zealand'. There are similar joint committees of Australia and the United States of America and of Australia, Papua–New Guinea and Indonesia. Veterinary Attaches located in Brussels and Washington also regularly consult with officials of importing countries.

Australia also has obligations to the World Health Organisation and the International Health Regulations which cover the welfare of those travelling by sea or air.

FIRST PORTS OF ENTRY

Vessels arriving in Australia are required to enter at particular ports; sixteen ports for aircraft and 64 for boats. In the 10 years before mid 1983, 15,970,000 passengers entered Australia through these ports; more than the entire Australian population. Countless tonnes of cargo, personal effects and baggage are also checked by quarantine staff every year.

CUSTOMS

At the 'ports of entry', passengers, their baggage and cargo are controlled by a barrier system maintained by the Customs and Immigration authorities. Passengers are inspected at a 'primary health line', then Quarantine Declarations are made to officers at the Customs and Immigration barrier.

Persons are selected for quarantine examination on the basis of their Declaration and the country from which they have come, on other information and experience, and in addition a random sample is examined. Primary searches are made by Customs Officers, and Animal and Plant Quarantine Officers make supplementary searches and take decisions on individual items of agricultural interest.

The proportion of restricted goods found by this means has been assessed by a total re-examination of some previously examined passengers. The results of these tests have not been published, for obvious reasons. However they have shown that, quarantine examinations are effective, though not totally. It is estimated that, to recover 50% of restricted items not detected by the existing system, 90% of passengers would have to be searched; this is clearly impracticable and could well be counterproductive to Australia's broader interests.

PENALTIES

Penalties for offences under the Quarantine Act have been increased recently to a maximum of $100,000 or 10 years gaol for individuals or $200,000 for corporations.

CARGO

The 'containerisation' makes it difficult to inspect cargo. As a result containers now enter Australia only through approved container depots where they can be properly inspected. Containers are often infested with insects, therefore Plant Quarantine, which provides the entomological expertise in the Service, together with Customs, is primarily responsible for examining containers, which, like passengers, are systematically sampled. This is done by links with the Customs computer entry systems and by random and biased selection of cargoes for detailed inspection.

INSECT CONTROL

Aircraft and certain vessels are freed of insects using methods approved by the World Health Organisation. Regular surveys of the results of these actions confirm that this is an essential procedure especially for aircraft arriving from certain countries.

It is necessary to prevent the entry of exotic insects which are pests or which might be vectors of diseases, especially of animals and man, such as malaria, haemorrhagic, dengue and yellow fever. It is considered particularly important to exclude the mosquito, *Aedes aegypti*, that is able to transmit yellow fever flavivirus. Also Rift Valley fever, which is caused by a bunyavirus, is considered to be an important threat to animals and man as it might find a receptive environment in Australia; the cattle disease akabane, which is caused by a related virus, is endemic in Australia.

QUARANTINE FACILITIES

Originally the Quarantine Stations at the major Australian ports were used to isolate boat passengers in order to exclude smallpox and the haemorrhagic fevers of man from the country. Recently these Stations have been replaced by the National High Security Quarantine Unit at Fairfield Hospital, to which potentially infectious patients can be transported in mobile isolators.

Plant Quarantine facilities in capital cities and other strategic locations now have glasshouses, where imported plant materials can be grown in isolation, and also facilities for fumigating plant material.

Animal Quarantine stations are located in Brisbane (soon to be closed for economic reasons), Sydney, Melbourne, Adelaide and Perth. A quarantine station for holding animals in isolation *en route* to Australia is located in the Cocos (Keeling) Islands.

RODENT CONTROL

Traditional methods for stopping rodents entering Australia from ships are at present being reinforced because Korean haemorrhagic fever, a virus disease spread by rodents, has emerged as an important human disease. Rodents are being trapped in the vicinity of wharves to assess the potential danger.

QUARANTINE PUBLICITY

It has become clear that most Australians, and more particularly most travellers, are not aware of the consequences of a failure of quarantine. As a result, a major quarantine publicity programme has recently been developed. The program has included a nationwide television programme featuring the 'environmentalist', Harry Butler, and more specific publicty at agricultural shows and schools, and also the distribution of multilingual publicity material at airports and sea ports, and with visas.

COASTAL SURVEILLANCE

Although Australia's territorial waters are patrolled by boats from the Royal Australian Navy and aircraft of the Royal Australian Air Force, the arrival of large numbers of refugees in boats from South-East Asia during the late 1970s pointed to the need to detect and check boats landing without authority or control.

Boats can bring unhealthy occupants, wood borers and other undesirable migrants in the foods and ballast that they carry. Many refugee vessels of the 1970s arrived in very poor conditions, and the lives of the occupants were at risk, for, unless detected quickly by coastal surveillance, they could have been stranded for long periods in inhospitable areas of northern Australia. As a result, since the early 1980s, a detailed daily search of the coastline between Townsville and Geraldton in Western Australia has been made by chartered aircraft carrying trained observers and reporting to the Australian Coastal Surveillance Centre in Canberra. This program is expensive ($10.5m), and so it was completely reviewed in 1983 and all the risks re-assessed. It was decided to concentrate resources in areas of maximum risk (Cairns to Karratha), and to investigate all unauthorised landings; a responsible team consisting of Quarantine, Customs and other officials as appropriate visit the site, clean the area and, if necessary, impound the vessels under Quarantine or Fisheries legislation.

TORRES STRAIT

The Torres Strait Treaty, which is about to be formalised, between Australia and Papua–New Guinea provides PNG access to fishing and other sea bed resources in the Torres Strait. The Treaty allows movement by traditional people by traditional means for traditional purposes. The Islands of Boygu and Saibai, though Australian, are very close to the PNG coast. This makes it impossible to enforce quarantine by traditional barrier methods. However daily surveillance flights, helicopter visits and the assistance of local native authorities on individual islands, effectively provides quarantine. A team located on Thursday Island oversees activities in the area and receives reports from 'head men' on the Islands. Frequent visits are made to check animal and plant health, and insects are trapped to check for the screw worm fly, the oriental fruit fly and certain mosquitoes.

REGULATORY ACTIVITIES

The 'barrier service' at 'ports of entry' is the most visible form of activity. However, other controls play an equally important part in quarantine protection, these include:

- assessment and control of the entry and use of biological materials for medicine and agriculture;
- biological control programs using imported agents;
- the Security Assessment Group which monitors the microbiological security of the Australian National Animal Health Laboratory;
- screening of mail and parcels at Mail Exchanges;
- aid projects in developing countries;
- control of potential disease vectors around airports;
- control of waste food material from visiting boats and aircraft; acknowledged as major means of international transmission of animal and plant diseases;
- quarantine inspection of foods of animal and plant origin;
- imports of plant genetic material, as specimens for growth in quarantine as tissue culture;
- research into transmission of disease through plant material;
- control and inspection of imports of semen and ova of animals as well as live zoo specimens, livestock, laboratory animals and pets.

OVERVIEW

The activities of the australian quarantine service are very wide and varied, and I hope the above outline does justice to the Service. As was noted by the Senate Standing Committee on National Resources, this service is provided at a 'bargain basement price', $41.3m in 1983–84. This sum included approximately $10.5m for coastal surveillance and $18m for payments to the States for field duties, quarantine waste disposal, ship clearances, etc.

The necessity for this expenditure is well illustrated by the estimate that an outbreak of foot and mouth disease (probably the most infectious disease known) would cause a drop in export earnings of about $1500m in the first year!

REFERENCES

Doyle, K.A. (1979). *Proc. 2nd Int. Symp. Vet. Epid. and Econ.* (Canberra), 279.
Gee, R.W. (1980). *Proc. Exotic Animal Disease Emergency Seminar.* (Macedon).
Ilbery, P.L.T. (1984). *J. Commonwealth Dept. Health* 1, 5.
Johnston, J. (1980). *Exotic Animal disease Emergencies in the Australian Grazing Sector — An Economic Study.* Report for Bureau of Animal Health and the Australian Wool Corporation.
Longmire, J.L. *et al* (1980). Twenty Fourth Annual Conference of the Australian Agricultural Economics Society, Adelaide. BAE Working Paper 80.
Navaratnam, S.J. (1982). *Proc. Int. Conf. Pl. Prot. in Tropics,* 605.
Senate Standing Committee on National Resources (1979). Report to Parliament on "The Adequacy of Quarantine".

Chapter Six

EXOTIC DISEASE EMERGENCY PLANS
R.W. Campbell

In Australia we aim to control exotic diseases by eradicating them. To do this involves planning the legal and financial framework to accomplish this in the shortest possible time. Although some exotic diseases, such as bluetongue or vesicular virus diseases in feral animals, may be extremely difficult to eradicate, the long term aim is eradication. Cost/benefit studies (see Chapter 8) support this objective.

Planning involves several iterative steps; plan formulation, action programs, review, plan reformulation etc. This must be a continuing process to take account of advancing technology, community expectations, changes in legal and financial arrangements, and ecological considerations.

A PLETHORA OF PLANS

It has become obvious in recent years that exotic disease eradication planning involves a wider range of issues than just the veterinary ones. Interestingly there is still considerable confusion about what is a 'plan' and it might be useful to consider that point first. Animal health authorities in Australia have, over many years, prepared 'model control plans' that define the disease control principles and practices to be used when there is an exotic disease outbreak. These are also the basis on which cost-sharing arrangements between the Commonwealth and the States were developed, whereby the Commonwealth and all States contribute to the eradication costs of an exotic disease outbreak no matter in which state or states it may occur. But these model control plans are insufficiently detailed to mobilize the resources needed to control an exotic disease outbreak, and are much more correctly defined as manuals of veterinary procedures.

The nature and extent of a true exotic disease plan became apparent at a seminar held at the Australian Counter Disaster College, Macedon, four years ago. This seminar was attended by Commonwealth, State and livestock industry people representing those likely to be involved in an exotic animal disease eradication program. It was resolved at the seminar that an exotic disease plan was essentially a short statement outlining arrangements for the efficient and effective use of the community's total resources, including definition of the agreed responsibilities of the agencies concerned, procedures for the activation of these separate agencies and mechanisms for ensuring that tasks to be undertaken were compatible with the community's general counter-disaster arrangements.

Within this framework, more detailed sub-plans, manuals of procedures and job cards are developed as necessary to detail the variety of tasks that are to be done by various individuals and organisations.

With this concept of an exotic disease emergency plan, it is more apparent why there must be separate (but obviously related) State, Commonwealth and Federal/National plans. Each State and the Commonwealth needs a plan consistent with its individual constitutional responsibility and the resources at its disposal. Similarly, for the States and Commonwealth collectively to mount an efficient and effective Federal or National campaign, there must be an agreed Federal/National plan that ensures that the total resources of the country can be marshalled and priorities assessed.

With these things in mind, it is useful to recall that effective planning requires several types of information:
1. an assessment of the extent of an outbreak;
2. an assessment of the resources that will be required to control it;
3. an assessment of the information needed by the organisations involved and the public.

Armed with this knowledge, it is possible to begin the planning process.

EDUCATION

Most Australian veterinarians, farmers, police, state emergency service personnel and politicians have no experience of exotic animal diseases. This ignorance will hamper planning because the people involved will be trying to prepare for something that is totally outside their experience. Even those veterinarians and others, who have helped eradicate livestock diseases overseas, have to guess what would happen in similar circumstances in Australia. Given the importance of livestock exports to our economy, it is hard to understand why it is so difficult to convince Governments of the value of sending Australian veterinarians overseas to train in the management of exotic disease outbreaks, when the likely benefits to our exotic disease planning are so apparent.

The 1980 Exotic Animal Disease Emergency seminar at Macedon highlighted the value of education in preparing exotic disease eradication plans; an aspect which had previously been largely ignored. Education involves not only the public relations and news media arrangements, but discussion of the effects on the community of a downturn in the economy, and all the problems that would be faced in the rural industries, by farmers, by abattoir and dairy factory operators, by manufacturers of livestock products, and in the event of an outbreak of rabies, by millions of owners of dogs and cats. There is also need to consider the disruption of the lives of people who would be responsible for controlling the disease outbreak. However, careful preparation should minimize the disruptive effects of an outbreak.

PLAN EVALUATION

It is obvious that plans must be tested. In Victoria, we are fortunate that our 'vesicular disease plans' have only been tested over the past 100 years by mock outbreaks. However these simulated outbreaks, despite their obvious limitations, are a useful means of testing basic procedures and identifying planning weaknesses. With the completion of ANAHL at Geelong, test exercises and exotic disease training in Australia will take a further step. In future we plan also to involve private veterinary practitioners, laboratory personnel and farmer representatives in training programmes.

Three non-technical areas of exotic disease planning that are receiving considerable attention at present are:

1. information systems;
2. public relations and news media requirements;
3. financial arrangements.

The use of electronic devices to handle the vast amount of data generated has given veterinary managers the means to make better-informed decisions than ever. The Animal Health Emergency Information System (ANEMIS) sponsored by the Bureau of Animal Health has proved to be a very good system for collecting and collating data and providing the necessary activity reports. Minor modifications are proposed but the intrinsic worth of the system for adoption Australia-wide augurs well for the future.

PUBLIC RELATIONS

The area with greatest potential for exacerbating planning problems in the event of an exotic disease outbreak is the matter of public relations and news media liaison. The Legana incident in Tasmania demonstrated the destructive effects of a poorly-informed press, and this incident is still having ramifications. Conversely, a similar disease incident in New Zealand, with as many, if not more, problems, was hailed internationally as an outstanding success simply because there was a more favourable news media coverage. Planning arrangements for the news media in the event of a vesicular disease outbreak in Victoria were considerably aided during the recent FMD 'Extortion Threat' and quickly allayed fears that we may have been inadequately prepared for such an eventuality. Considerable effort must be directed to this area, as experience overseas indicates that the news media is invariably critical of mistakes, and some inevitably occur. Similarly, public support for any control program is paramount to its successful conclusion, particularly when the high costs associated with eradication procedures compete for community finances.

THE COST

People likely to be affected by an exotic disease outbreak are becoming increasingly aware of the disastrous effect that loss of profit and loss of income could have on their businesses. Insurance is available but few take advantage of it. The possible role of industry-sponsored insurance is being considered as is the oft-discussed subject of the individual's legitimate financial responsibility versus that of the Government.

SUMMARY

Planning for an exotic disease emergency is a daunting task. The range of diseases to be considered and the range of outbreak circumstances that may occur are legion. The people who are charged with such planning have limited experience on which to proceed and they work against a backdrop of potential economic disaster.

However, such planning is proceeding with considerable enthusiasm and test-exercises are being used to minimise the likelihood of unforeseen events. Considerable co-operation is being received from the large variety of government, semi-government, and industry agencies involved and there is room for cautious optimism for the future.

REFERENCES
Report of Proceedings of the Exotic Animal Disease Emergency Seminar held at the Australian Counter Disaster College, Macedon, 2–7 March, 1980.

Chapter Seven

EXOTIC DISEASES AND THE LAW

Douglas J. Whalan

THE CONSTITUTIONAL DIMENSION

When we Australian lawyers see difficulties in dealing with problems that are potentially Australia-wide, we tend to blame our Constitution. However, it is a scientific fact, understood even by a lawyer, that exotic diseases do not automatically stop at State boundaries. But, rather than complaining about our Founding Fathers, we must devise practical ways of dealing with exotic disease problems. Although it is satisfying to note just how much has been done, there are still several legal powers that have not been used to their fullest extent. I also believe that much that has been done by administrative means has been accomplished in spite of the law or despite the absence of law. Nevertheless, although there is much amicable administrative co-operation between the Commonwealth and the States, we must recognize that the law ultimately obtrudes. Most constitutional power in this area of exotic disease prevention, control and compensation, lies with the States. Nevertheless, in theory at least, there are many Commonwealth powers, in addition to those actually used, that could be used.

I will first discuss powers that are potentially available to the Commonwealth, then see how much Commonwealth legislation is currently in place that could be used to deal with exotic disease problems, and then turn to look at State legislation.

COMMONWEALTH CONSTITUTIONAL POWERS

Potential Power

The range of possible Commonwealth powers is substantial. The powers in Section 51 of our Australian Constitution could be used by the Commonwealth. Possible powers are those over trade and commerce, defence, quarantine, insurance, corporations, pharmaceutical sickness benefits and medical services, immigration and emigration, internal affairs and the incidental power. In fact the range is so wide, that the only powers that I rule out with any confidence are the Commonwealth powers over lighthouses, lightships, beacons and buoys! However, as in so many cases, the politics and reality of Federal-State relations probably require that many potential powers not be used.

Quarantine Power

One very powerful weapon that can be used in the battle against an exotic disease invasion is the quarantine power in Section 51(ix). The Commonwealth *Quarantine Act 1908* contains very strong powers that can be used to help to control any human, animal or plant outbreak. Where the Governor-General[1]:

" ... is satisfied that an epidemic caused by a quarantinable disease or danger of such an epidemic exists in a part of the Commonwealth, the Governor-General may, by proclamation, declare the existence in that part of the Commonwealth of that epidemic or of the danger of that epidemic. ... Upon the issue of a proclamation the Minister may, during the period the proclamation remains in force, give such directions and take such action as he thinks necessary to control and eradicate the epidemic, or to remove the danger of the epidemic, by quarantine measures or measures incidental to quarantine."

"Quarantinable disease" covers those diseases mentioned in the Act itself and any disease proclaimed by the Governor-General to be a quarantinable disease[2]. The present list is smallpox, plague, cholera, yellow fever, typhus fever, leprosy, rabies, Marburg fever, Lassa fever and Ebola fever.

There is also a very wide power given to the Minister administering the Act to deal with an emergency situation. Where in the Minister's opinion[3]:

" ... an emergency has arisen *which requires the taking of action not otherwise authorized by this Act,* he may take such quarantine measures, or measures incidental to quarantine, as he thinks necessary or desirable for the diagnosis, prevention and treatment of any quarantinable disease."

This is a most unusually wide legal power, as it gives the Minister a power that is not even determined in any detail in the legislation. But there is an obverse side to that coin: Because it *is* so wide, acts done under it would be looked at very carefully if they were challenged in Court.

Although quarantine powers are generally exercised at points of entry into a country, power is also given to the Governor-General in the *Quarantine Act* to prevent the removal of animals, plants or goods from one part of the Commonwealth to any other part of the Commonwealth, and to declare as a quarantine area any part of the Commonwealth where there is, or there is suspected to be, any quarantinable disease or any disease or pest affecting animals or plants[4]. These powers can only be exercised if the Governor-General is satisfied that their exercise is necessary for preventing the spread of a quarantinable disease, or a disease or pest affecting animals or plants[5].

These powers are so extensive that there could be some problems in working out their legal relationship with the State legislation that covers some of the same areas.

As the quarantine power is a Commonwealth power under the Constitution, Commonwealth legislation would be held by the Courts to be paramount. Indeed, in an emergency situation the Courts need not be approached as the *Quarantine Act* gives power to the Governor-General in such circumstances to over-ride State laws by proclamation[6]. Although Commonwealth power in this area is paramount, in practice, of course, there is a great deal of co-operation between Federal and State authorities to try to ensure that harmful diseases do not gain entry into Australia.

THE STATE POWERS — "ANIMAL, VEGETABLE OR ... HUMAN"

I now look at State legislation which could be effective and, in doing so, I play the old game of "animal, vegetable or .... Human".

All six States, the A.C.T. and the Northern Territory have relevant legislation. All eight jurisdictions have health or public health Acts[7], animal or stock diseases legislation[8] and plant diseases provisions[9].

Whether the legislation covers human outbreaks, outbreaks in animals or in plants we find there are somewhat similar patterns in the three areas. Almost all provide for notification of outbreaks of disease[10], most have quarantine provisions aimed at preventing the spread of disease[11], and the diseases are usually specified in the legislation[12].

In most the legislation provides for the addition of a further disease to the list by proclamation or regulation[13]. This means that quick action can be taken without waiting for Parliamentary approval if there was an out-break of an unlisted disease. However, there are problems that arise because the legislation is not uniform. For instance, there are gaps in the quarantine or quarantine-like provisions and therefore there is no uniformity of *legal* approach that can be activated if a national emergency situation arises. Administratively, no doubt, a plan can be worked out, but it is not yet lawyer-proof. There are also some difficulties in integrating the State and Federal legislation. As we have seen already, quarantine is a Federal responsibility and thus aspects of State legislation that are strictly quarantine matters could be over-ridden. However, there could be difficulty in deciding when State legislation ceases to be quarantine legislation and becomes merely public health legislation, or when it becomes merely animal disease legislation or plant disease legislation.

COMPENSATION AND LEGAL LIABILITY

Compensation for Loss

In all eight jurisdictions there are provisions for compensation for some losses. The provisions are not well developed in the case of human or plant diseases, but the provisions relating to stock diseases are more extensive and much better integrated between jurisdictions. Even in the animal diseases area there are still omissions or doubts that could cause legal difficulties in some jurisdictions.

Where human disease is involved some jurisdictions give compensation if property is destroyed to stop the spread of disease[14], but often there are provisions which permit the recovery of costs or expenses from people affected by disease[15].

In the plant diseases legislation there are many provisions which allow recovery of costs or expenses from a person[16]. There are some that provide no compensation for loss unless there is negligence or malice on the part of officials[17], and in just a few situations compensation will be paid for loss[18].

The animal diseases compensation provisions are the best organized. In 1955 a Commonwealth-State Cost-Sharing Arrangement was entered into to share the costs of eradication of foot-and-mouth disease, should an outbreak occur[19].

It has subsequently been extended to cover all vesicular diseases, as well as rinderpest, swine fever, African swine fever, rabies, Newcastle disease, fowl plague and bluetongue. Under the Arrangement the Commonwealth pays half the cost of any programme and the States and Territories the other half. It is truly co-operative in that, if an outbreak occurs in one State, that State does not pay half the costs involved and the

Commonwealth the other half, but *all* the States contribute to the States' half in proportions that are agreed in the Arrangement.

In the animal area too, costs and expenses can sometimes be recovered. However, all jurisdictions give compensation for loss of stock and usually other property as well; compensation in most cases is now at market or current value unless there is contributory negligence on the stock or property owner's part or the owner is convicted of an offence in relation to an eradication programme; and some, but not all, jurisdictions cover a wider range of exotic diseases than is covered by the Arrangement[20]. It is my view that all jurisdictions should amend their legislation to cover the full range of exotic diseases. Most jurisdictions expressly exclude compensation for loss of profit or use[21] and, because the compensation provisions of other jurisdictions provide only for specific cases that will be compensated, the law is probably the same in all jurisdictions.

The move towards uniformity in animal disease legislation is to be welcomed and I suggest that in all areas uniform legislation should be developed to ensure that there is uniform treatment of all innocent victims of an outbreak of an exotic disease.

Legal Liability

There is another side to the legal liability coin. As we know from the illustration of efforts to control Paterson's Curse (Chapter 24), where an injunction has been granted to restrain the release of a biological control agent, there are legal inhibitions on the use of some weapons to overcome outbreaks of a disease.

Legal liability could well flow from the use of an agent for destroying an exotic disease. For instance, it could be a spray[22] or a biological control agent. Furthermore, just as an exotic disease will not recognize State boundaries, neither would a bug such as is available to attack Paterson's Curse. If it is decided that it is in the public interest that such a bug should be used, there is very little likelihood that the Commonwealth alone would have legislative power to cover the whole area. If policy decrees that anyone releasing or using the bug should be relieved of potential legal liability, then here, too, it is my view that uniform interlocking State and Territory legislation would be appropriate. I realize that along the way will lie delay and difficulty, but it seems to me the only secure way to go[23].

CONCLUSIONS

Much has been done administratively to aid those guarding against outbreaks of exotic diseases and Commonwealth-State co-operation is substantial. However, ultimate action and responsibility must rest on legal controls and remedies.

Both Commonwealth and State authorities would wish to act as swiftly as possible in the face of a particular outbreak of an exotic disease which threatened either the public health of our community or economic base. But quick action can, in the "worst case" (when someone makes a legal challenge), be taken only if laws are in place to back up the administrative machinery. Unless adequate laws are in place, any administrative action can, quite properly, be challenged by a person who is disadvantaged by the action or any proposed action. People, plants or animals cannot be subject to restraint or action unless that action rests on a solid legal basis.

It is my contention that, in many areas, Australia does not have adequate legal means for very rapid control, and in some areas provisions for compensation for loss are not adequate. If there were legal challenges then, even in an emergency situation, much of the amicable co-operation that occurs may not be legally sustainable. Of course law is only the starting point, but I argue that there is a strong case, both in justice and for Australia-wide efficiency, for uniform laws to be developed to help us in any fight against any exotic disease invasion.

REFERENCES
1. *Quarantine Act 1908* s.2B.
2. Id s.5. Diseases in relation to animals and plants are also defined in s.5 and they, too, may be declared by Governor-General's proclamation.
3. Id s.12A.
4. Id s.13(1) (g) and (h).
5. Id s.13(3).
6. Id s.2A.
7. A.C.T. *Public Health Ordinance 1928* and *Public Health (Infectious and Notifiable Diseases) Regulations;* N.S.W. *Public Health Act 1902;* N.T. *Public Health Act;* Qld. *Health Act 1937–1982;* S.A. *Health Act 1935;* Tas. *Public Health Act 1962* and *Quarantine Act 1881;* Vic. *Health Act 1958;* W.A. *Health Act 1911–1982.*
8. A.C.T. *Stock Diseases Ordinance 1933;* N.S.W. *Stock Diseases Act 1923;* N.T. *Stock Diseases Act;* Qld. *Stock Acts 1915–1979* and *Exotic Diseases in Animals Act 1981–1982;* S.A. *Stock Diseases Act 1934;* Tas. *Stock Act 1932;* Vic. *Stock Diseases Act 1968;* W.A. *Stock Diseases (Regulations) Act 1969–1978* and *Exotic Stock Diseases (Eradication Fund) Act 1969–1983.*

9. A.C.T. *Plant Diseases Ordinance 1934;* N.S.W. *Plant Diseases Act 1924;* N.T. *Plant Diseases Control Act;* Qld. *Diseases in Plants Act 1929–1972* and *Diseases in Timber Act 1975;* S.A. *Fruit and Plant Protection Act 1968;* Tas. *Plant Diseases Act 1930;* Vic. *Vegetation and Vine Diseases Act 1958;* W.A. *Plant Diseases Act 1914–1981.*

10. In the following footnotes the sections cited without the name of an Act or Regulation refer to the relevant Act or Regulation cited in full in footnotes 7–9. That is, when discussing human disease, animal disease or plant disease, the references are to the relevant public health, stock or plant legislation.
    Human diseases. A.C.T. Regs. 4, 4A, 4B and 15; N.S.W. ss.29 and 50D; N.T. (there seems to be no specific provision); Qld. ss.30 and 51; S.A. ss.73, 127, 128 and 135; Tas. ss.38, 142(1A) and Reg. 3 of *Public Health (Notifiable Diseases) Regulations 1967;* Vic. ss.137, 138, 141, 162A and 342; W.A. ss.152 and 276–289. Also Commonwealth *Quarantine Act 1908* ss.22.
    Animal diseases
    A.C.T. ss.7 *and 8 and Commonwealth Livestock Diseases Act 1978* s.6; N.S.W. s.9; N.T. ss.35 and 36; Qld. s.23 and *Exotic Diseases in Animals Act 1981–82* s.8; S.A. s.19 and *Health Act 1935* s.107; Tas. s.10; Vic. ss.11 and 12; W.A. s.10(2)(c).
    Plant diseases
    N.S.W. s.10; N.T. s.12; Qld. s.16 and *Diseases in Timber Act 1975* s.10; S.A. s.8; Tas. s.9; Vic. ss.5A and 43; W.A. s.10.

11 *Human diseases*
    A.C.T. Regs. 5, 7–10 and 17; N.S.W. ss.30A and 32A and *Public Health (Amendment) Act 1937* ss.4 and 6; N.T. (no specific provision); Qld. ss.33, 34A, 35–38, 50, 53 and 56; S.A. s.144; Tas. ss.16, 17, 20 and *Quarantine Act 1881.* Also Commonwealth *Quarantine Act 1908* s.5.
    Animal diseases
    A.C.T. s.10 and Commonwealth *Livestock Diseases Act 1978* s.6; N.S.W. ss.3(1) and 10–15; N.T. ss.12–26; Qld. ss.3 and 13–14A and *Exotic Diseases in Animals Act 1981–1982* ss.9–19; S.A. ss.5(1), 6–8b and 13; Tas. ss.9 and 12; Vic. ss.3, 4 and 28; W.A. s.10.
    Plant diseases
    A.C.T. s.6; N.S.W. ss.5, 6 and 8; N.T. ss.9–11; Qld. ss.4 and 15 and *Diseases in Timber Act 1975* s.4; S.A. ss.3 and 5–7; Tas. ss.4, 8 and 13; Vic. ss.3, 5, 5A, 25, 31A and 33; W.A. ss.5, 6, 12 and 15.

12 *Human diseases*
    A.C.T. Reg.3; N.S.W. ss.28 and 50C; N.T. s.10; Qld. ss.29 and 32; S.A. s.4 and Second and Third Schedules; Tas. ss.3 and 13 (declared by Governor); Vic. s.3; W.A. s.3.
    Animal diseases
    A.C.T. ss.3 and 4 (declared by Minister); N.S.W. s.3(1); N.T. ss.5 and 6 and *Exotic Diseases (Animals) Compensation Act* s.3 and Schedule; Qld. s.3 and Exotic Diseases in Animals Act 1981–1982; S.A. s.5(1), *Health Act 1935* s.106 and *Foot and Mouth Disease Eradication Fund Act 1958* ss.3 and 4 (declaration by Governor); Tas. s.3(1) (proclamation); Vic. ss.3 and 4 (declaration by Governor), 20 and 38; W.A. s.6 and *Exotic Stock Diseases (Eradication Fund) Act 1969–1983* s.4.
    Plant diseases
    A.C.T. s.5 (declared by Minister); N.S.W. ss.3 and 4 (proclamation by Governor); N.T. ss.6 and 12 (notice by Minister); Qld. ss.4 and 5 and *Diseases in Timber Act 1975* ss.3 and 4 (proclamation by Governor); S.A. ss.3, 8 and 8a (proclamation by Governor); Tas. ss.2, 4 and 13 (proclamation by Governor or by regulation); Vic. ss.3, 12A and 32 (proclamation by Governor); W.A. s.4 (by proclamation).

13 *Human diseases*
    A.C.T. Reg. 3; N.S.W. ss.28 and 50C and *Public Health (Amendment) Act 1937* s.3; N.T. s.10; Qld. s.29; S.A. ss.5 and 5A; Tas. ss.3 and 13; Vic. s.3; W.A. ss.3 and 248.
    Animal diseases
    A.C.T. ss.3 and 4; N.S.W. s.4; N.T. ss.5 and 6; Qld. s.3 and *Exotic Diseases in Animals Act 1981–82* s.5; S.A. ss.5(1) and 8a and *Foot and Mouth Disease Eradication Fund Act 1958* ss.3 and 4; Tas. s.3(1); Vic. ss.3, 4 and 38; W.A. ss.6 and 7 and *Exotic Stock Diseases (Eradication Fund) Act 1969–1983* ss.4 and 5.
    Plant diseases
    A.C.T. s.5; N.S.W. ss.3 and 4; N.T. ss.6 and 12; Qld. ss.4 and 5 and *Diseases in Timber Act 1975* ss.3 and 4; S.A. ss.3 and 8; Tas. ss.2, 4 and 13; Vic. ss.3 and 12A; W.A. s.4.

14 N.S.W. ss.36 and 50A; Qld. ss.40, 161 and 162; Tas. s.24; Vic. s.134; W.A. s.259.
15 A.C.T. s.10; N.S.W. s.34; S.A. ss.66–69 and 145; Tas. s.26.
16 A.C.T. s.13; N.S.W. ss.16 and 27; N.T. ss.20 and 24; Qld. ss.10, 11 and 13; S.A. s.13; Vic. ss.11 and 29; W.A. ss.14, 18, 22 and 24.
17 A.C.T. s.13A; N.S.W. s.25; N.T. s.22; S.A. s.14; Tas. s.11; Vic. s.22 (but there are some provisions for compensation); W.A. s.32.

18 Qld. s.21(2) and *Diseases in Timber Act 1975* s.11; S.A. *Phylloxera Act 1936* ss.40–42; Vic. ss.22, 28 and 37–40.
19 The relevant Commonwealth legislation is now the *Livestock Diseases Act 1978*.
20 A.C.T. s.8 and Commonwealth *Livestock Diseases Act 1978* ss.11–17; N.S.W. ss.17B–17HA; N.T. *Exotic Diseases (Animals) Compensation Act;* Qld. ss.15, 15A and 25B and *Exotic Diseases in Animals Act 1981–1982* ss. 25–37; S.A. ss.13, 26, 27 and 43a, *Swine Compensation Act 1936, Cattle Compensation Act 1939* and *Foot and Mouth Disease Eradication Fund Act 1958;* Tas. ss.13, 14 and 19A and *Public Health Act 1962* s.24; Vic. ss.42–47; W.A. s.14 and *Exotic Stock Diseases (Eradication Fund) Act 1969–1983* ss.9–14.
21 N.S.W. s.17F; N.T. s.11; Qld. s.32; S.A. s.13; Vic. s.45; W.A. s.12.
22 Although it involved a spraying programme for the control of grasshoppers and not an exotic disease, a Council was held liable for the death of cattle where grass had been sprayed: *Buffier* v. *Warrah Shire Council* (1938) 56 W.N. (N.S.W.) 9. This was an application of the ordinary principles of negligence and nuisance; what was unusual was that the Council was under a statutory duty to spray, but did it negligently.
23 Since this paper was delivered the *Biological Control Bill 1984* has been introduced into the Federal Parliament. It follows the pattern outlined here. It will negate legal liability where an eradication organism is released after a complex inquiry and approval process has been followed. It will also apply only to the A.C.T., but it is intended to be a template for uniform legislation throughout Australia.

Chapter Eight
ECONOMIC EFFECTS OF AN EXOTIC DISEASE OUTBREAK IN AUSTRALIA

Joe Johnston

INDIVIDUAL ACTIONS VERSUS COMMUNITY RESPONSE

The reactions of individuals to a disease invasion are not always, in aggregate, the best for society as a whole. In past emergencies, individuals have shown understandable but countersocial behaviour, sometimes in complete ignorance of the hazards created for the community. For example:

1. people have gone ahead with disease control action of their own, without coordinating with their neighbours even though this might have given the best results;
2. individuals have sought to rid themselves of infected material without proper precautions;
3. people, who are ignorant of diseases, have kept well-loved, personally much needed, but infected plants or animals.

Individual actions such as these impose costs or benefits on others, but are not subject to reward or, often, to penalty. Because such reactions have effects which extend beyond the affairs of the individual decision maker, economists have labelled such behaviour as 'externalising'. Johnston (Jim) and McInnes (1983) have recently reviewed many of these issues as they relate to animal health and pest control. Efficient control of such behaviour is very important in public policy relating to health, environmental and developmental issues.

Governments sometimes must intervene and invest in disease control, because the often conflicting interests of people as individuals, as opposed to special interest groups, such as industries, might lead to market failure. The solution is for governments to introduce forms of legislation or to select policies, which will give people the incentive to operate nearer to what is best for all.

Depending upon the nature and seriousness of a disease outbreak, the areas where governments could become involved are:

1. helping disadvantaged groups;
2. judging whether compensation should be paid and how it should be administered, and;
3. establishing what help should be given to reconstruct or restructure disrupted societies and industries.

INFORMATION SOURCES FOR POLICY MAKING

Following an exotic disease invasion, government advisors and decision makers may be only slightly better informed than individuals. Information upon which to decide on, for example, appropriate levels of incentives or compensation would probably be lacking because of the very 'newness' of the disease discovered and the lack of relevant economic research. In such a situation, the role of government would be to provide welfare, to assist the affected industries and individuals, to provide any available information, and to maintain community standards on the use of drugs, chemicals and medicines.

Economic decisions with a long-term effect, such as whether to attempt to combat the disease, or to allow it to 'run its natural course', and possibly become endemic, are more difficult because the epidemiological characteristics of a disease in a new environment are probably unknown.

The costs of providing information rapidly would probably be great as it might involve diverting specialists from other high priority tasks. However for an urgent and important emergency, this might be justified, though if a particular disease was known to be a real threat, planning can be done well in advance.

An example of this 'strategy of preparedness' is the study done on the possible entry into Australia of screwworm fly (Australian Bureau of Animal Health, 1979). This work led to a number of policies being initiated that better insured the livestock industries against the pest. Another study (Johnston, 1982), done for the wool industry and government, examined the likely impact on the Australian economy of the entry of foot-and-mouth disease (FMD) or Rift Valley fever (RVF), and outlined some policies, which could help to minimise the adverse effects that would follow.

Contingency plans, the availability of trained personnel, and the established Commonwealth/State disaster cost-sharing arrangement which form a normal part of the Australian system of handling exotic disease emergencies, provide an infra-structure well capable of dealing with all but the most serious disease intrusions. Recently it has been decided to strengthen this capability by increasing the supply of information, 'combat resources' and trained expertise by a program of contingency research and training within a more general program of research. This is part of the rationale of the new high security animal health laboratories recently commissioned in both Australia and the Republic of South Africa.

IMPACTS

Exotic disease in any nation can affect communities in a bewildering number of ways. The case histories included in this volume provide a wide range of carefully selected examples which may be examined against the socio-economic framework given here.

Disease can affect people directly, as would be the case with a newly introduced human disease. But its influence also can be indirect through social, physical and economic mechanisms.

Impacts on individuals

The impact can be direct, for example, in effects on:

— human health and activity;
— food supplies, e.g. in subsistence farming systems;
— the availability of animals for draft and transport;
— availability of fuel and building materials.

Consequential effects, sometimes requiring some time to appear, might be noticed in:

— the work and leisure environment;
— family and community lifestyles;
— the availability of employment;
— food prices and availability;
— prices received for produce;
— taxes, costs of social and health services;
— the standard of living;
— the natural heritage;
— the value of personal or business assets;
— investment options;
— the ability to borrow and repay.

Impacts on industry

The reasons for the consequential effects on individuals would usually be a result of the new disease having an impact through commerce, industry, or the provision of government services. The impacts of a new disease can be relatively direct, it might affect, for example:

— output quantity i.e. yields;
— costs of production, of distribution;
— quality of output and harvestability.

But, the complexities and interactions of commerce render more and more groups vulnerable through business association, in the following ways:

— by controls on the distribution of commodities;
— by affecting associated sectors e.g. transport/processing;
— by affecting regional resource use opportunities;
— by changing domestic and export returns;
— by restricting reinvestment possibilities;
— by eroding competitive capacity;
— industry indebtedness and debt repayment capacity

Economic impacts

All these forces add up to a national economic impact, whereby the following could be affected:

— balance of trade;
— gross domestic product;
— banking flows;
— export revenue;
— taxation revenue;

— employment;
— savings/interest rates;
— retail sales;
— investment.

IMPLICATIONS FOR AUSTRALIANS

Epidemiologists have observed that the increasing levels of intensity of agricultural, forestry and animal industry throughout the world render those industries and interdependent communities more susceptible to exotic diseases and their effects (Ellis, 1980). Increased world travel increases pressures on quarantine services, increasing the chance of pest and disease penetration. On the other hand the concentration of some intensive industry installations into high security areas can be an advantage in protection against disease entry and in control following an outbreak.

Trends in the technological developments of Australian rural industries, and the participation of its community in international travel growth, are similar to those of many other nations. As a consequence, the people of Australia and the Australian economy face many of the same problems as those of other 'westernised' nations if there was an outbreak of an exotic disease.

The impact of exotic disease on resources or industries in Australia which are oriented towards the domestic market can mostly be predicted with precision. Recently, there has been considerable industry involvement in undertaking studies of possible strategies to increase preparedness for a range of possible disease problems.

Models for examining the likely effects, and testing the economic merit of different strategies in domestically closed economies, have been demonstrated, for example, for eradication strategies for FMD (Power and Harris, 1973; McCauly et al, 1979), and also for screw-worm fly (Australian Bureau of Animal Health, 1979). The main requirement for such predictive work and exploration of policy options is some knowledge of the likely effect of the disease on the technical and economic performance of the industry under study.

In outbreaks, it has been found that producers are affected by cost rises associated with movement controls, increased health inputs, destruction of stocks and facilities, the cost of idle capital, and lost sales. Consumers are affected by price rises of the commodity involved; due to increased scarcity.

Of course, there are those that benefit, too. Producers of commodities which can substitute for the one affected get a windfall opportunity to expand their enterprise, and producers not involved in the outbreak have scope to use any surplus production capacity to supply the unexpected demand. These factors have been satisfactorily modelled in economic terms to provide guidance in contingency studies.

The effects of outbreaks of avian disease on domestic poultry (Roe, 1984); of disease outbreaks in pig production installations, and problems with the health of forestry and plantation crops (all examples of industries based on a specialised infrastructure) can be very serious for those directly concerned in production. Substitution effects and other adjustments can absorb these impacts as they flow through the economy. This means that these types of outbreak are rarely sufficiently powerful to be observed to have any effect in national economic indicators.

Because the outbreak of exotic disease in industries with internal markets poses similar socio-economic problems in Australia, US American and European countries (McCauly et al, 1979; Power and Harris, 1973), considerable benefits are derived from international exchange of experiences.

Human disease outbreaks appear to create similar social, welfare, and economic problems irrespective of the country in which they occur.

TRADE IMPLICATIONS

It is not generally realised that for some nations, including Australia and New Zealand, severe economic effects would result more from the discovery of the disease agent than from its pathological effects. The discovery of a disease which affects. The discovery of a disease which affects, or is thought to affect, the international saleability of produce can cause the official disease status of a nation to be changed. For export-dependent countries this is an extremely important issue, because change in status can threaten the export possibilities for the whole nation, and not just one commodity but a range of commodities (Doyle, 1980). Producers and processors involved with that range, irrespective of whether they have the disease or not, irrespective of whether it is affecting animals or plants in the immediate district or one 2000 kilometres away, are then put in a position where they have to reconsider at short notice their future production and marketing options.

Australia has experienced a number of potentially serious trade setbacks in recent years, fortunately of relatively short duration. One case, in the late seventies, involved the discovery of a bluetongue virus in northern Australia. This was proved by subsequent investigations to be a harmless relative of the bluetongue orbiviruses that cause disease in sheep or other parts of the world. Other instances have involved the suspected presence of vesicular disease agents, which on investigation have been found to be false alarms.

The pattern of contribution of the major sectors of the economy towards the 'gross domestic product' (GDP) in Australia is currently, for agriculture, mining, manufacturing and tertiary industries about 6%, 5%, 20% and

69% respectively. The value of total exports, with about one third originating from each of the agricultural, mining and manufacturing sectors, is about $20,000 million, 15% of the GDP. The economy of New Zealand has similar characteristics, with a greater proportion depending on exports of agricultural and forestry products. A substantial change in output or exports could significantly affect these economies.

The export items from the rural industry which could substantially affect the economy are wheat, wool and meat. Disruption of the exports of any one of these could be sufficiently influential to affect the exchange rate. The study by Johnston (1982) of the effect of an outbreak of trade-disrupting foot-and-mouth disease in Australia indicated that net income losses in the rural sector, without counting any effects carrying through to other sectors, would amount to at least $2,000 million in the first year following an outbreak. The implications of such a large 'shock' to the established agricultural community is obvious.

Thus some types of exotic disease would affect the international trading of animal and plant products by Australia and New Zealand quite differently from many other countries. As a consequence, despite the technical problems being similar to those likely to be faced overseas, the main priority after an outbreak of disease would be to reinstate trade as rapidly as possible, if expensive industry restructuring is to be avoided. Where rapid eradication of the disease is not possible, or that access to markets overseas cannot be quickly regained, industries would have to adjust to the new situation. If the industry affected is capable of adjusting to produce alternative commodities, the effects need not be too serious. Young (1984) emphasized that much of the Australian rural industry has been able to adjust to severe economic pressures resulting from changing overseas demand for our produce over recent decades. Such adjustment has required investment capital, the adoption of new techniques, and changed management, but some of the changes have taken place surprisingly quickly. The Johnston study of the possible impact of a prolonged FMD emergency concluded that if producers realized that the emergency was to be longlived, producers with the capacity to adjust would make the best of new conditions and salvage a substantial portion of the net income they lost initially. However some highly specialised industries, using land and/or an infrastructure for which no economic alternatives exist, might have to close down. Industries likely to be vulnerable to this problem are the pastoral sheep and cattle industries of northern and inland Australia.

With several hundreds of millions of dollars at stake, and the possibility of saving trade by adopting stringent and vigorous measures to rid the nation of trade-impeding disease, Australians must decide whether they would support the strong veterinary actions, and financial incentives such as compensation measures, required to achieve fast reinstatement of trade. Economic analysis indicates that in a major epidemic of FMD it could be economically justified to slaughter up to one quarter of the national herd and flock to eradicate the disease promptly, rather than employing a strategy of prolonged vaccination. However the financial implications of compensation are immense, as would be the sheer scale of veterinary operations. By contrast, the conventional wisdom in countries such as the USA, where domestic supply and demand are the most important factors, is to replace a failing slaughter campaign by the slower vaccination policies after only one percent of the population of livestock has been eliminated.

It must be stressed that such emphatic and stringent policies would only be justified in relatively few instances, such as the outbreak of FMD in our cattle and sheep industries, or the discovery of a disease such as "Karnal bunt" in our wheat industry. Disease affecting our own industries internally would largely be dealt with by existing arrangements and infrastructure.

FUTURE DEVELOPMENTS

There is increasing interest in identifying ways to cope with exotic disease outbreaks in Australia. The approach taken by governments is through contingency planning, agreed cost-sharing arrangements between State and Federal authorities, and planning of counter-disaster arrangements (Australian Bureau of Animal Health/Department of Defence, 1980). Efforts of governments are also directed towards providing information (e.g. French and Geering, 1979) and training to maintain skills to recognise disease and take appropriate action. In recent years industry bodies involved in disease issues have increased the level of their participation in the debate on exotic disease policies. The question of possible trade embargoes and the issues surrounding compensation are highly relevant to industry.

Plans and policies are constantly under review and development so that they move with changes in the structure of the economy and incorporate new ideas and information as they become available. The process of review is soon to concentrate on the Commonwealth/States cost-sharing agreement. As part of this process it is likely that the relative roles of governments, industries and individual producers in disease control and eradication will be reexamined.

The question of compensation payments and government help to persons or groups affected by significant disease influences will probably be examined as part of the review. Should Australia ever have to face the issue of administering compensation payments in a large and prolonged outbreak of a trade-disrupting disease such as FMD, it is likely that adherence to currently laid down valuation and administration concepts would not achieve the dual objectives of equitably compensating those most affected and encouraging the best allocation of resources.

Basically, compensation is applied to cases where the community requires private resources to be destroyed. In an emergency environment characterised by extremely unstable and falling market prices, payment of any compensation to operator having diseased stock would advantage them over their counterparts having non-diseased stock. The administration of government assistance, be it compensation, restructuring loans, welfare assistance or other forms of help, should also aim to acheive an appropriate restructuring of the industry. The majority of people concerned will probably consider that what existed before the outbreak is the desirable norm which should be restored, however this may not be the best form of the industry for the future. Compensation can never be successfully applied as a measure to restore an industry to its original status following an emergency affecting industry structure. These issues have been reviewed against current economic theory by Johnston (Jim) and McInnes (1983).

In national disasters or where the whole community is affected, it is perhaps best if, in the early stages of a compaign, administrators merely attend to the welfare issues. To attempt to administer once-and-for-all compensation payouts in this phase is unworkable. Far better, to wait until new market forces have emerged and indicate the optimal post-invasion industry structure, then appropriate assistance and welfare could be allocated to optimize the value to society as a whole.

REFERENCES

Australian Bureau of Animal Health (1979). *Screw-Worm Fly — possible prevention and eradication policies for Australia* Australian Government Publishing Service, Canberra.

Australian Bureau of Animal Health/Department of Defence (1980). *Exotic Animal Disease Seminar — report of proceedings* Australian Counter Disaster College, Macedon.

Doyle, K.A. (1980). *Proceedings of the Second International Symposium on Veterinary Epidemiology and Economics* (W.A. Geering, R.T. Roe and L.A. Chapman, eds) p. 279 Australian Government Printing Service, Canberra.

Ellis, P.R. (1980). *Proceedings of the Second International Symposium on Veterinary Epidemiology and Economics* (W.A. Geering, R.T. Roe and L.A. Chapman, eds) p. 11 Australian Government Printing Service, Canberra.

French, E.L. and Geering, W.A. (eds) (1978). *Exotic Diseases of Animals — a manual for diagnosis* Service Publication (Animal Quarantine) Number 11, Department of Health/Australian Government Printing Service, Canberra.

Johnston, Joe (1982). *Exotic Animal Disease Emergencies in the Australian Grazing Sector — an economic study* Australian Bureau of Animal Health, Canberra.

Johnston, Jim and McInnes, K. (1983). Compensation planning in public pest control. Paper at 27th *Conference of Australian Agricultural Economists Society*, Brisbane.

McCauly, E.H., Aulaqi, N.A., New, J.C., Sundquist, W.B. and Miller, W.M., *A Study of the Potential Economic Impact of Foot and Mouth Disease in the USA* University of Minnesota, 1979.

Power, A.P. and Harris, S.A. (1973). *Journal of Agricultural Economics* 14, 573.

Roe, R.T. (1984). Economic consequences of an avian exotic disease outbreak in Australia Paper at *Poultry Industry Association Seminar*.

Young, R. (1984). in *Economic Development in East and South East Asia — implications for Australian agriculture in the 1980's* p. 19. Bureau of Agricultural Economics, Australian Government Publishing Service, Canberra.

Chapter Nine

THE AUSTRALIAN RESPONSE TO THE THREAT OF AN EXOTIC DISEASE INCURSION: THE ESTABLISHMENT OF ANAHL

Ron Johnston and Pam Scott

The establishment of the Australian National Animal Health Laboratory has, for many years, been presented as the major element in the Australian armoury against an exotic disease incursion.

This paper examines the way in which the series of decisions to construct the $160 million facility were made. In particular, we review the arguments and evidence used, and interests involved in establishing:

1. the need for a maximum security animal health laboratory;
2. the economic justification of the expenditure; and
3. the value of the proposed functions of diagnosis, training, research and vaccine production.

The debate about the need and nature of ANAHL was conducted almost entirely among institutions and individuals committed to the laboratory, and their perception of need and judgements of value largely determined the decisions reached. The changing scientific and political climate led to a continuing shift in the arguments and evidence offered, but these were all designed to maintain the basis for the initial concept of ANAHL. These shifts, however, eventually served to highlight the uncertainties and value judgements in the arguments, and to an erosion of the credibility of the decision-making institutions and the proponents of ANAHL.

Now established, ANAHL may make a useful contribution to minimising the harmful effects of exotic diseases on Australian agriculture. However, as an exercise in science policy decision-making, it demonstrates the weakness of a system where scientific proponents dominate the roles of counsel, judge, and political decision-maker as well. It also highlights the inadequacies of major science policy decisions taken on an *ad hoc* basis, in a climate of special pleading, and in the absence of a general guiding policy.

INTRODUCTION

There has been a long history of concern in Australia over the invasion of exotic diseases. the combination of isolation from other countries and their various diseases, and the different climatic conditions provide what has long been recognised as considerable advantage to Australian agriculture, and these advantages have been jealously guarded.

Since at least 1964, the establishment of a maximum security laboratory has been held by many to be one of the most important means for protecting this continent against the effects of exotic diseases, in particular foot and mouth disease (FMD). This means has finally been realised, amidst considerable controversy, in the form of the Australian National Animal Health Laboratory (ANAHL).

The purpose of this paper is not to evaluate the appropriateness or effectiveness of ANAHL as a protection against exotic disease invasion. Still less is it concerned with determining responsibility or attributing blame, or credit, for the outcome. Rather, it is concerned with evaluating the way in which a major science policy decision was made, the arguments and evidence that were brought to bear, and the adequacy of the procedures used.

A brief history of ANAHL decision-making
The first initiative in the establishment of ANAHL was a request to Dr E.A. Eichorn, a senior official in the U.N. Food and Agricultural Organisation with experience in foot-and-mouth disease (FMD) eradication, to visit Australia "to draw up a plan to protect this continent from the disease"[1] The outcome was a

recommendation that Australia establish its own maximum security laboratory to provide diagnostic, and vaccine testing and production facilities, in the event of an outbreak of exotic disease.

In response to this Report, the Commonwealth States Veterinary Committee formed a Working Party whose terms of reference included an investigation of the desirability of, and need for, a maximum security animal health laboratory, its functions, siting and staffing. Its recommendations, supporting the establishment of such a facility, were presented to the Australian Agricultural Council (AAC) in February, 1970.

Independently, the Commonwealth Minister of Health set up an inter-departmental Committee made up of members from the Departments of Health, Primary Industry and Treasury and the CSIRO. Its recommendations which were in general agreement with those of the Veterinary Committee, were presented to the AAC in July, 1970.

At the July meeting of the Council, the Standing Committee and senior officials recommended that a panel should be formed, comprising senior representatives of the States to consult with the Commonwealth. This Panel met in August, formed an eleven-person Advisory Proposal Committee which in turn formed a Proposal Evaluation Team (PET). Between October 1970 and December, 1970, the members of PET visited fifteen overseas laboratories. The PET Report was published in 1972 and this report formed the basis of a joint submission by the Ministers for Education and Science, Health, and Primary Industry to the Commonwealth Government. In October 1972 the Commonwealth Government agreed in principle to the establishment of a maximum security animal health laboratory.

A further joint submission to the Government was made in 1973 by the Ministers for Science, Health, Primary Industry and Northern Development. This submission was accompanied by the CSIRO Proposal for a National Animal Health Laboratory Report and an Environmental Impact Study produced by the CSIRO and the Department of Works. In 1974 the Government approved the selection of the Geelong site, and the proposal was referred to the Parliamentary Standing Committee on Public Works (PWC) by the House of Representatives.

A public enquiry was held at Geelong in September 1974 by the PWC. The hearing was dominated by the CSIRO, the Department of Health, the Department of Housing and Construction and the Bureau of Agricultural Economics. The Australian National Cattle Council and the Australian Veterinary Association sent written submissions only. The recommendations and conclusions contained in its report to Parliament were:

1. There is a need to establish a maximum security Animal Health Laboratory to ensure the prompt and reliable diagnosis of exotic animal diseases.
2. The proposal is economically justified.
3. The "box within a box" principle of design of the laboratory will ensure microbiological security.
4. The proposed functions of the laboratory are appropriate.
5. The precautions taken to prevent the escape of infectious disease viruses will be an improved version of those which have been successful in a number of similar laboratories overseas.
6. After a suitable proving period the laboratory should be authorised to handle foot-and-mouth disease virus before an outbreak of the disease occurred in this country.
7. The site selected is suitable.
8. The Committee recommended the construction of the laboratory as proposed.
9. The Committee considered that the construction and establishment of the laboratory should proceed as a matter of urgency[2].

This report, including the recommendation that construction should proceed as a matter of urgency, was accepted in its entirety when submitted to the Parliament in late 1974. However, building did not commence until 1978, apparently in response to the isolation of a bluetongue orbivirus in northern Australia in November, 1977, and the presentation of a submission by the four major primary producer organisations, calling for the immediate commencement of construction of ANAHL[3].

In December 1979, the Government decided to accelerate construction by one year at an additional cost of $7 million. This followed an exotic disease scare at Legana in Tasmania, where pigs developed symptoms of a vesicular disease[4].

In 1978 the ANAHL Consultative Committee was established by the Ministers for Health, Primary Industry and Science and Technology, to advise the CSIRO Executive on all matters pertaining to the program and operations of ANAHL. In 1979 this Committee recommended that ANAHL should undertake research on the development of FMD vaccines, which would involve the introduction of FMD virus prior to an outbreak of the disease in Australia.

This proposal was put to the Animal Health Committee (formerly the CSVC) and eventually the Agricultural Council. In February 1980, the AAC endorsed the principle of allowing live FMD virus to be imported before an outbreak occurred.

The CSIRO Executive then advised the Minister for Science and the Environment that it would be in the national interest for ANAHL to have access to FMD virus in advance to an outbreak. The Minister for Science and the Environment sought the support of the Ministers for Primary Industry and Health in July 1980, and together they approached the Prime Minister, who endorsed this recommendation in November, 1980.

This decision heralded the start of much greater public involvement and controversy. Although triggered by the decision to import live FMD virus before an outbreak, the controversy extended to a questioning of the need for, and the functions of ANAHL. The events and resolution of the controversy is beyond the scope of this paper. However, it included a Forum at Geelong in August, 1982, reports by the Australian Science and Technology Council and the Australian Academy of Sciences, a decision by the Government not to permit the importation of live FMD virus for five years, and a review of the functions of ANAHL. In summary, these events were:

Year	Month	Event
1964		Visit by Dr Eichhorn of FAO.
1970	February	Commonwealth–States Veterinary Committee recommendations to Australian Agriculture Council
	July	Inter-Departmental Committee recommendations to Australian Agricultural Council
	August	Australian Agricultural Council Panel forms an Advisory Proposal Committee which in turn forms a Proposal Evaluation Team (PET)
	October–December	Proposal Evaluation Team visits overseas laboratories
1972	October	Proposal Evaluation Team Report published Joint submission to Government by Ministers of Education and Science, Health and Primary Industry; Construction of ANAHL approved in principle by Government
1973	May	CSIRO Report "A Proposal for a National Animal Health Laboratory"
	October	Environmental Impact Study of Geelong Site
1974	April	Geelong Rifle Range selected as site
	July	Proposal for ANAHL referred to Parliamentary Standing Committee on Public Works (PWC) by House of Representatives
	September	PWC Public Enquiry
1977	November	Primary industry organizations press "The Urgent Case".
1978	March	Building commences
1980	February	Australian Agricultural Council endorses importation of live foot-and-mouth disease virus
	November	Endorsed by Prime Minister
1981	April	Professor B. Morris critical at Annual Conference of Cattle Council of Australia
1982	August	ANAHL Forum at Geelong
1983	May	ASTEC and Australian Academy of Science reports produced. Importation of live FMDV subsequently banned by Government for five years.

The Public Works Committee Inquiry provides a basis for the structure of this paper. Three of the major conclusions of that Inquiry were:

1. There is a need to establish a maximum security animal health laboratory;
2. The proposal is economically justifiable;
3. The proposed functions are appropriate.

We have selected these as the main decisions for analysis. However our analysis is not restricted to the findings of the PWC Inquiry. Rather, it examines the ways that arguments and evidence were used and changed over time in support of the achievement of particular scientific and political objectives.

1. THE NEED FOR ANAHL

Although the basic question of how Australia should be protected against exotic disease would, *a priori*, appear capable of a number of different answers, that Dr Eichorn recommended a maximum security laboratory appears to have served to discount all possible alternatives. Nor should his recommendation be regarded as very surprising, given his position as officer-in-charge of the Palo Alto Diagnostic and Vaccine Production Institute before joining the FAO.

The Inter-Departmental Committee, arranged by the Department of Health and the Veterinary Committee, considered only the desirability and need for the laboratory and not whether there were alternative means of achieving the desired objectives. By the time the Agricultural Council Panel was formed in 1970, the need for ANAHL had apparently become beyond question, and the terms of reference of the Project Evaluation Team were:

1. to determine the feasibility of establishing within Australia a research, diagnostic and vaccine safety and potency testing laboratory together with a unit for producing FMD vaccine;
2. to estimate the approximate cost of establishing and running such a laboratory[5].

One of the terms of reference for the Public Works Committee Inquiry was to investigate the need for a maximum security laboratory. But by then, not only had there been ten years of assumption of the need, but the Government, just five months prior to the public hearing, had:

1. approved the establishment of ANAHL on the Geelong Rifle Range Site;
2. agreed to the formation of a Consultative Committee to assist in the management of ANAHL;
3. noted that the recurrent cost would be additional to the CSIRO's budget requirements; and
4. approved that the Department of Housing and Construction should document the project to the point where reference could be made to the PWC[6].

The case for the need for ANAHL at the PWC Inquiry was largely made by representatives from CSIRO, but the same line was followed by the submissions of the Department of Health, the Australian National Cattlemen's Council, and the Geelong Regional Planning Authority. The arguments can be divided into two types.

The first argument focussed on the value of the livestock industry, its economic importance to Australia, and the likely cost of an exotic disease outbreak. In addition to estimates of the local production, the CSIRO representative devoted much of his evidence to the costs and impact of several types of animal diseases in various countries, and concluded "the reasons why Australia needs a maximum security laboratory are apparent from what has already been said"[7]. But a close examination of this evidence revealed no explicit argument of the need for a maximum security laboratory. In fact, a recurring and unquestioned assumption apprears to be that the value of export trade and the cost of disease outbreaks overseas are evident justification for ANAHL. To the committed, the connection is obvious. To the critical, the evidence could be interpreted in exactly the opposite direction, as most of the countries in the cited examples already possessed their own laboratories at the time of the outbreak.

The second type of argument concentrated on the inadequacy of quarantine to protect Australia and the increasing risk of an outbreak. Dr Pierce of CSIRO stated: "The quarantine service operated by the Australian Department of Health has so far proved an effective barrier against the accidental introduction of these diseases, but no quarantine service, however efficient, can hope to provide an absolute guarantee against their entry"[8]. With the advent of fast air travel and as more people travel to Australia "the risk of exotic diseases penetrating our quarantine barrier inevitably becomes greater"[9].

During cross-examination, Senator Poyser (PWC Member) questioned this assertion, indicating surprise that this was included in the CSIRO submission rather than the Department of Health submission, and asked, "Is this an assessment of CSIRO in relation to this matter?" Dr Allen for the CSIRO replied: "It does not represent a formal assessment on our part, but it is a fact which I think most people are aware of and it is fairly widely quoted"[10]. On the basis of this "widely quoted" observation, Dr Pierce stated in his submission that there is a "real and growing probability that an exotic virus will, sooner or later, penetrate Australia's quarantine barriers"[11]. However, the Quarantine Division of the Department of Health, not unnaturally, took a rather more positive view of the effectiveness of measures to exclude exotic diseases. The First Assistant Director General, Mr Bill Gee, in response to a question from Mr Garrick about the sort of extra precautions required, replied:

"I think that our precautions at the moment are adequate. We get periodic problems through human error, admittedly, which are very quickly brought to our attention, but I think as air traffic increases that we will require more staff to be able to handle the people and their goods . . . The main answer I think will be people to deal with the increased number of passengers and increased numbers of aeroplanes rather than involved and expensive equipment or research"[12].

When Mr Keogh suggested that the CSIRO evidence made it appear that the entry of FMD disease would be inevitable, Mr Gee replied:

"No, I do not think that it is inevitable, but I certainly would not be prepared to say there is absolutely no risk. There is a continuing although small, risk in my opinion of FMD being imported into the country. I think the risk is very low while we do not permit the import of live animals or livestock products from FMD infected countries . . . the probability of the disease being transmitted by shoes and clothing to animals in a dose sufficiently high to set up the infection and to initiate an outbreak, I think is low . . . I do not believe there is an extremely high risk of the introduction of FMD"[13].

It should be remembered that Australia has not had any probable outbreaks of FMD since 1872, rabies since 1867, or rinderpest since 1923. Contagious bovine pleuro-pneumonia, which entered Australia on 1858 was eradicated in 1967[14]. The Consultative Committee of the AAC, which meets only in the face of a suspected disease situation has met only four times in the last twenty years; in 1966 in regard to a "very mild strain of Newcastle Disease"[15], in 1973 for what was incorrectly thought to be a disease in horses, in 1977 when the avirulent bluetongue virus was isolated in northern Australia, and in 1982 for the Legana Pig incident in Tasmania.

The only argument offered at the PWC Hearing in support of the likelihood of an exotic disease incursion was the increased speed and volume of air travel. However, it was this argument that was wholeheartedly endorsed by the PWC Report, its argument relying on, and quoting extensively from, the CSIRO submission. Since that hearing, a number of scientists have questioned whether the risk of introduction and survival of FMD has been substantially increased by passenger air travel.

Nevertheless, the argument of increased risk appears to have achieved the standing of dogma, such that in the 'Urgent Case' of 1977, the claim was made that:

> "we believe the risks of a serious outbreak of an exotic disease have increased to such a degree that it is not now a case of 'if' there is a serious outbreak in this country, but rather, 'when'"[16].

Yet the only new argument added, was the risk associated with the inflow of Vietnamese refugees.

There can be no dispute that the economic value of the Australian livestock industry provides grounds for action to protect it, or that the outbreak of foot and mouth disease would have a very serious financial impact. However, these facts alone do not provide the justification for a maximum security laboratory apparently accepted throughout the various stages leading to the establishment of ANAHL.

The limited role of a maximum security laboratory was apparently accepted by the CSIRO:

> "An Australian National Animal Health Laboratory with maximum security facility cannot keep exotic disease out of Australia; that is not its purpose. Rather, the laboratory is an insurance against the day an outbreak of an exotic disease occurs. When that happens, the laboratory will be a vital factor in minimising the impact that such an outbreak could have on the Australian economy"[17].

However, the rhetoric frequently suggested that ANAHL could prevent an outbreak occurring, through the use of phrases such as, ANAHL is "essential to safeguard"[18] the livestock industry, and "to maintain its position with meat-importing countries as a disease free area"[19].

Hence, the *need* for a maximum security laboratory, or its appropriateness for the Australian context appears almost never to have been the subject of debate. The proponents of the facility have been able to use the economic importance of the livestock industry, and the unsubstantiated threat posed by increased air travel (an increase which has now ceased), to provide support for their scheme. Alternatives were given little consideration. Nor, as shown later in this paper, were very strong arguments mounted as to the nature and extent of the contributions a maximum security facility could make to 'protecting' the livestock industry.

2. THE ECONOMIC JUSTIFICATION FOR ANAHL

It is customary in considering an 'insurance policy' to weigh the costs of the policy against the risks to be insured and the benefits offered. While the risks remained uncertain, an analysis of the costs and benefits was undertaken by the Bureau of Agricultural Economics in June 1974 at the request of Dr K. Kesteven, consultant to the Department of Agriculture, and presented to the PWC Hearing.

No evaluation had been previously made of the economics of the proposal, other than the oft repeated reference to the value of the meat export industry. Indeed, at the PWC Hearing, neither the CSIRO nor Department of Health included a statement of the cost of the facility. Mr Kelly (PWC Member) remarked:

> "this is the first time I have seen such a submission without any mention of the question of the cost of it. There is a certain coyness in the evidence. No one mentions such a mundane matter as money, but one of the things you most do when you have to evaluate the need for it is put it alongside the cost of it. What is also notable is that there has been no discussion about the operating costs of it[20].

It was on the basis of the BAE study alone that the PWC concluded that the proposal to build ANAHL was economically justified. In its Report the PWC stated:

> "In an examination of the economic aspects of the Animal Health Laboratory proposal, the Bureau of Agricultural Economics concluded that it could be a viable proposition as a result of the expected benefits arising from research programs alone and if these benefits were combined with a disease outbreak situation, there seems to be little doubt regarding the economic viability of the proposal"[21].

In subsequent reports and discussions, the BAE analysis of the economic viability of ANAHL and the PWC's acceptance of it become the authoritative basis for favourable judgement and decisions. The approach taken by the BAE was to attempt to calculate the magnitude of the benefits needed to equal the estimated costs of building and running ANAHL, and it was assumed that tangible benefits would result and that these benefits could be measured in monetary terms. The report admitted the analysis was extremely approximate:

> "There is a large degree of uncertainty associated with the type of benefits, their possible magnitude and time of occurrence. Virtually no data were available on any of these aspects"[22].

However, three sources of benefits were identified on the basis of "various descriptions of the objectives and functions"[23] of ANAHL and were listed as:

1. the prevention of substantial losses in export revenues earned from livestock product sales;
2. the reduction in production losses and livestock slaughterings that might be considered necessary without an ANAHL; and
3. the benefits which might be expected to be derived from programs of research at the laboratory[24].

In order to give an estimate of these benefits in monetary terms twelve assumptions were adopted by the BAE[25]. These assumptions can be divided into two types. The first type included estimates of the number of livestock and the value of livestock and livestock products along with estimates of the cost of ANAHL and the time of construction, and these could be described as scale assumptions. While there is room for debate over the precise value assigned to these assumptions, the validity of the item itself is not in dispute. The second type of assumption is more problematic. This group involves assumptions about events that may occur, about the period of time over which the laboratory will be active and about the benefits resulting from ANAHL, and these could be described as substantive assumptions. For example, the basis for the selection of forty years as the effective life of the laboratory was given by Mr Miller (Acting Director, BAE) in answer to a question posed by Mr McVeigh:

Mr McVeigh: "You have just taken this as a period of forty years for the purpose of your exercise, not that because some expert from another department said that this is going to last forty or fifty years? You have just taken this?"

Mr Miller: "That is correct"[26].

The assumption made regarding the annual operating costs contains both scale and substantive assumptions. The cost under a quiescent disease situation is an assumption of scale and was estimated at $2.8 million per annum. The cost of operation in an outbreak situation would involve substantive assumptions but although the BAE recognised that these costs would "increase substantially" during an outbreak "no assumption as to the magnitude of the possible increase was made"[27]. This means that the potential benefits were calculated on the basis of an outbreak of FMD but the BAE "considered it unnecessary"[28] to estimate the true costs to ANAHL of handling a FMD outbreak.

Each of the three identified sources of benefits was considered separately, and using the remaining assumptions an

"attempt (was made) to demonstrate the feasibility of obtaining the magnitude of benefits during the period from year 11 to 50 which would at least equal the magnitude of cost of the ANAHL"[29].

The first benefit considered was the prevention of substantial losses in export sales. According to the discounting assumptions, it was calculated that if an outbreak of FMD occurred in year 11, $120 million of export revenue would have to be saved. If the outbreak occurred in year 46, the complete export revenue of about $2,000 million would need to be saved. In that case ANAHL would not only have to "reduce export losses" but prevent them altogether, an assumption that is not made explicit anywhere in the report and indeed one which is contrary to any evidence presented in any of the submissions. An alternative calculation in this example shows that if ANAHL prevented losses of $10 million each year from years 11 to 50 this would equal the investment in the laboratory.

In concluding, the BAE stated that these examples "provide a clear indication of the order of magnitude of benefits which *might be expected* in the event of an exotic disease outbreak occurring"[30]. However, this appears to be a deliberate inversion of the argument. As noted previously, the calculations are based on the magnitude of benefits which would be *required* to equal the costs of ANAHL with no discussion or indication of the feasibility of achieving this result.

The second example used by the BAE looked at the benefits from a reduction in production and slaughter losses that might be considered necessary without an ANAHL. Again the calculations were undertaken to indicate the order of magnitude of the benefits, in this case the reduced slaughter and production losses, *required* to equal the costs of ANAHL. And again, the conclusion formed was that these required benefits were, in fact, benefits that might be expected from the operation of ANAHL, without any examination of the ways in which this might be achieved.

In order to achieve the required magnitude of benefit it was assumed that 0.6 percent of Australia's livestock was affected by an exotic disease outbreak each year and that the presence of ANAHL would reduce the effects of the disease to a nine month period each year and that it would save half the value of stock slaughtered. However, the CSIRO evidence claimed that in the event of an outbreak all stock in the area would be slaughtered "whether showing signs of infection or not"[31]. There appears to be no reasonable basis for assuming that ANAHL could reduce the time period of the disease from twelve months to nine months and the basis for assuming 0.6 percent disease rate seems to be that it gives the correct answer.

"It was discovered that if 0.6 percent of Australia's sheep and cattle population is affected by an exotic disease outbreak each year, the annual benefits from savings in production and disease control slaughterings combined are $10.97 million, which exceed the annuity required for the ANAHL project to break even"[32].

The third example considered the benefits resulting from research at ANAHL. Although the BAE admitted that "estimates of the monetary returns which might be expected from research programs are subject to a very large degree of uncertainty",[33] they were able to give the following "simple examples":

"Assume a research breakthrough occurred in year 11. If the value of that technological advance added $10 million or 0.29 percent to the gross value of $3,384 million of livestock production in that year, and no further increase was achieved through the research efforts of the ANAHL staff, the project would break even in economic terms, i.e., from years 11 to 50 the annual gross value of livestock production in 1972–73 prices would be $3,394 million.

Alternatively a contribution from research at the ANAHL to the annual gross value of livestock production ($3,384 million) equivalent in value to a compound rate of gain of 0.0007 percent per year from year 11 to 50 would be required for the project to break even. Even if its is assumed that a 10 year period elapses between a research discovery and adoption of the practice, the necessary compound rate of gain in value terms for the project to break even rises from 0.007 percent over 40 years to 0.02 percent from year 21 to 50"[34].

From these examples, and apparently despite the uncertainty surrounding research, the BAE was able to conclude that ANAHL "could be viable proposition as a result of the expected benefits arising from research programs alone"[35].

No attempt was made to determine the likelihood of a research breakthrough which would add $10 million to the value of livestock production. Indeed, there was no attempt to address the particular area at all. Rather, this is an abstract calculation which could be used to justify the value of a $50 million investment in research in *any field*. The returns on research were apparently considered so self-evident that they needed no justification. Indeed one might wonder why the logic of the argument was not turned on its head and a case made that the increase in the value of production might well be $20 million, so the Government should invest $100 million of public money in ANAHL.

Although the BAE maintained that ANAHL could be viable on the basis of research alone, it added that "if these benefits were combined with a disease outbreak situation there seems to be little doubt about the economic viability of the ANAHL proposal"[36]. This could be taken as suggesting that the ANAHL might not be viable on the basis of prevention of substantial losses in export revenue, production and slaughtering alone, a suggestion that would seem to run counter to the much repeated justification of ANAHL.

The authority of the institutions and individuals, together with the mystification of numbers and scientific method, apparently precluded any serious examination of the validity of these arguments. As Mr Keogh (PWC member) said: "If somebody such as Mr Miller (Acting Director, BAE) and others are convinced of the economic viability of it, with the CSIRO running it, we are not in disagreement with the question"[38].

This was in spite of the admission (during questioning) of the extremely weak basis of the calculations:

"The data simply does not exist for the type of analysis that would need to be done before one could say professionally that this laboratory is definitely an economic proposition. What we have done is to, if you like, do some scribbling on the back of a used envelope, pulled some figures out of the hat and tried to provide the inter-departmental committee and this Committee with some figures which will help to form a judgement as to whether an insurance policy of this type is likely to be beneficial"[38].

Not only does this statement appear to have been overlooked in assessing the evidence, but is quite inconsistent with the authoritative and scientific tone of the BAE Report.

In discussion of the economic analytical framework used, the BAE claimed that "ideally the economic criteria should be used to compare a number of projects in order to select those with the greatest expected net benefits and to determine how a limited supply of capital should be allocated to the projects in the most economically efficient way"[39]. Since there were no other alternatives presented for controlling outbreaks of exotic disease in Australia, indeed the consideration of alternatives had been systematically avoided, the analysis compared the "magnitude of estimated livestock product losses without a NAHL with the possible saving of a proportion of those losses if the NAHL was built"[40]. Such a comparison could give only one answer.

It would be difficult not to argue that the cost-benefit analysis was framed and conducted with only one possible outcome in mind — the provision of a particular kind of authoritative support for the construction of ANAHL.

3. THE APPROPRIATENESS OF THE FUNCTIONS OF ANAHL

The Eichorn report recommended diagnosis and vaccine testing and production as the appropriate functions for a NAHL. However, the subsequent inquiries and reports successively broadened the range of functions, such that the list proposed to the Public Works Committee hearing by CSIRO, and subsequently adopted in full by them, consisted of:

1. to provide a diagnostic service to support the control and eradication of exotic diseases of livestock should they be introduced into Australia and to ensure that livestock imported into an offshore quarantine station are free of exotic;
2. to undertake research into indigenous and exotic diseases of livestock;
3. to train laboratory and field personnel in the diagnosis and control of exotic diseases; and

4. to provide facilities for producing 200,000 doses of FMD vaccine per month[41].

In other words, the functions of quarantine testing, research, and training have been added.

It would appear the various proponents of ANAHL recognised the inappropriateness of having an expensive maximum safety facility established and staffed, waiting for an emergency that may, or possibly may never, occur. Indeed, the involvement of CSIRO from the earliest stages of proposal development, not only as administrators and operators of the ANAHL, but as active proponents, suggest that research and training were always seen, at least by them, as major functions. That, after all, was their business.

The implicit argument that was pressed throughout the decision-making process was that a maximum security facility was necessary to perform the diagnostic function, and once you had such a facility you needed to use it effectively by performing research and providing training within it. Moreover, staff of appropriate quality would only be attracted under these conditions. This is the model of operation of the facilities at Pirbright, U.K. and Plum Island, U.S.A.

Presented in this way, such an argument appears entirely reasonable. But at no stage was there an evaluation of the effectiveness of investing in research and training appropriate to a maximum security facility, with using that money to do other work. It was the concern about this lack of explicit comparison, and more particularly the possible effect on resources available to support other animal research, that was the basis for early criticism of the ANAHL proposal from scientists within CSIRO. For reasons of time and space, only the arguments in support of the central diagnostic function will be presented here.

The diagnostic function The major arguments used in support of the diagnostic function for ANAHL — the function widely accepted as its major purpose — were the inadequacies of relying on foreign laboratories and the advantages conferred by a national facility. Primary producer organisations argued that the size of the livestock industry justified Australia becoming self-reliant in diagnosing exotic diseases:

"It is to be deplored that Australia — a country that relies so much on livestock production — must use other countries' facilities for testing possible major threats to its valuable primary industry"[42].

The basis for this argument is little more than one of nationalism — that we are a big country, so we should have our own facility and not be forced to rely on foreigners.

However, in addition to arguments of nationalism, three specific arguments were propounded:

1. the volume of testing required in the second phase of control and eradication, after an outbreak had initially been diagnosed, is too large to be effectively carried out overseas;
2. overseas laboratories were unavailable, even for initial testing;
3. a local laboratory would reduce the eradication time, and hence lost trade.

Follow-up testing — the arguments presented apparently substantiated this claim. Overseas laboratories would be unlikely to be available to perform follow-up testing on the scale required, and even if they were, it would not be a convenient efficient way to conduct an eradication program. Dr Snowdon argued: "No overseas laboratory could be expected to undertake the large numbers of tests required during the control and eradication of an exotic disease"[43]. This claim was also made by Dr Pierce in his evidence: "Overseas laboratories could not be expected to carry out tests on anything like the scale that may be required", and, he continued, "Moreover, the need to send material continually overseas for diagnosis would severely limit the speed and effectiveness of the many decisions that would have to be made during the course of an eradication campaign"[44].

Initial diagnosis — in seeking to establish the inadequacy of reliance on overseas laboratories, the proponents of ANAHL went to considerable lengths to identify programs. One argument was that the services of overseas laboratories might not always be available. Mr Gee, in his evidence, went on to state: "Contact with some of these laboratories is infrequent and there is always the possibility that co-operation may be affected by changing circumstances of, e.g. politics, war or communications"[45]. To substantiate his claim that overseas laboratories may be unreliable, Mr Gee pointed out that in a routine reconfirmation of co-operation only six of the ten co-operating laboratories replied. However, of the remaining four, two were Plum Island and the Wellcome Research Institute in England, with which close collaboration existed, and the others were specialist centres for diseases of little relevance to Australia[46].

In contrast to this position, however, during questioning, Dr Pierce of the CSIRO stated, "they (Pirbright) are a world reference centre ... their responsibility is to establish a diagnosis"[47]. In an article in Rural Research the CSIRO stated, "reference laboratories, whether or not funded by the Food and Agriculture Organisation of the United Nations are only committed to a primary diagnosis"[48].

Another argument against reliance on overseas laboratories was the uncertainties involved in transporting specimens:

"The procedures required to obtain diagnoses in overseas laboratories are accompanied by uncertainties and delays. For instance, material may be lost in transit or exposed to conditions which could destroy disease agents which it may contain, and unless Australian authorities are in constant contact with the overseas laboratories that have agreed to carry out the diagnostic tests, there can be misunderstandings resulting in failure to obtain a diagnosis"[49].

Whilst this is undoubtedly possible, it should not be regarded as the regular pattern. During cross-examination, Dr Snowdon replied that Pirbright would give a diagnosis, "possibly within two or three to four hours of receipt of the sample, although, if they have to grow it up in tissue culture, which is the second procedure, I would say that would take 48 hours, but possibly 24 if everything works well"[50].

This time frame was confirmed by the events at Humpty Doo, Northern Territory, where on Friday 26th May, 1983, two pigs were suspected of having a vescicular disease. These animals were examined by a Northern Territory pathologist on Saturday 27th May, and were subsequently killed and autopsied. Tissue samples were sent to Pirbright and these arrived on Sunday 29th May and testing commenced immediately[51]. By Monday 30th May, the Minister for Primary Industry, John Kerin, was able to announce that preliminary tests were negative and that a final clearance should be available within two to three days[52].

Reduction in eradication time — the argument was presented in the form that:

"As mentioned previously, much of our overseas trade in meat and livestock products would cease overnight if Australia had an outbreak of FMD. In the absence of any other evidence, those nations consituting our major overseas market for meat would normally require that Australia remain free of FMD for some time after a declaration of successful eradication and before allowing a resumption of meat imports. On the other hand, if Australia had diagnostic facilities, these countries might be expected to reduce this period considerably as the result of the provision of laboratory diagnostic tests showing negative results. A reduction in lost trading time of even a few months could represent a gain of millions of dollars from agriculture exports"[53].

The same argument was echoed by the submission of the Australian National Cattlemen's Council:

"The presence of the ANAHL would mean that in the event of an outbreak of exotic disease, an eradication program would commence more rapidly than if no such laboratory was available, and it was necessary to rely on overseas facilities. Each month's delay in having Australia declared free after such an outbreak, is worth approximately $5.5 million in beef exports alone"[53].

However, little evidence was presented that ANAHL could reduce the export ban by any period approaching a month. The importance of early diagnosis has always been addressed as the critical element in control. "The essence of containment of an introduced exotic disease is early diagnosis"[55] and "of vital importance in any control and eradication campaign is the early detection and diagnosis of the exotic disease"[56].

But this argument would seem to point to the need for the training of primary producers and veterinarians in the recognition of exotic diseases rather than the need for a diagnostic laboratory. The greatest potential time loss is likely to be in the initial response of the farmer and veterinary officer. The incident at Humpty Doo in the Northern Territory, referred to previously, confirms this view. The BAH report stated the lesions on the suspect pigs were one to two weeks old. Results from analysis conducted by the Pirbright Laboratories were available only three days after examination by pathologist in the Northern Territory. The difference in time between sending samples to Geelong or Pirbright were almost irrelevant in comparison with delays in recognition.

So although it has been claimed that ANAHL will reduce the time taken to obtain diagnosis, it may in fact reduce minimally only one part of a chain of events, all filled with uncertainty. Only in the area of follow-up testing are the benefits evident.

CONCLUSIONS

The decision making processes with regard to the establishment of ANAHL appear to have been almost monopolised by proponents of the scheme. After the Eichorn Report there was almost no independent evaluation of the need for a maximum security facility, and the focus of the inquiry processes became the best means of achieving this pre-given end. The reports themselves became vehicles of advocacy rather than assessment. The central question implicitly became posed in terms of doing something to help protect the valuable livestock industry, or doing nothing. Once the argument became posed in these terms, it became virtually impossible to produce grounds to oppose the construction of the ANAHL facility, regardless of either cost or potential benefits.

No alternative to the establishment of a maximum security animal health laboratory was considered in the reports prior to the PWC and the evidence submitted to the PWC was presented in such a way as to suggest that there was no feasible alternative to ANAHL. In this way a choice between alternatives, and the political and value judgements that this implied was avoided.

In the arguments justifying the proposed functions of ANAHL, alternatives were given, but only to highlight their inadequacies. The PWC were told that Pirbright may be unreliable for primary diagnosis and would be unable to perform the necessary follow-up testing after an outbreak was confirmed. It was also claimed the Pirbright was the only safe source of vaccine but that there were many drawbacks in relying on this source. Although current training involved attendance at overseas courses and outbreaks, this was said to be unsatisfactory because of the limited numbers who could attend and the cost involved. So the PWC were not given any real alternatives to ANAHL performing the various functions. The submissions suggested that all alternatives had been investigated but rejected as unsuitable, and this was accepted by the PWC.

Throughout all the arguments justifying the need for ANAHL and the need to perform its proposed functions, was the suggestion of the economic advantages that would result. However, the BAE cost benefit study, which was prepared for the PWC Inquiry, provided the first and only formal study of the economic viability of ANAHL. On the basis of this study, the PWC concluded that ANAHL was economically justified. However, an examination of this study revealed that it was based on questionable assumptions, used questionable methods of calculation and examples and contained a high degree of uncertainty.

Furthermore, an examination of the Minutes of Evidence revealed that the PWC did not investigate or assess the report itself but accepted its validity on the basis of the credibility and reputation of the institution who prepared the report. The claim that ANAHL was economically justified became an authoritative statement which was not challenged.

The justifications for ANAHL were presented to the PWC as factual and objective evidence and accepted as such, and the many value judgements, uncertainties and assumptions were overlooked or concealed.

The arguments of need relied heavily on the assertion that there was an increasing risk of an outbreak of an exotic disease. However, the perception of risk and the assessment of the degree of risk are not objective 'scientific' activities, but depend on interpretations and value judgements.

The proponents of ANAHL, and in particular CSIRO relied heavily on the authority of their scientific expertise as the basis for judgement on matters of value. Some of the arguments they presented relied on the authority of these earlier reports, but they also held themselves out as experts in areas of quarantine, economics and risk safety assessment. The PWC accepted the CSIRO arguments about the risk of an exotic disease entering Australia rather than the Quarantine Department's assessment; they accepted their arguments that the various functions could provide economic advantages, and that the site was suitable and the microbiological security was adequate. Thus it would appear that the CSIRO had defined a wide range of areas 'scientific' and had defined themselves as experts in these areas and these definitions were accepted by the PWC without question.

Wynne[57] has argued that the increasing complexity of technology and its consequences has led to a focus of concern and evaluation on the institutions which appear to control technology, rather than the technologies themselves. If decision-makers and the general public are unable to evaluate the technology itself, then the credibility and trustworthiness of the decision-making processes and institutions is evaluated instead: "Questions of public trust, credibility, openness and, significantly, the past track record in these respects, become the key factors in framing social attitudes"[58]. "Expertise and authority can be 'negotiated' by concealing uncertainties and value judgements and by claiming that decisions are based on objective scientific evidence"[59]. Although this situation may facilitate decision-making in the short term, Wynne maintains that eventually it is counter productive when the falsification is revealed; "attempts to gain authority on specific decisions and issues by using spurious images of certainty, gives rise to a greater and more general loss of authority by the institutions as a whole when that image is eventually punctured, as they nearly always are"[60]. The changing scientific and political climate led to a continuing shift in the arguments and evidence offered, but these were all designed to maintain the basis for the initial conception of ANAHL. These shifts, however eventually served to highlight the uncertainties and value judgements in the arguments, and to an erosion of the credibility of the decision-making institutions and the proponents on ANAHL.

As an example of 'Big Science' decision-making, there is much to learn from the case of ANAHL. Recent experience with other major technological decisions, (e.g. 'Starlab') suggest that the same approaches are still being adopted.

Three particular suggestions I would like to offer are:

1. Independent and critical evaluation and decision-making is unlikely in a situation where proponents of a particular proposal dominate the various processes and levels of decision-making. Assessment inevitably gives way to advocacy. But the system of appointment of expert committees and IDC's based on interested Departments almost always ensures that proponents are well represented. Nor is it unknown for proponents to encourage and assist particular community interest groups to lobby appropriately.

    In matters of science and technology, a Public Works Committee of Parliamentarians can too easily be mislead, persuaded or pressured by the authority of scientific expertise and institutions. Moreover, it would be extremely difficult for them to over-ride the scientific arguments which have determined need, and hence confined themselves to matters of detail of operation and costing. Hence: **there is a need for every major technological proposal to be evaluated by an entirely independent set of evaluators, with adequate scientific expertise but without direct representation of proponents.**

2. The fact that 'Big Science' decisions always occur as one-off decisions lends great pressure for an *ad hoc* approach and for a favourable decision. Inevitably, once a case is mustered it is presented in terms of doing something for which substantial potential economic benefits can be seen (e.g. helping to protect the $1500 million livestock industry) or doing nothing. The latter action clearly involves major

political disadvantages ("The Minister threatens the future of . . . etc. etc.) The only way to overcome this bias is to ensure that Big Science decisions are not made on an *ad hoc* basis, but rather in a systematic way within the framework of the overall policy and budget in support of research.

In addition to a critical evaluation of costs and potential benefits, there is a need also for consideration of the opportunity costs of a particular disease and whether investment in that area offers more promise than investing in a range of alternative scientific approaches.

3. In both of these processes there is a need for careful exposition of relevant factual knowledge and its limitations, and an admission of areas where opinion and judgement, including judgement of values, predominate. (For example, the question was never asked as to whether the livestock industry was more in need of support than some other industry.) Advocacy has its place, but where advocates, cloaked in scientific authority, dominate the decision-making processes, judgements of value are likely to be concealed, the political process usurped, and decisions made on the basis of sectional interests rather than the national good.

REFERENCES
1. *Canberra Times*, 20/10/82, P. 2.
2. Parliamentary Standing Committee on Public Works, *Animal Health Laboratory, Geelong, Victoria: Report*, 1974, p. 12.
3. Primary Industry Submission. *The Urgent Case*, November 1977.
4. CSIRO. *A Review of the Division of Animal Health.* November 1982, p.72.
5. CSIRO. *A Summary of the History of the Australian National Animal Health Laboratory, Its Development and Future Operation.* 1982, p. 2.
6. Parliamentary Standing Committee on Public Works. *Animal Health Laboratory Minutes of Evidence.* September 1974, p. 6.
7. *ibid.*, p. 5, A4.1.
8. *ibid.*, p. 2, A2.1.
9. *ibid*.
10 *ibid*, p. 29.
11 *ibid.*, p. 7, A4.24.
12. *ibid.*, p. 70.
13. *ibid*.
14. ASTEC Report. *ANAHL: The Use of Live Exotic Animal Pathogens*, 1982, p. 2.
15. Parliamentary Standing Committee on Public Works. *Animal Health Laboratory: Minutes of Evidence, op. cit.*, p. 66.
16. Primary Industry Submission, *op. cit.*, p. 1.
17. Parliamentary Standing Committee on Public Works. *Animal Health Laboratory: Minutes of Evidence, op. cit.*, p. 7. A4.25.
18. *ibid.*, p. 82.
19. Primary Industry Submission. *op. cit.*, p. 3.
20. Parliamentary Standing Committee on Public Works. *Animal Health Laboratory: Minutes of Evidence, op. cit.*, p. 20.
21. Parliamentary Standing Committee on Public Works. *Animal Health Laboratory, Geelong, Victoria. Report, op. cit.*, p. 4, para 33.
22. Parliamentary Standing Committee on Public Works. *Animal Health Laboratory: Minutes of Evidence, op. cit.*, p. 98.
23. *ibid*.
24. *ibid*.
25. *ibid.*, p. 99.
26. *ibid.*, p. 103.
27. *ibid.*, p. 99.
28. *ibid*.
29. *ibid.*, p. 100.
30. *ibid.*, p. 98.
31. *ibid.*, p. 4, A3.9.
32. *ibid.*, p. 101.
33. *ibid*.
34. *ibid*.
35. *ibid.*, p. 98.
36. *ibid*.
37. *ibid.*, p. 103.
38. *ibid.*, p. 102.

39. *ibid.*, p. 99.
40. *ibid.*
41. *ibid.*, p. 14–15.
42. Primary Industry Submission, *op. cit.*, p. 1.
43. Parliamentary Standing Committee on Public Works. *Animal Health Laboratory: Minutes of Evidence, op. cit.*, p. 19.
44. *ibid.*, p. 5–6.
45. *ibid.*, p. 60.
46. *ibid.*, p. 67.
47. *ibid.*, p. 23.
48. CSIRO, *Rural Research Quarterly No. 113.* December 1981, p. 5.
49. Parliamentary Standing Committee on Public Works. *Animal Health Laboratory: Minutes of Evidence, op. cit.*, p. 18, C7.1.
50. *ibid.*, p. 25.
51. Bureau of Animal Health, *Bulletin No. 15.* 2/6/83.
52. Primary Industry Media Release, PI83/91. 30/5/83.
53. Parliamentary Standing Committee on Public Works. *Animal Health Laboratory: Minutes of Evidence, op. cit.*, p. 6, A4.6.
54. *ibid.*, p. 132.
55. *ibid.*, p. 61.
56. Australian Bureau of Animal Health. *ANAHL and Exotic Disease Control.* September 1982, p. 4.
57. B. Wynne, "Technology, Risk and Participation: On the Social Treatment of Uncertainty". In: J. Conrad (ed.), *Society, Technology and Risk Assessment,* Academic Press, 1980.
58. *ibid.*, p. 176.
59. *ibid.*, p. 195.
60. *ibid.*, p. 196.

Chapter Ten

VIRAL PATHOGEN IMPORTATION — THE RISKS AND BENEFITS

A.J. Della-Porta and R.I.B. Francki

Australia is an island continent which has been spared the ravages of many exotic virus diseases of animals and plants. However, the increased importation of animal and plant genetic stock and the international travel to Australia, has seen the introduction of a number of exotic diseases over the last 20 years (Chapter 3).

This paper will concentrate on the benefits of importing some exotic animal and plant viruses for diagnostic and research purposes and an assessment of the risks associated with such importations. As described by other contributors, considerable losses could be caused by the accidental introduction of such agents into Australia.

SOME ADVANTAGES OF IMPORTING VIRAL PATHOGENS

There seem to be four main advantages of importing viral pathogens. We believe that, given proper containment and handling facilities, the benefits of importation of some exotic viruses far outweigh the risks. The benefits are:

1. *Rapid and accurate diagnosis of viruses causing disease outbreaks*
   One of the most important reasons for importing exotic animal and plant viruses is to aid in the rapid and accurate diagnosis of possible exotic disease outbreaks in Australia. The sooner that a disease can be diagnosed, the greater the chances of being able to eradicate it and hence prevent economic loss. Further, a misdiagnosis may be as expensive as an exotic disease invasion and many of the incursions that we have had show the inadequacy of diagnosis based on a single test.

   Rabies is probably one of the greatest threats to Australia. It is found close to our northern borders in Indonesia (Thaib, 1982). If it should enter Australia and become established in our feral and wild animals (Murray and Snowdon, 1976), it would be very difficult to eradicate. It would have a major impact on the lives of all Australians, with compulsory vaccination of cats and dogs and severe controls on dog movements. Live rabies virus is at present required for the World Health Organisation's fluorescent antibody test, being an essential positive control for the rapid detection of the virus (Dean and Abelseth, 1973). Attenuated fixed rabies virus is at present used at the Commonwealth Serum Laboratories in Melbourne for this work, which will eventually be transferred to the Australian National Animal Health Laboratory in Geelong, as part of the Australian Government's policy to locate all exotic animal viruses at that laboratory.

   There was an outbreak of fowl plague in Victoria in 1976 (Turner, 1976). Rapid control, with a small financial loss, was possible because a senior veterinarian of the Victorian Department of Agriculture had just attended an exotic diseases course at the Plum Island Animal Disease Center in the USA where he had seen fowl plague. Unfortunately, very few overseas people can attend this course. The availability of exotic agents in Australia could enable many more field veterinarians to be trained to recognise the clinical signs of exotic diseases and hence assist in their rapid diagnosis.

   Outbreaks of new plant virus diseases can often be eradicated if appropriate action is taken quickly. this is also usually dependent on prompt and accurate diagnosis of the virus responsible. Rapid and conclusive virus identification is dependent not only on the availability of reagents such as appropriate antisera, but also on the availability of a culture of the type virus for comparison and use as a positive control. In the past there have been numerous instances of serious difficulties and delays in the identification of plant viruses encountered in Australia for the first time. In most cases these problems could have been avoided if appropriate virus type cultures had been available.

A good example of such an instance is when a virus was isolated from gladioli which after preliminary testing was suspected of being tobacco ringspot virus (TRSV). However, when it was checked against an imported antiserum to TRSV, the tests proved negative. Further extensive studies, which took several months, continued to indicate that the virus was indeed TRSV and it was finally concluded that the available antiserum may have been faulty. This suspicion was confirmed when another antiserum to TRSV was obtained from colleagues overseas (Randles and Francki, 1965).

Several years later in the same laboratory, a virus was isolated from *Vicia faba* grown from seed imported in bulk. From a number of its properties it was suspected that the virus was either broad bean stain virus (BBSV) or true broad bean mosaic virus (TBBMV), neither of which had previously been reported in Australia. Antisera to both viruses were imported but tests showed that the unknown virus reacted with both antisera although other tests indicated that the virus was not a mixture of BBSV and TBBMV. Eventually the antisera were tested by a colleague overseas against their homologous viruses which established that whereas the anti-BBSV antiserum was true to its label, the other contained antibodies to both BBSV and TBBMV. When this was revealed it become clear that the virus was BBSV (Moghal and Francki, 1974). Problems with non-reacting, incorrect or poor reference antisera have occurred with attempts to diagnose both exotic plant and animal viruses. Such situations could lead either to a failure to diagnose an exotic disease incursion or, just as seriously, to an incorrect diagnosis. The identity of the plant viruses in both the above examples took several months to be established at a disproportionate investment of time and resources. All the viruses could have been rapidly identified if type cultures and their respective antisera had been available in Australia.

2. *Characterisation of indigenous viruses*

Indigenous viruses may cross-react in serological tests with exotic viruses and cause an unnecessary disease scare if they were avirulent forms with no potential to cause disease.

Bluetongue virus was first isolated in Australia in 1975 and, since then, many other related orbiviruses have been detected. Extensive overseas studies, causing considerable delay and much expense, have been necessary because exotic bluetongue viruses were not available in Australia (Della-Porta, et al., 1981). Evidence now suggests that the low virulence Australian isolates may be significantly different from the virulent South African bluetongue isolates (Chapter 20; Gorman, et. al., 1983). Furthermore, a number of Australian orbiviruses cross-react with bluetongue diagnostic reagents in serological tests, emphasising the problems of interpretation of single procedures and the need for better tests (Della-Porta, et al., 1984).

There is a low virulence Newcastle disease virus (V4) in Australian poultry (Simmons, 1967). If we had available medium and high virulence isolates of the virus we could use them to develop tests for differentiating our avirulent strain from highly virulent isolates. Also we could use them in cross-protection studies to determine whether the V4 strain would protect Australian poultry against virulent Newcastle disease virus. Such studies have at present to be done overseas (Spadbrow, et. al., 1978; Westbury, 1984; Westbury, et. al., 1984).

A number of rhabdoviruses that have been isolated in Australia, including bovine ephemeral fever, Kimberley, Kununurra, Parry Creek and CSIRO 368 viruses (St. George and Kay, 1982), which present potential problems in the diagnosis of rabies and vesicular stomatitis. For example, biochemical studies done at the Animal Virus Research Institute, Pirbright, United Kingdom, indicate that the proteins of bovine ephemeral fever virus resemble those of rabies related viruses (Della-Porta and Brown, 1979). Also, a recent report from the USA indicates that bovine ephemeral fever virus may be antigenically related to Lagos bat virus, a rabies related virus (Tesh, et. al., 1983). Unfortunately, these workers could only use bovine ephemeral fever virus antiserum and could not handle Lagos virus, thus only a one-way comparison, without controls, was possible. There is no evidence that this virus causes a rabies-like disease (Della-Porta and Snowdon, 1980). We know little about the relationships of the other rhabdoviruses with those reported elsewhere.

There are a number of viruses in Australia which have been characterized locally but have not been compared in detail to obviously related exotic viruses and hence the degree of relationship remains unknown. In some instances the Australian virus has been named as a new virus in spite of the possibility that it could be synonymous with a virus described and named elsewhere. For example, there are now serological data (J.W. Ashby, personal communication) indicating that subterranean clover red leaf virus, which is an important pathogen of subterranean clover in Australia (Kellock, 1981), is closely related to soybean dwarf virus described in Japan (Tamada and Kojima, 1977). Nevertheless, the question whether the two viruses are closely related but distinct or whether they are isolates of the same virus cannot be resolved until they are both available for detailed studies in the same laboratory. There are a numerous other situations where the exact taxonomic position of a virus can only be determined by direct comparisons with type virus cultures.

3. *Genetic engineering*

There is currently considerable basic research effort going into developing a technology for the introduction of genetic material into plants (Howell, 1982). In the search for suitable vectors to achieve this, some plant viruses are being considered as candidates (Hull and Davies, 1983). At present most attention is being paid to viruses with DNA genomes; those in the caulimovirus and geminivirus groups. However, in the future it is almost certain that viruses with RNA genomes will also be considered. There are well in excess of 400 known plant viruses world-wide (Matthews, 1982) of which no more than 150 have been found to be present in Australia. thus it seems that research into the potential use of plant viruses as genetic vectors here may be frustrated unless some exotic viruses can be imported.

4. *Biological control agents*
The use of exotic pathogens as biological control agents is discussed in Chapter 18. An excellent example is the use of myxomavirus to control the European rabbit in Australia (Chapter 23). There are also many attractive possibilities of using pathogenic viruses for the control of insects (Falcon, 1982). At present, plant viruses are not popular subjects of research as potential biological control agents. However, there must be some situations where they offer possibilities for use in weed control. Also, low virulence strains of some plant viruses could be used to protect against infection by virulent strains. It may well be that in the future, situations will emerge where importation of exotic viruses for biological control purposes could be of considerable advantage.

CONTAINMENT AND SAFE HANDLING OF EXOTIC VIRUSES

Plant viruses
Unfortunately, plant viruses cannot be maintained in tissue culture which would present them with the least chances of escape. However, most viruses grown in plants can be kept under secure isolation conditions in appropriately designed glasshouses or growth chambers. When building such secure facilities the properties of the viruses to be maintained must be considered carefully. It must be borne in mind that some viruses have arthropod, nematode or fungal vectors, some can be transmitted through seed or carried by pollen and some have very stable particles which can persist in soil or water for very long period.

Some moderately secure facilities exist in most States that enable quarantine and testing of imported plant breeding stocks. However, at present there is not a high security laboratory for handling high risk plant viruses.

Animal viruses
With animal viruses there are probably three main categories that pose containment problems in the laboratory. The high risk agents (such as foot-and-mouth disease virus) present problems because of their high infectivity and their transmission by aerosols. A second category of high to medium risk agents comprise those that infect man as well as animals, the zoonotic viruses, such as rabies virus. The low risk agents have low transmissibility, arthropod vectors, or direct cell-to-cell transmission. The majority of exotic animal viruses fall into the low risk group, when properly handled within secure facilities.

The exotic animal viruses, there exists a facility for safe handling, the Australian National Animal Health Laboratory (ANAHL) at Geelong (Figure 1). This is a C2 laboratory that also contains C3 facilities. The building is an airtight concrete structure, with air-locks between laboratory areas and the outside. The higher risk agents are handled in laboratories maintained at lower pressure than the surrounding laboratories which, in turn, are held at a lower pressure than the outside. The staff enter, with a change of clothes, through an airlock. Inside the laboratory they handle all infectious viruses in biological safety cabinets (Figure 2). They do not come in contact with the live virus. On leaving the laboratory, they shower. All air is filtered, once entering the laboratory and twice on leaving it, through High Efficiency Particulate Air (HEPA) filters. All liquid waste is sterilized before leaving the building and all solid wastes are incinerated. In the C3 laboratory, viruses are contained within glove-ported class III safety cabinets, similar to those used at Fairfield Hospital for handling exotic human viruses (Chapter 2) or with staff fully enclosed in a plastic 'space suit' (Figure 3), with their own independent HEPA filtered air supply. The suits are decontaminated in a chemical shower. Air from the C3 facility can be incinerated as well as filtered. Large and small animals can be handled under similar conditions, with suit facilities being available for high risk or zoonotic agents.

ANAHL is probably the world's most advanced microbiologically secure laboratory specifically built to handle exotic animal viruses. It will be able to accommodate exotic pathogens in 1985 after its microbiological security and operating procedures have been certified by an independent body, the ANAHL Security Assessment Group (Chapter 5). A wide consultative process has been agreed upon before approval will be granted to import viral pathogens. ANAHL is a facility of national and international importance that should be fully utilized for the benefits of Australia and the agricultural industry. Collaborative research projects have always been envisaged and visiting scientists from Australia and overseas can be accommodated.

Fig. 1. Australian National Animal Health Laboratory, Geelong, Victoria. A maximum security facility for safe handling of exotic animal pathogens.

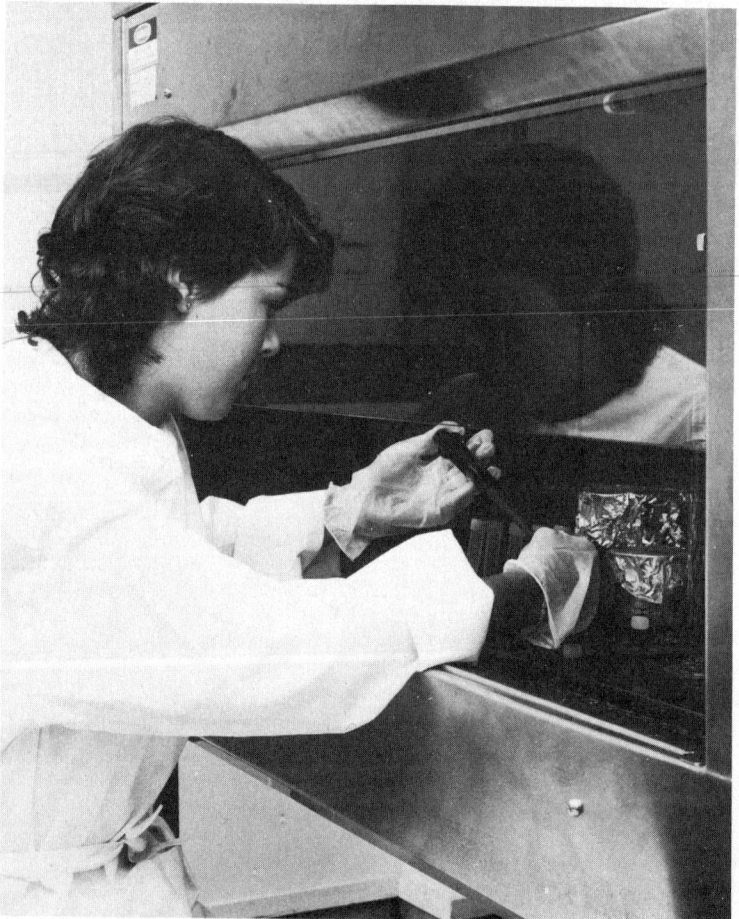

Fig. 2. A worker handling virus in a class II biological safety cabinet.

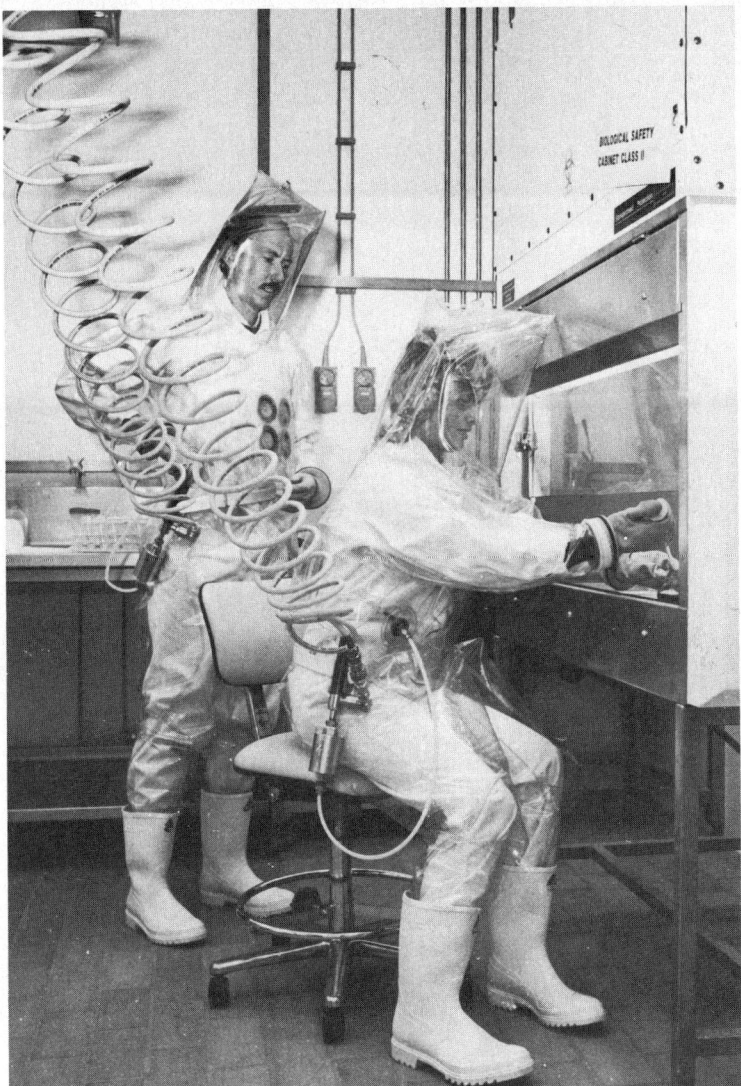

Fig. 3. A worker handling high risk agents in a plastic 'space' suit. The suit is decontaminated in a chemical shower.

CONCLUSIONS

It is doubtful whether any exotic plant or animal disease entering Australia could be fully diagnosed by a single simple test. The assumption that any exotic plant or animal disease entering Australia could be readily diagnosed by a single test needs to be seriously challenged. The complex inter-relationships among strains of bluetongue, Newcastle disease and fowl plague viruses differing in virulence, illustrate some of the pitfalls in their identification for disease control purposes; pitfalls that may result in overreaction with costly consequences to agriculture. Provided that adequate secure facilities and operating procedures are adopted, it should be possible to handle exotic plant and animal viruses safely. The benefits, to Australia's export income, in rapid diagnosis and the avoidance of misdiagnosis cannot be over-stressed. Agriculture is a major export income generator and the proper resources are needed for its protection.

REFERENCES

Dean, D.J. and Abelseth, M.K. (1973). In: *Laboratory Techniques in Rabies*. (M.M. Kaplan and H. Koprowski, eds) 3rd edition, p. 73, World Health Organisation, Geneva.
Della-Porta, A.J. and Brown, F. (1979). *J. Gen. Virol.* **44**, 99.

Della-Porta, A.J., Herniman, K.A.J. and Sellers, R.F. (1981a). *Vet. Microbiol. 6*, 9.
Della-Porta, A.J., McPhee, D.A., Wark, M.C., St. George, T.D. and Cybinski, D.H. (1981b). *Vet. Microbiol. 6*, 233.
Della-Porta, A.J., Parsonson, I.M. and McPhee, D.A. (1984). In: *International Symposium on Bluetongue and Related Orbiviruses.* (T.L. Barber and M.M. Jochim, eds) in press, Alan R. Liss, New York.
Della-Porta, A.J., Sellers, R.F., Herniman, K.A.J., Littlejohns, I.R., Cybinski, D.H., St. George, T.D., McPhee, D.A., Snowdon, W.A., Campbell, J., Cargill, C., Corbould, A., Chung, Y.S. and Smith, V.W. (1983). *Vet. Microbiol. 8*, 147.
Della-Porta, A.J. and Snowdon, W.A. (1980). In: *Rhabdoviruses* (D.H.L. Bishop, ed.) Vol. 3, p. 167, CRC Press, Boca Raton, U.S.A.
Falcon, L.A. (1982). In: *Population Biology of Infectious Disease Agents.* (R.M. Anderson and R.M. May, eds) P. 191. Report of the Dahlen workshop on population of infectious disease agents, Berlin 1982. Springer-Verlag, Berlin.
Gorman, B.M., Taylor, J. and Walker, P.J. (1983). In: *The Reoviridae* (W.K. Joklik, ed.) p. 287. Plenum, New York.
Howell, S.H. (1982). *Ann. Rev. Plant. Physiol. 33*, 609.
Hull, R. and Davies, J.W. (1983). *Adv. Virus. Res. 26*, 1.
Kellock, A.W. (1981). *Aust. J. Agric. Res. 22*, 615.
Matthews, R.E.F. (1982). *Intervirology. 17*, 1.
Moghal, S.M. and Francki, R.I.B. (1974). *Aust. J. Biol. Sci. 27*, 341.
Murray, M.D. and Snowdon, W.A. (1976). *Aust. Vet. J. 52*, 547.
Randles, J.W. and Francki, R.I. (1965). *Aust. J. Biol. Sci. 18*, 979.
St. George, T.D. and Kay, B.H., eds. (1982). *Proc. 3rd Symp. Arbovirus Res.* p. 249. CSIRO and Queensland Institute of Medical Research, Brisbane.
Simmons, G.C. (1967). *Aust. Vet. J. 43*, 29.
Spradbrow, P.B., Ibrahim, A.L., Mustaffa-Babjee, A. and Kim, S.J. (1978). *Avian Diseases 22*, 329.
Tamada, T. and Kojima, M. (1977). *Commonwealth Mycological Institute and Association of Applied Biologists. Descriptions of Plant Viruses No. 179.*
Tesh, R.B., Travassos da Rosa, A.P.A. and Travassos da Rosa, J.S. (1983). *J. Gen. Virol. 64*, 169.
Thaib, S. (1982). In: *Viral Diseases in South-East Asia and the Western Pacific* (J.S. Mackenzie, ed.) p. 236. Academic Press, Sydney.
Turner, A.J. (1976). *Aust. Vet. J. 52*, 384.
Westbury, H.A. (1984). *Aust. Vet J. 61*, 5.
Westbury, H.A., Parsons, G. and Allan, W.H. (1984). *Aust. Vet. J. 61*, 10.

Section C

Study of pests, parasites and their hosts

This section of the book describes various aspects of the scientific study of pests and parasites; their isolation and characterization by traditional and new methods (Chapters 11, 12, 13 and 14), and the wars and alliances of the genes, the 'hidden agenda' of the struggle between pests, parasites and their hosts (Chapters 14, 15 and 16).

The final two chapters are about modern approaches to ways of controlling unwanted pests and parasites. First, new ways of producing vaccines, including the use of 'genetic engineering' (the current revolutionary 'intellectual epidemic' of biological research!), and finally those exotic disease outbreaks we attempt to deliberately produce in order to control unwanted insect pests.

Chapter Eleven

AGENT DETECTION AND IDENTIFICATION: NEW AND TRADITIONAL TECHNIQUES

G.W. Burgess

This paper will summarize the conventional techniques which have been used for the detection and identification of animal viruses, although some of the principles apply equally well to other agents. Some techniques that are more experimental will be outlined in the following two papers.

Viruses are only able to replicate in living cells. Therefore when a sample is assayed for the presence of viruses, its infectivity is tested by inoculating a variety of host systems including the natural host, laboratory animals, embryonated eggs, cell cultures or organ cultures.

The growth of virus in the natural host may be recognised by the presence of clinical signs or pathological changes. Samples may be collected for further tests such as immunofluorescence, or serological tests which detect the appearance in the blood serum of proteins called antibodies, which will react with particles of the virus.

There are severe limitations to the use of whole animals. They have an intact immune system. Susceptibility may also be modified by genetic variation, by age or by an immune response resulting from previous exposure to the virus being tested. Facilities needed to contain infected large animals may be elaborate and expensive (see Chapter 9).

Smaller laboratory animals, such as mice, may be used as an alternative host. They suffer from many of the same limitations as the natural host, and in addition it may be necessary to 'passage' the virus through several individuals before well adapted mutants become selected and produce clinical signs.

Embryonated eggs have often been used in the past for growing mammalian viruses such as influenza and for most of the avian viruses, and indeed still are used for some purposes, such as growing viruses for vaccines. The presence of the virus may be indicated by changes such as death of the embryo, stunting of the embryo, haemorrhages, pocks or plaques on the chorio-allantoic membrane or the presence of a haemagglutinin in the allantoic fluid. The range of viruses which will grow in eggs is limited and unless the eggs are derived from specific pathogen-free flocks, contaminating viruses may complicate the picture.

Cultured cells provide the scientist with a relatively well defined host system devoid of an immune response. The presence of viruses may be indicated by death or changes in the cells (a cytopathic effect; CPE), presence of viral antigen within the cells or in the medium around them (as indicated by procedures such as immunofluorescence or haemagglutination) or the presence of viral particles when lysed cells or the medium is examined in an electron microscope.

'Organ cultures', small pieces of tissue maintained in culture medium, are useful for growing some fastidious viruses which do not readily grow in cell cultures. The viral growth may be indicated by various changes, for example if the tissue is a piece of ciliated trachea, virus infection may stop cilial movement. Virus may be demonstrated by immunofluorescent staining or growth in an alternative host system.

All of the systems discussed above have serious limitations including the long time taken to confirm the presence of an agent, insensitivity, lack of specificity, ability to support the growth of a limited range of viruses as well as an inability to recognise inactivated viruses. Thus the use of a whole range of techniques is the basis of the art of virus diagnosis.

They include observation of clinical signs, pathological changes and epidemiology of disease, the growth of virus in various host systems, identification of the virus and measurement of immune responses. Viral diagnosis may also rely on the direct demonstration of intact virus, viral antigens or even the viral genome.

Viral identification

Once an unknown virus has been isolated, it must be identified and characterized. The basis of this process is to obtain information about a range of characters of the virus, compare this information with that of previously described viruses. If it is found that the unknown organism is possibly the same or similar to one that is already known, this is confirmed by very specific tests such as serological tests or those involving nucleic acid hybridization. The range of characters to be investigated include the following though not necessarily in the following order:

1. The ability of the virus to grow in various host systems.
2. The type of change produced in the host systems.
3. The site of growth within host cells, and the presence of inclusion bodies in infected cells.
4. The morphology and size of its particles, including the presence of a lipid envelope around the particles.
5. Presence of haemagglutinin and others antigens in the particles.
6. The composition and type of nucleic acid in its particles.
7. The electrophoretic mobility of viral polypeptides.
8. The stability of the infectivity and the effect on it of heat, pH extremes, trypsin and other agents.

DEMONSTRATION OF VIRUS AND VIRAL ANTIGENS

In an attempt to overcome some of the difficulties in obtaining information of the characters outlined above, methods for directly searching for virus particles or viral antigens have been developed. The most commonly used procedures are immunofluorescence, complement fixation and electron microscopy.

Immunofluorescence has allowed the rapid detection of virus particles in separated cells, such as desquamated epithelial cells, tissue smears and tissue sections. The complement fixation serological technique may be used to identify antigens present in tissue suspensions or vesicular fluid. Antigens are tested against a library of antibodies.

Electron microscopy has allowed the recognition of a range of formerly undiscovered viruses which fail to grow or do not produce changes in traditional host systems. the sensitivity and specificity of the test may be improved by the addition of specific antibody to form aggregates of virus. The morphology of viruses producing similar diseases may be quite different and electron microscopy may provide a rapid indication of the family of viruses responsible for a lesion.

The methods of direct demonstration of viruses and viral antigens are usually insensitive and unspecific, especially when polyclonal antibodies are used. These techniques are rapidly being replaced by contemporary procedures, such as those outlined in the following three chapters, with advantages such as greater sensitivity, specificity, speed or portability. The enzyme labelled antibody techniques are becoming increasingly versatile.

Those wishing to learn more of the conventional techniques should read reviews following.

REFERENCES

Andrews, C., Perreira, H.G. and Wildy, P. (1982). *Viruses of Vertebrates.* Fourth edition, Bailliere Tindall, London.

Freshney, I.R. (1983). *Culture of Animal Cells.* A Manual of Basic Techniques. Alan R. Liss Inc., New York.

Kruse, P.F. and Patterson, M.K. (1973). *Tissue Custure, Methods and Applications.* Academic Press, New York.

Lennette, E.H. and Schmidt, N.J. (1979). *Diagnostic Procedures for: Viral and Rickettsial and Chlamydial Infections,* Fifth Edition. American Public Health Association, Washington.

McNulty, M.S. and McFerran, J.B. (1984). *Recent Advances in Virus Diagnosis.* Current Topics in Veterinary Medicine and Animal Science, 29. Martinus Nijhoff.

Osterman, L.A. (1984). *Methods of Protein and Nucleic Acid Research.* 1 Electrophoresis. Isoelectric Focusing Ultracentrifugation. Springer-Verlag, Berlin.

Voller, A., Bartlett, A. and Bidwell, D. (1981). *Immunoassays for the '80s.* MTP Press Ltd, Lancaster, England.

Wardley, R.C. and Crowther, J.R. (1982). *The Elisa: Enzyme-linked Immunosorbent Assay in Veterinary Research and Diagnosis.* Martinus Nijhoff.

Chapter Twelve

NEW DEVELOPMENTS IN THE USE OF DNA PROBES FOR THE RAPID DETECTION OF VIRAL PATHOGENS

R.H. Symons

DNA probes labelled with radioactive phosphorus (^{32}P) are widely used in molecular biology research for the specific detection of a nucleic acid by a technique often called molecular hybridization analysis. In this technique, the ^{32}P-DNA probe hybridizes to form a double-strand nucleic acid only with a nucleic acid (DNA or RNA) which is exactly complementary to the DNA probe, whereas no such double-strand hybrid is formed with other nucleic acids. There are various ways of detecting the presence and stability of this DNA probe-nucleic acid hybrid.

Hybridization analysis is being increasingly applied as a diagnostic procedure for specifically detecting viral pathogens of man, animals and plants. For widespread use in all types of laboratories, however, the use of ^{32}P poses many problems. Recent developments in the preparation of both radioactive and non-radioactive probes are considered here.

INTRODUCTION

Most specific diagnostic methods for the detection of viral pathogens fall into two main classes. The first of these is an immunological approach which relies on the use of antibodies prepared against the viral coat protein and a variety of assays have been developed. The other involves the use of molecular hybridization analysis in which radioactive complementary DNA (cDNA) prepared from the purified viral nucleic acid or, preferably, from a recombinant DNA clone of that nucleic acid, is incubated with appropriate test extracts to test for the presence of viral nucleic acid by the formation of a highly specific cDNA:viral nucleic acid complex. Only the second approach is considered here.

There is a rapidly growing interest in the potential commercial use of DNA probes for the specific and rapid diagnosis of many diseases (Klausner and Wilson, 1983). Traditionally, DNA probes have been labelled with such radioisotopes as ^{3}H, ^{32}P or ^{125}I, with ^{32}P being most frequently used. Although isotopic labelling offers great sensitivity, there are unfavourable cost, safety and stability factors inherent in the use of radioactive materials and this has stimulated interest in the development of cheaper and more easily managed non-radioactive labelling techniques. Emphasis here is on the development of these new types of probes, rather than on detailed aspects of their potential use.

PRINCIPLE OF MOLECULAR HYBRIDIZATION ANALYSIS

The biological properties of naturally occurring DNA and RNA molecules are determined by the sequence of the four component nucleotides. Since the nucleotide sequence of a viral nucleic acid is essentially constant and since different viral nucleic acids differ in their nucleotide sequences, any procedure which distinguishes between nucleotide sequences in nucleic acids could be of potential use in the specific detection of a particular viral nucleic acid. A powerful and widely used technique which can do this is called molecular hybridization analysis.

When a cDNA probe and its complementary viral nucleic acid (DNA or RNA) are incubated under appropriate conditions, usually for 10 to 20 hours, the two nucleic acids anneal to form a DNA–RNA or DNA–DNA hybrid (hence the term hybridization) such that A, C, G and T residues in DNA faithfully base pair, respectively, with U, G, C and A residues in RNA or T, G, C and A residues in DNA. No such hybrid forms when the sequences are not complementary. Methods have been developed, one of which is described below, which measure the extent of hybrid formation, and hence the detection of a specific viral nucleic acid.

Preparation of radioactive DNA probes

Techniques have been well established for preparing recombinant DNA clones of viral nucleic acids in both double-strand and single-strand bacterial vectors (Thompson, 1983) and these methods are not considered here. Such DNA clones are the starting material for the preparation of radioactive DNA probes.

A widely used method to prepare ^{32}p-labelled DNA is 'nick translation' (Rigby et al, 1977) which requires double-strand DNA. The procedure which uses a cloned viral insert in the bacterial plasmid vector pBR322 is outlined in Figure 1A. Random nicks are introduced into the DNA clone by incubating it with a small amount of deoxyribonuclease I (DNase I). In the presence of this nicked DNA and the four dNTPs, one or more of which is labelled with ^{32}p (^{32}p-dNTP), *Escherichia coli* DNA polymerase I binds to each nick and uses the 3'hydroxyl group as a primer to start the synthesis of a DNA strand which is complementary to the opposite or template strand. While this synthesis occurs, this remarkable enzyme also catalyses removal of the single-strand DNA in its path so that the old strand in front of the enzyme is continually being hydrolysed while it is being replaced with a new, and hence radioactive, strand behind it. The position of each nick in the DNA therefore moves along with the enzyme. The end result is usually a population of DNA molecules in which 20% to 50% of the old DNA of both strands is replaced by newly synthesized ^{32}p-DNA. The total DNA is then used as the ^{32}p-DNA probe.

When viral nucleic acids are cloned in the single-stranded genome of the bacteriophage M13, the 'vector', the procedure of Figure 1B (Hu and Messing, 1982) is a simple and rapid method for the preparation of ^{32}p-DNA probes. An M13 clone with a minus insert (i.e. an insert which is complementary to the viral nucleic acid) is incubated with a specific synthetic oligodeoxynucleotide 15 residues long (15-mer) which hybridizes to the M13 vector on the 'upstream' (5'-terminal) side of the cloned insert. The primer is then extended by a variant of *E. coli* DNA polymerase I (Klenow Fragment) in the presence of ^{32}p-dNTPs to give a population of molecules with varying lengths of newly synthesised DNA hybridized to the template strand. In practice, synthesis extends only part of the way around the 7,000 residue circular genome of the M13 phage. The product is then purified from enzyme and substrates by simple column chromatography and is ready for use.

Dot-blot hybridization, a procedure for detecting viral nucleic acids

A technique which is widely used for detecting viral nucleic acids by hybridization with ^{32}p-DNA probes (Alwine et. al., 1979; Thomas, 1983) is called the dot-blot procedure and is outlined in Figure 2. Test samples, which are usually crude or partly purified nucleic acid extracts in $5\mu l$, are 'spotted' onto a thin sheet of cellulose nitrate and then baked *in vacuo* at 80°C for two hours, nucleic acids become irreversibly bound to the paper by an unknown mechanism. The cellulose nitrate sheet is incubated in a hybridization solution, the ^{32}p-DNA probe is added and incubation continued, usually overnight, to allow the ^{32}p-DNA probe to hybridize to any viral nucleic acids bound to the filter. The excess probe is removed and the filter is exposed to X-ray film for 20–48 hours. A dark spot on the film indicates the presence of a ^{32}p-DNA probe:viral nucleic acid hybrid. An example of a dot-blot obtained in the detection of the RNA of a plant viroid with a specific ^{32}p-DNA, prepared as described in Figure 1B, is given in Figure 3.

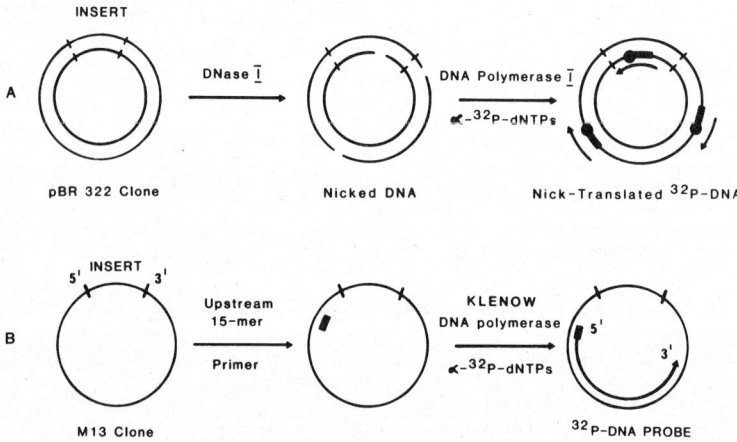

Fig. 1. Outline of two procedures for the preparation of ^{32}p-DNA probes.

a. Nick translation (Rigby et al., 1977). The arrows give the direction of synthesis of ^{32}p-DNA (thick line) by *E. coli* DNA polymerase I (filled circles) acting on a cloned insert in the plasmid vector pBR322 after nicking by DNase I.

b. Use of a single-strand bacteriophage M13 DNA clone and a synthetic 15-mer primer. Further details given in text.

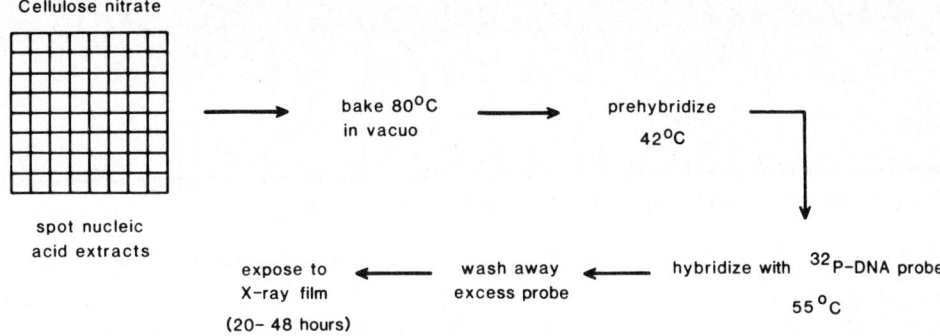

Fig. 2. Outline of dot-blot procedure using ^{32}p-DNA probes.

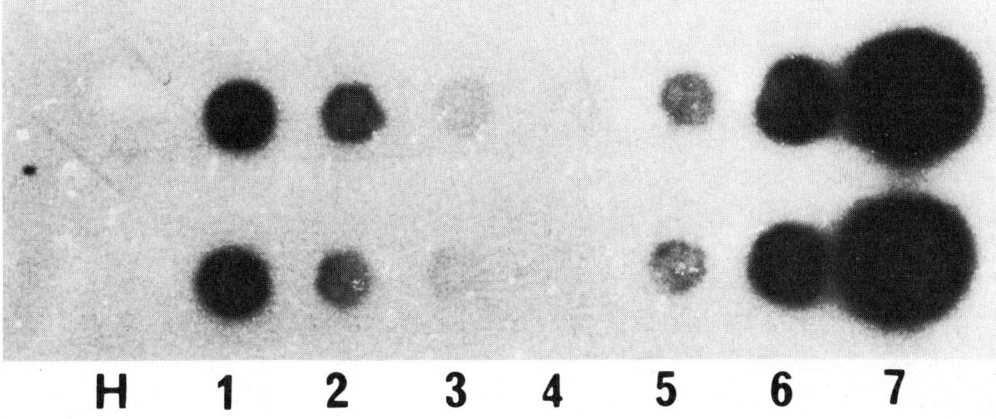

Fig. 3. Dot-blot assay for the detection of avocado sunblotch viroid in extracts of avocado leaves using a ^{32}p-DNA probe prepared as in Fig. 1b. Duplicate samples spotted. H — healthy extract. Samples 1 to 4 — 50pg, 20pg, 5pg and 2pg purified avocado sunblotch viroid mixed with a healthy leaf extract. Sample 5 — 5pg purified viroid in 0.1 mM EDTA. Samples 6 and 7 — extracts from leaves of two avocado trees infected with avocado sunblotch viroid.

Development of non-radioactive DNA probes

There is a clear need for a detection system for cDNA:nucleic acid hybrids which is simple and approximately as sensitive as that obtained with radioisotopes. An approach initiated by Langer et al., (1981) makes use of analogues of dTTP in which the methyl group on the 5-position of the pyrimidine ring to dTTP is replaced by a biotin molecule covalently attached through a linker arm (Figure 4; Langer et al., 1981). Such nucleotides, called bio-dUTP, are efficient substrates for several DNA polymerases when substituted for dTPP. Thus, Leary et al., (1983) prepared biotinylated DNA probes in the standard nick translation reaction of Figure 1B by replacing the ^{32}p-dTTP with bio-dUTP. As the biotin is attached through the 5-position of the pyrimidine ring it does not, as originally predicted, interfere significantly with the base pairing reaction of the two carbonyl groups of the pyrimidine ring with the complementary adenine residue (Langer et al., 1981).

Biotin-labelled DNA probes have been successfully used in a variation of the dot-blot assay of Figure 2. The ^{32}p-DNA probe was replaced by biotinylated DNA probe and hybridized to DNA or RNA immobilised on nitrocellulose filters in exactly the same way. After washing away the excess probe, residual bound probe can be detected in two related ways (Figure 5). In the first method, the filters are incubated for several minutes with a covalent avidin-enzyme complex in order to allow the highly specific and very tight binding of the protein avidin to the filter-bound biotin. The unbound avidin-enzyme reagent is then washed away and an appropriate soluble enzyme substrate added; the substrate is chosen such that the reaction product is completely insoluble and coloured and is precipitated *in situ*. Hence, the presence of bound biotin-DNA probe is detected by the appearance of an insoluble, coloured product. Either of two enzymes can be used, alkaline phosphatase or horse radish peroxidase (Figure 5). In the second detection method, enzyme coupled to antibodies prepared against biotin is used but all other aspects of the assay are the same.

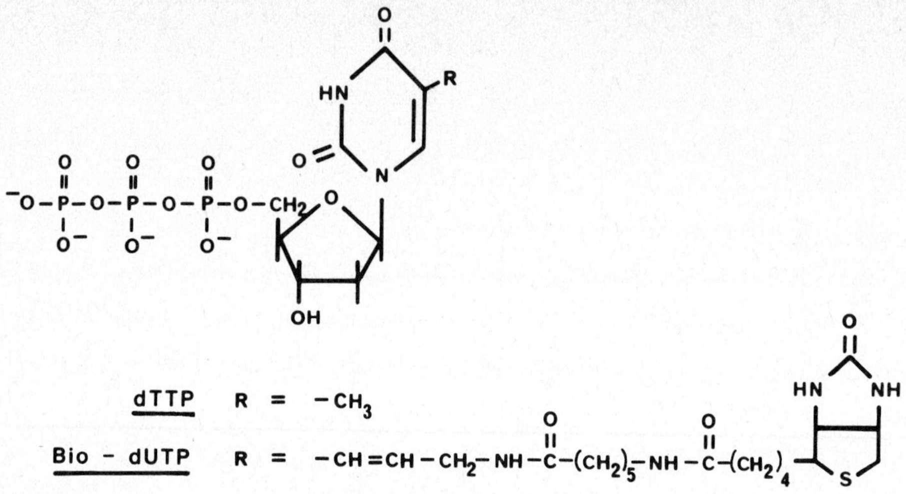

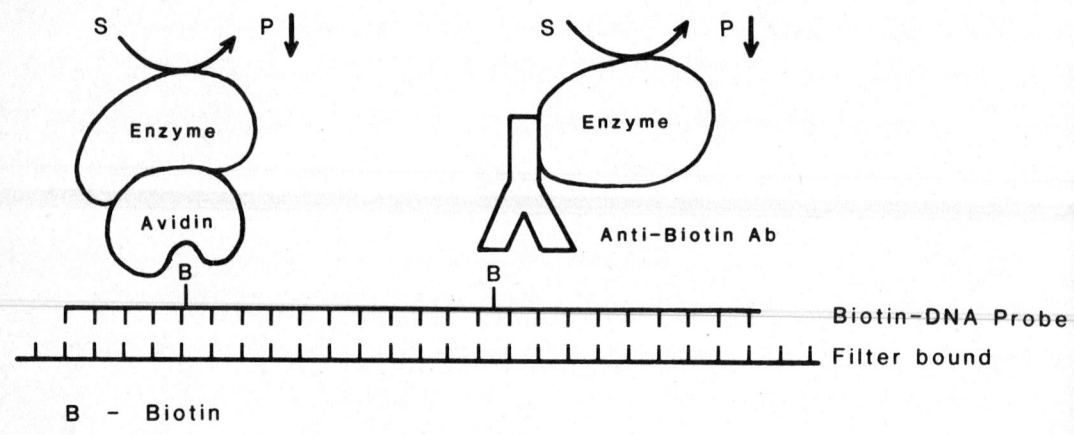

Fig. 4. Structure of dTTP and its analogue, bio-dUTP.

Fig. 5. Representation of way in which biotin-DNA probe hybridized to cellulose nitrate-bound viral nucleic acid using either an avidin-enzyme or an (antibiotin-antibody)-enzyme complex.

Variations of this basic method have been developed by Leary *et. al.* (1983) to increase the amplification of the enzyme detection system. These are not considered here but they do allow the specific detection of 1 to 10pg of target nucleic acids bound to cellulose nitrate. Hence, it appears that the sensitivity of this non-radioactive assay is about the same as that usually achieved with ^{32}p-DNA probes. Although reagent kits incorporating the use of bio-dUTP are now available commercially, there have been no published reports so far on the use of non-radioactive DNA probes for the routine detection of viral nucleic acids.

Alternative approaches to the preparation of non-radioactive DNA probes
    A practical disadvantage of the method of Leary *et. al.* (1983) is the requirement for an enzymic reaction to incorporate biotin into the DNA probe. With nick translation, a double-strand template is required and it requires large amounts of expensive enzymes and reagents to prepare milligram quantities of biotinylated DNA probe. We have adopted an alternative approach by preparing a synthetic reagent which allows the simple chemical coupling of biotin to any nucleic acid (Forster *et. al.*, 1985).

This reagent, called photobiotin (Figure 6), consists of biotin attached by a linker arm to a photoactivatable group. When, for example, a single-strand recombinant DNA clone of a viral nucleic acid in a bacteriophage M13 vector is mixed with photobiotin and irradiated with visible light for 10 to 20 minutes under defined conditions (Figure 7), the azido group is converted to a very reactive nitrene which allows the formation of a very stable bond to various components of the nucleic acid. After removing excess photolysed photobiotin, the biotin-DNA probe is ready for use. Reaction conditions are adjusted so that about one biotin is coupled per 100 to 200 nucleotide residues; this corresponds to about 35–70 biotins per molecule of M13 vector which consists of about 7,000 residues. Two major advantages of this reagent are that it can be used with either single- or double-strand DNA or RNA and that milligram quantities of biotin probe can be rapidly and economically prepared.

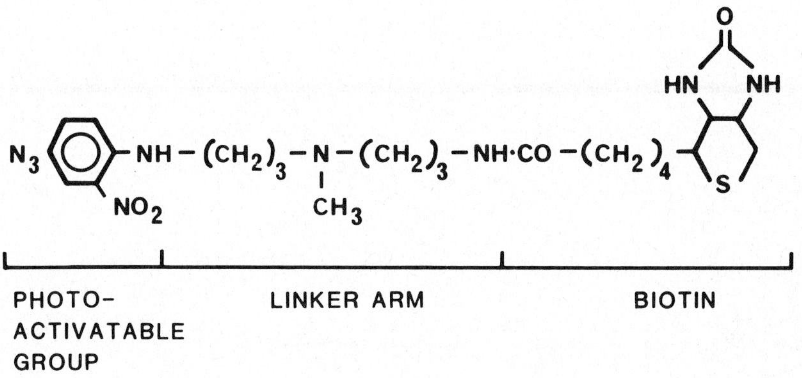

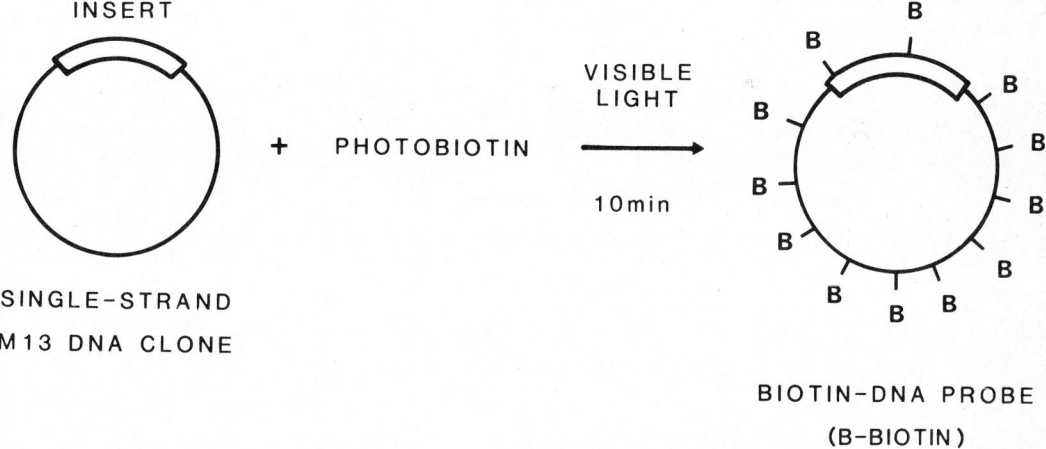

Fig. 6. Structure of photobiotin.

Fig. 7. Preparation of a non-radioactive DNA probe by photolysis of photobiotin with a single-strand M13 DNA clone.

Initial trials are underway to test the usefulness and sensitivity of biotin-DNA probes prepared in this way for the rapid detection of viral nucleic acids by the dot-blot procedure. A diagrammatical representation of the complex formed on addition of the avidin-enzyme reagent to the filter-bound biotin-DNA probe is presented in Figure 8.

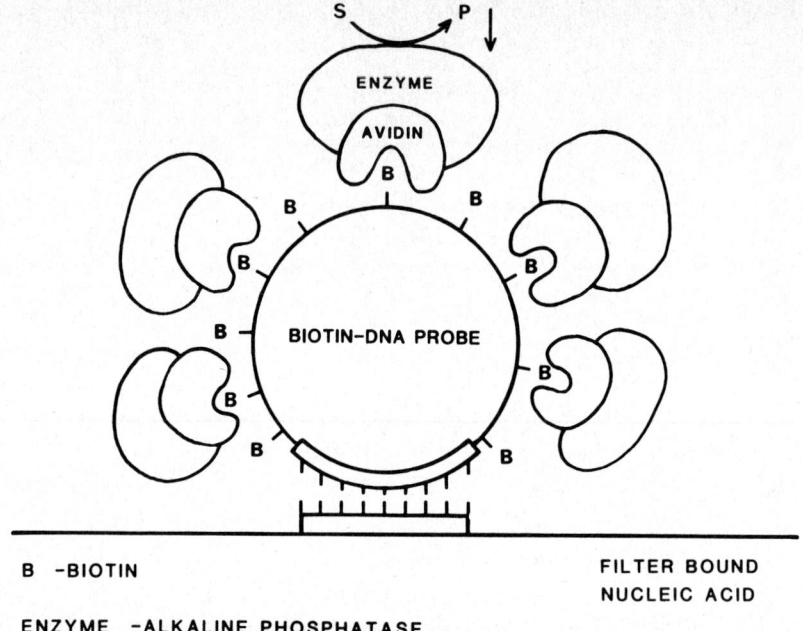

Fig. 8. Representation of way in which chemically biotinylated M13 DNA probe which is hybridized to cellulose nitrate-bound nucleic acid.

CONCLUSIONS

These is nowadays intense interest in developing new diagnostic procedures for rapidly detecting viral and other pathogens of man, animals and plants. Non-radioactive nucleic acid probes and methods for their use are likely to be continually improved to the stage where they will compete effectively with other methods now in routine use because of their speed, specificity, economy and practicality.

REFERENCES

Alwine, J.C., Kemp, D.J., Parker, B.A., Reiser, J., Renart, J., Stark, G.R. and Wahl, G.M. (1979). *Methods Enzymol. 68*, 220.
Forster, A.C., McInnes, J.L., Skingle, D.C. and Symons, R.H. (1985). *Nucleic Acids Res. 13*, 745.
Hu, N. and Messing, J. (1982). *Gene 17*, 271.
Klausner, A. and Wilson, T. (1983). *Biotech. Aug..* p. 471.
Langer, P.R., Waldrop, A.A. and Ward, D.C. (1981). *Proc. Natl. Acad. Sci. USA, 78*, 6633.
Leary, J.L., Brigati, D.J., and Ward, D.C. (1983). *Proc. Natl. Acad. Sci. USA, 80*, 4045.
Rigby, P.W.J., Dieckmann, M., Rhodes, C. and Berg, P. (1977). *J. Mol. Biol. 133*, 237.
Thomas, P.S. (1983). *Methods Enzymol. 100*, 255.
Thompson, R. (1980). In: *Genetic Engineering, Vol. 3* (R. Williamson, ed.) p. 1, Academic Press, London.

Chapter Thirteen

AGENT DETECTION AND IDENTIFICATION — NEW IMMUNOLOGICAL TECHNIQUES

M.R. Brandon and H.G. Bults

The immunological detection of disease agents in animal blood and tissues depends on the availability of antisera. Until recently antiserum was produced by injecting a disease agent or agent preparations into an animal and collecting serum, and antibodies in these sera specifically reacted with the disease agent. Polyclonal antisera made in this way are a complex mix of antibodies and difficulties arise with their limited supply, weak titre and with false reactions.

Recent advances in cell biology have resulted in the production of antibodies in tissue culture, the antibody being the product of the progeny of a single cell and being of a single specificity. The single cell can be immortalised as a tumour cell line and the product of the cell line is known as a monoclonal antibody. Monoclonal antibodies can be produced in unlimited amounts and problems associated with conventional antisera eliminated. Monoclonal antibodies have the potential of revolutionising immunological disease diagnosis.

NEW IMMUNOLOGICAL TECHNIQUES

The Enzyme Linked Immunosorbent Assay (ELISA)

Over the past ten years there has been an increase in the number of immunological tests used for agent detection. One of the reasons for this has been the development of methods which use labelled antigens or antibodies, resulting in tests with very high levels of sensitivity and specificity. Fluorescent labels have been attached to antibodies and these conjugates have proved useful for the rapid identification of organisms responsible for infectious diseases as well as the measurement of antibody levels in infectious and autoimmune diseases. Radioisotopes have also been found to be useful, especially as labels on antigens (for instance, disease agents), and radioimmunoassay (RIA) has become the most widely used method for sensitive assays of both large and small molecular weight substances, particularly for steroid hormones. However both immunofluorescence and RIA have limitations. Immunofluorescence is time consuming and not easily automated and requires sophisticated equipment for detection. The RIA is particularly suitable for large scale operations but the short shelf life of the reagents, the expensive equipment required for measuring radioactivity and the strict regulatory control on the use of isotopes limits the use of RIA to laboratories. This has led to a search for alternative labels for antibodies and antigens. The most promising new labels are enzymes. These can be linked to antibodies and antigens and the resulting complexes have both immunological and enzymatic activity. The quantitation of antigen or antibody relies on the measurement of colour changes and this is particularly suited to development of assays suitable for use outside of laboratories.

Disease Agent Detection by ELISA

Two methods are available for detecting the presence of a disease agent (the antigen) by ELISA. The competitive ELISA test involves mixing an enzyme labelled disease agent (antigen) with a test sample containing the suspect disease agent, which competes for a limited amount of antibody against the disease agent. The test antibody is absorbed to a 'solid phase' which is usually a particulate material such as a test tube, bead or disc made of polyvinyl, polyproplene, polycarbonate, glass or silicone rubber. The reacted (bound) antigen is then separated from the free material and its enzyme activity is estimated by the addition of substrate to develop a colour reaction.

The most commonly used method for detecting disease agents is the double antibody sandwich technique shown in Figure 1. In this method a solid phase is coated with specific antibody. This is then reacted with the next sample containing the disease agent, then enzyme labelled specific antibody is added, followed by the enzyme substrate to develop the colour reaction.

Detection of Disease Agent Antibody by ELISA

Often the best method of diagnosing an infectious disease is to use ELISA to detect antibody produced by infected animals against the disease agent. The indirect method of ELISA has proved particularly useful for this. In the indirect method the antigen is either immobilised by passive adsorption to the solid phase or anchored by the use of an antibody as shown in Figure 2. Test sera are then incubated with the solid phase and any antibody in the test sera becomes attached to the disease agent secured on the solid phase. After washing to remove unreacted serum components, an anti-antibody enzyme conjugate is added which attaches to any antibody already fixed to the antigen. Washing again removes unreacted material and finally the enzyme substrate is added. The colour change is a measure of the amount of the conjugate fixed, which is itself proportional to the antibody level in the test sample.

Application of ELISA to the Diagnosis of Foot and Mouth Disease Virus

Many papers have been published on the use of ELISA for the detection of FMDV (Abu Elzen and Crowther, 1978, 1982; Crowther and Abu Elzen, 1979, 1980). The indirect ELISA developed by Abu Elzen and Crowther (1982), illustrated in Figure 3, yields similar results to the complement fixation (CF) test but offers two main advantages in that antibody from any animal species may be used to establish relationships by ELISA by using different anti-species conjugates; this is not possible using routine CF tests. The ELISA has also proved to be *100* times more sensitive than a CF test, so that a greater economy of reagents is achieved. For example, in a study by Crowther and Abu Elzen (1979), it was demonstrated that all 19 positive samples in a study were types by ELISA from original material, whereas the CF test typed only 8 samples and further tissue culture passage was necessary before CF typing was obtained. The ELISA method was not affected by anti-complementary factors as was seen for 5 samples using the CF test before passage.

In practical terms the test is simple to perform, bovine antisera to all seven serotypes are readily available and as only small amounts of specific antibody are needed to coat typing plates, a single solution of antibody provides enough material for several years work. These new techniques based on conventional antisera are quicker, hundreds of times more sensitive than the established techniques presently used and can detect either FMDV antigen or antibody in infected animals and do not require live virus as a positive control.

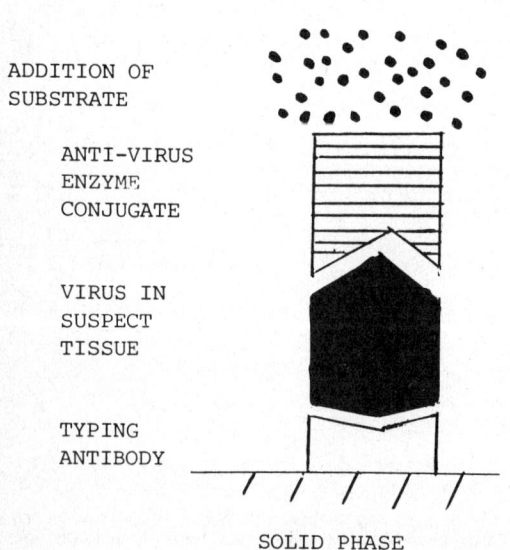

Fig. 1. Detection of antigen of a disease agent by ELISA.

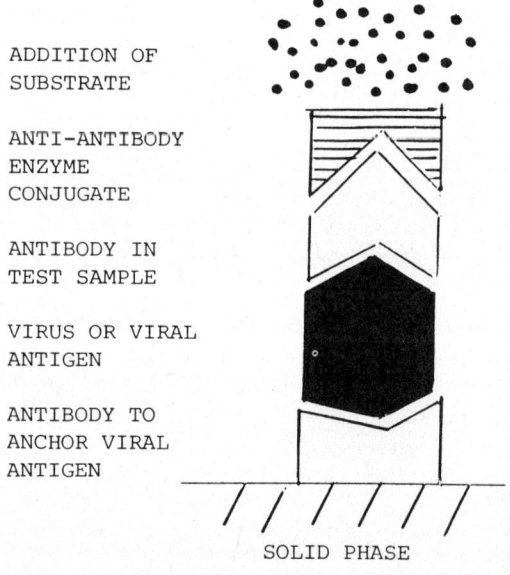

Fig. 2. Detection of disease agent antibody by ELISA.

APPLICATION OF ELISA TO DETECTION OF FMDV

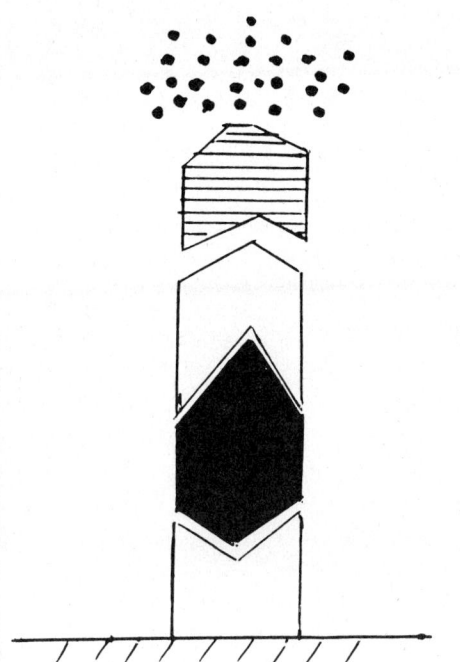

Fig. 3. Application of ELISA to detection of foot-and-mouth disease virus.

MONOCLONAL ANTIBODIES

When an animal is injected with an immunising agent (antigen or disease agent) it responds by making antibodies against different antigenic molecules on the surface of the antigen and against different determinants on a single antigen, and even different antibodies that fix, more or less well a single determinant. As it is extremely difficult to separate the various antibodies, conventional antisera contain mixtures of antibodies and the mixtures vary from animal to animal.

Each antibody however is made by a different clone of lymphocytes. If each clone could be grown separately it would produce a monoclonal antibody. Unfortunately antibody-secreting cells cannot be maintained in culture media. However there are malignant tumours of the immune system called myelomas whose rapidly proliferating cells produce large amounts of abnormal immunoglobulins called myeloma proteins. A tumour is an immortal clone of cells descended from a single progenitor and so myeloma cells can be cultured indefinitely and all the immunoglobulins they secrete are identical in chemical structure.

In 1975, Kohler and Milstein fused mouse myeloma cells with lymphocytes from the spleen of mice immunised with a particular antigen. The resulting hybrid-myeloma or 'hybridoma' cells expressed both the ability of the lymphocytes to produce a single type of antibody, and the immortality of the myeloma cells. Such hybrid cells can be manipulated by the techniques applicable to animal cells in permanent culture. Individual hybrid cells can be cloned, and each clone may produce large amounts of identical antibody to a single antigenic determinant. The individual clone can be maintained indefinitely and at any time samples can be grown in culture and injected into animals for the large scale production of monoclonal antibody. Very specific monoclonal antibodies produced by this general method are proving to be extremely valuable and versatile biochemical probes.

Use of monoclonal antibodies for diagnosis of myxoma poxvirus

In 1979 studies were started to attempt to distinguish between three strains of myxoma virus and to measure their incidence in the Australian rabbit population. Previous studies using polyclonal antisera produced by immunising rabbits with whole myxoma virus failed to detect antigenic differences between strains. However antigenic differences were detected in polyconal antisera obtained from rabbits immunised with enriched preparations of soluble antigens isolated from the myxoma strains Glenfield, Lausanne and Urana Field. The Glenfield and Lausanne strains were shown to be different and the Urana Field strain was probably a variant of the Lausanne strain (Bults and Brandon, 1982). As the polyclonal antisera are of limited supply and of low titre, experiments were done to produce monoclonal antibodies that could distinguish between the various strains of myxoma virus. Monoclonal antibodies were produced that distinguished between the Glenfield and Lausanne strains by both ELISA and RIA (Table 1). The monoclonals antibodies confirmed that the Urana Field strain is a variant of the Lausanne strain. By producing further monoclonal antibodies to other field strain isolates it should be possible to study the distribution of variants of myxoma virus in the Australian rabbit population and help select suitable virulent strains for distribution in regions where the rabbit population threatens agricultural production.

TABLE 1. *Reaction patterns of monoclonal antibodies against enriched preparations of soluble antigens isolated from the Glenfield, Lausanne and Urana Field strains of myxoma poxvirus*

specificity Antibody	Antigen source Myxoma strain		
	Glenfield	Lausanne	Urana Field
Glenfield strain	+	−	−
Lausanne strain	−	+	++
Urana Field strain	−	+	+

NEW SOLID PHASE ASSAYS USED IN CONJUNCTION WITH MONOCLONAL ANTIBODIES

Immuno-Western blotting

Complex mistures of antigens from disease agents can be quickly and easily separated by high resolution techniques such as electrophoresis in polyacrylamide gels using discontinuous detergent buffer systems. However once separated in this way it has been difficult to determine which of the separated species reacts with a given antiserum.

This problem has been overcome by electrophoretically transferring the separated proteins onto nitrocelulose sheets (Towbin, Staehlin and Gordon, 1979). Once attached to the nitrocellulose, the antigenicity of each of the separated components can be tested by treating the blot with antiserum and the bound antibody detected by ELISA or RIA. By using monoclonal antibodies with this technique different antigens of the disease agent can be detected (Figure 4) and the assay is so sensitive that picograms of protein may be detected.

Dot Immunobinding Assay. The recently developed dot immunobinding assay of Hawkes, Niday and Gordon (1982) is an alternative to the ELISA and possesses several distinct advantages. The dot immunobinding assay is of equal or greater sensitiveity than the ELISA and can function over a wide range of antigen/antibody ratios. The degree of nonspecific background reactivity can be directly compared to the specific immunoreactivity without assaying a duplicate sample.

In the assay, the test antigen from the disease agent is attached to nitrocellulose paper. Antiserum, or monoclonal antibody, is incubated with the nitrocellulose paper and the bound antibodies detected by ELISA or RIA. Using this assay the presence of less than 1 picogram of a glycoprotein from the cell surface of sheep lymphocytes can be detected. An example of a dot blot assay is shown in Figure 5. The presence of a particular antigen or antibody is shown as a positive colour reaction. This procedure could be readily adapted for on-farm testing of disease agents.

The use of the new immunological tests, with monoclonal antibodies, gives the potential for developing on-farm or in the field assays that could detect disease agents in the blood and tissues of animals with an accuracy and reliability previously only possible in research laboratories.

REFERENCES

Abu Elzen, E.M.E. and Crowther, J.R. (1978). *J. Hyg. Camb. 80*, 391.
Abu Elzen, E.M.E. and Crowther, J.R. (1982). *J. Virol. Meths. 3*, 355.
Bults, H.G. and Brandon, M.R. (1982). *J. Virol. Meth.* 5, 21.
Crowther, J.R. and Abu Elzen, E.M.E. (1979). *J. Hyg. Camb. 83*, 513.
Crowther, J.R. and Abu Elzen, E.M.E. (1980). *J. Immunol. Meth. 34*, 261.
Hawkes, R., Niday, E. and Gordon, J. (1982). *Anal. Biochem. 119*, 142.
Kohler, G. and Milstein, C. (1975). *Nature* 256, 495.
Towbin, H. Staehelin, T. and Grodoń, J. (1979). *Proc. Natl. Acad. Sci.* 76, 4350.

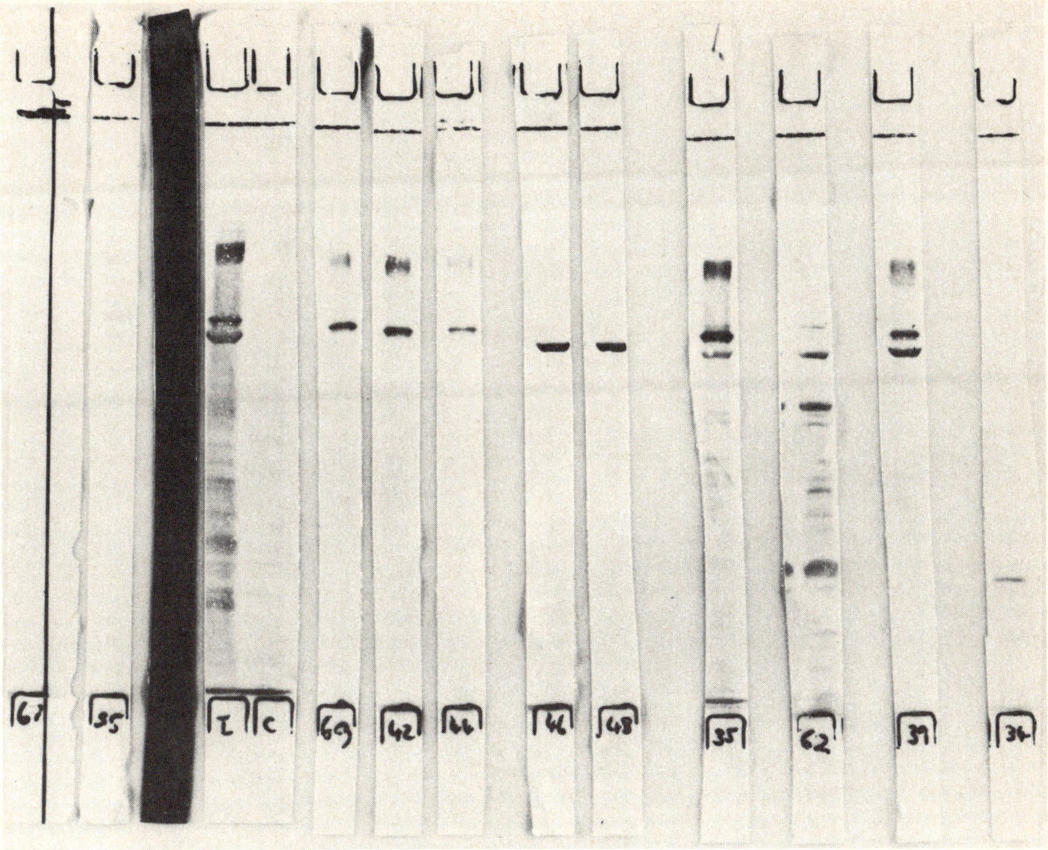

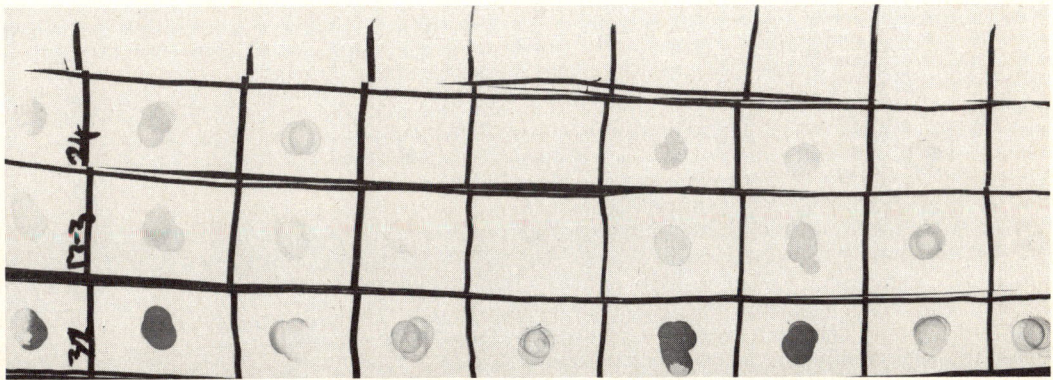

Fig. 4. Immuno-Western blot of an avian herpesvirus. Each lane represents the reaction of a different monoclonal antibody with nitrocellulose strips to which avian herpesvirus lysates have been blotted after electrophoresis in polyacrylamide gel (proteins solubilized in sodium dodecyl sulphate). It can be seen that the monoclonal antibodies detect numerous viral antigens. (Courtesy of Ms J. York, CSIRO Division of Animal Health, Parkville, Vic.).

Fig. 5. Dot immunobinding assay of a glycoprotein from the surface of sheep lymphocytes. ELISA using a monoclonal antibody detected as little as 1 picogram of the glycoprotein. A positive reaction is indicated by a colour change. The absence of colour shows the glycoprotein is not present in the cell extract being tested. (Courtesy of Ms J. Maddox, University of Melbourne, Parkville, Vic.).

Chapter Fourteen

THE STUDY OF GENETIC VARIATION — A BASIC TOOL FOR PEST AND PARASITE STUDIES

Adrian Gibbs and Rodney Mahon

Individual plants and animals differ from one another in appearance. Just by looking at two individuals we can often tell whether they are close relatives, members of different races or breeds of a single species, or come from different species.

Mankind has known for a long time that such differences are inherited, and that useful types of animals and plants can be obtained by selection and breeding; this process has been applied over the last few thousand years and has resulted in the domestication of many organisms. However, only recently has it been discovered how genetic information is inherited.

The first crucial experiments were done in a monastry garden over a century ago at Brno in Czeckoslovakia by a monk, Gregor Mendel. He experimented with garden pea varieties that had contrasting characters; tall versus short steams, wrinkled versus smooth seed, and green seed versus yellow. He found that when the flowers of tall plants were pollinated by those of short, or vice versa, the resulting seed produced only tall plants. When the flowers of these tall plants were pollinated among themselves and the resulting seed was sown, three quarters of the plants obtained were tall, but one quarter were short. Furthermore, when the tall plants were treated in the same way, seed from one third of them gave only tall plants, but from two thirds gave mixtures.

These experiments showed Mendel that some genetic traits dominate others in the same individual (i.e. tallness in peas dominates shortness), but those that are recessive are not lost or 'blended' with the dominant ones, they are merely obscured. Thus in sexually reproducing species, the genetic information is carried in pairs, at least, of discrete units, or genes, one inherited from each parents.

Over the past century, the line of research started by Mendel has blossomed into the science of genetics, the study of the behaviour of genes, and, during the past 30 years, into molecular genetics; the study of the molecules responsible for inheritance.

It has been found that the genetic information of organisms is carried in giant molecules called nucleic acids. All cellular organisms use deoxyribose nucleic acid (DNA) to store their genetic information, and various types of ribose nucleic acid (RNA), which convey that information and process it into the proteins that are the bricks, mortar and machinery of living cells. Certain microbes called viroids, and some viruses, use RNA to store their genetic information. In higher organisms, such as fungi, plants and animals, DNA is enclosed in organelles within each cell: the largest of these is the nucleus, and in it the DNA is organized into thread-like bodies called chromosomes, whereas in the organelles called chloroplasts and mitochondria, the DNA seems to be less organized, just like it is in the cells of bacteria.

Most DNA consists of two helically entwined chains made of alternating deoxyribose (a sugar) and phosphate molecules. Attached to each deoxyribose is one or other of the four bases, guanine (G), adenine (A), cytosine (C) and thymine (T). The sequence of these bases encodes the specific genetic information of every organism; thus the information is stored as a sort of genetic 'tape recording' (in one dimension), not as a 'blueprint' (in two dimensions), as is so frequently, and incorrectly, stated. Each base is attached to its deoxyribose by a strong bond, and to the nearest base in the partner chain by weak, but specific bonds, so that G in one chain only bonds to C in the other, and A only to T. Thus the sequence of bases in one chain is a 'mirror' or complement of that in the other; G to C, and A to T. This specificity indicates the way in which the cell copies the DNA to transfer the information accurately to progeny cells. Before each cell divides, the DNA partner chains separate and an enzyme moves along each chain 'reading' it and synthesizing a complementary chain. In this way two daughter molecules are produced that are identical to the parent molecule.

Some viruses, unlike all 'higher' organisms, do not use DNA to store their genetic information. Instead some use single-stranded DNA, others use single or double-stranded RNA. In RNA the sugar molecules in the chain are ribose, not deoxyribose as in DNA.

To store all the genetic information of a bacterium requires one to six million base pairs, whereas mammals, such as man, require around six thousand million base pairs, and the cells of salamanders and some plants contain even more! The DNA in a bacterial cells is usually in one large circular molecule and a few small ones; the large one is the bacterial chromosome and the small ones are known as plasmids. The total length of DNA in each bacterial cell is about one millimetre, so there is quite a packing problem involved in getting it into a bacterial cell that is only about one five-hundredth of a millimetre long. The packing problem is even greater for multicellular organisms; each human cell contains about two metres of DNA and this is packed into a nucleus that is only ten micrometres in diameter. In man, as in other organisms with nuclei, the problem is alleviated by dividing the DNA among organelles called chromosomes, of which there are 46 in most human cells. Within each chromosome the DNA is wound into coiled coils around special packing proteins called histones. Each cell contains two sets of chromosomes (Figure 1), one set inherited from its mother, the other from its father. However when egg or sperm cells are produced the chromosomes separate into two sets, so that egg or sperm cells contain only half as many chromsomes as a normal body cell.

The information 'tape recorded' in DNA is of two sorts. First there are control sequences which tell enzymes where to start 'reading' the DNA, where to pause and where to stop, just like the instructions stored on the more sophisticated magnetic recording tapes. These sorts of signals control the development of the organism, and only recently has much progress been made in understanding how they work. The other sort of sequence is much better understood. These are the ones that contain information for making the various proteins used by the cell. Proteins consist of large folded chains of amino acids. There are basically 20 different amino acids, and the order in which they are strung together during synthesis of a protein determines the structure and function of that protein. For the information in the DNAs to be converted into the structure of a particular protein, the DNA is first copied into a messenger RNA, which moves from the nucleus to the cytoplasm of the cell, where 'mini-organelles', called ribosomes, read the sequence of its bases and, at the same time, assemble the amino acids to form a protein. Remember that there are only four sorts of bases, but twenty amino acids. This potential problem is overcome by having a run of three bases (a triplet) specify each amino acid. There are

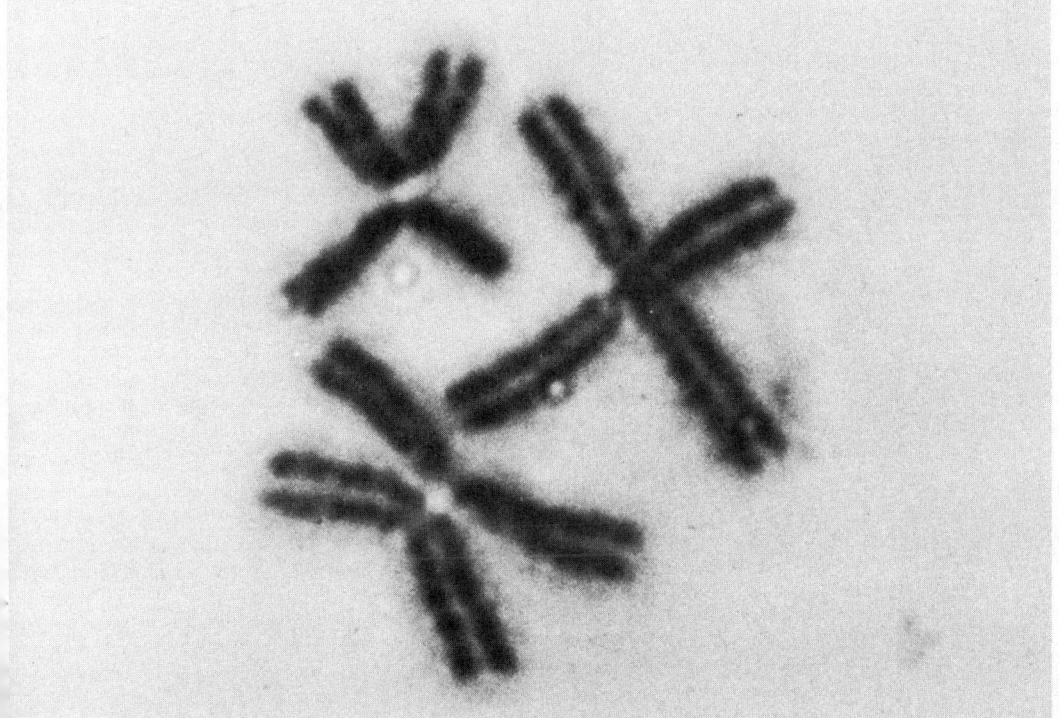

Fig. 1. Mitotic chromosomes of the mosquito *Aedes albopictus*. All species of mosquitoes have only three pairs of chromosomes.

$4 \times 4 \times 4 = 64$ possible triplet combinations of four different bases. Some code for stop and start signals, and some of the frequently used amino acids are specified by several triplets, for example CCT, CCC, CCA and CCG all code for the amino acid proline; note that the 'third position' of these triplets is redundant, as proline is specified irrespective of which base is in the third position.

So in biochemical terms, a gene is that segment of a nucleic acid responsible for specifying a particular trait, whether it is a 'control' gene or a 'structural' gene (i.e. specifies a particular protein), and the total gene set of an organism is called its genome. Inherited differences between related individuals merely indicate that they differ in the sequence of nucleotides in their genomes. The differences may be very few, even one nucleotide change can sometimes have a large effect, and genes control such seemingly complex traits as the rate of spread of a disease, or the ability of a weed to survive and spread in a particular locality, or the susceptibility of an organisms to poisons or parasites.

These differences may be studied in various ways and at different 'levels' of complexity. We can nowadays, for example, directly determine the sequence of nucleotides in a particular nucleic acid, or the sequence of amino acids in a protein that it specifies. Alternatively we can look at such differences indirectly by determining a property that is sequence dependent, such as the electrophoretic mobility of a protein, or the pattern or number of staining bands in chromsomes.

There are two most important reasons why these minor genetic differences are studied by scientists:

1. They may be used as subtle molecular labels or markers to obtain information about the present and past behaviour of organisms; their movement, mating behaviour, origins and evolution, etc.
2. They may show how the genes 'work' and why, for example, some individuals of a population resist a particular disease while others do not.

GENES AS MOLECULAR LABELS

Nucleic acids

The outline given above indicates what sort of genetic information may be used to characterize and compare individuals.

Sequence comparisons

The sequence of nucleotides of all the genes of an individual gives a complete genetic description of that individual. However nucleotide sequencing is a time consuming procedure, and is infrequently used to compare genes of individuals from 'macro-organisms'. It has been more frequently used to compare virus isolates (each a population of individuals), probably because viruses have small genomes and may be readily obtained from purified virus particles. For example sequencing studies has shown why influenza virus is able to cause an epidemic almost every year in the human population.

Influenza virus has rounded particles of irregular shape. Their surface is covered with protein 'knobs' of two sorts. One sort, the haemagglutinin, can stick red cells together. A human being who has recovered from an attack of influenza produces antibodies which circulate in their blood and which can combine with the haemagglutinin of the virus which infected them (or a closely related strain) and destroy its infectivity. At irregular intervals of 10–20 years an influenza virus with a novel haemagglutinin appears in the human population and, because few people, if any, have any antibodies against the haemagglutinin of that virus, it spreads rapidly and causes a major flu epidemic (e.g. Hong Kong flu, Asian flu, etc.). Between these major epidemics, there are minor annual epidemics and the haemagglutinin, though not novel, is found to undergo sequential minor changes in its antigenic sites (i.e. those parts with which the antibodies react). These minor changes are called 'antigenic drift' in contrast to the major changes which are called 'antigenic shift'. 'Shift' seems to result from a rare chance recombination between human and animal influenza viruses in a doubly infected individual; during such an encounter the human flu acquires the haemagglutinin gene of the animal flu and hence most, if not all, of the human population is susceptible to it. By contrast, 'drift' is the result of the sequential acquisition of mutations by the existing haemagglutinin gene principally in those parts that code for its antigenic determinants (Figure 2).

Molecular hybridization

Nucleotide sequences may be indirectly characterized and compared by various methods. When double-stranded nucleic acids are heated, the strands separate, and when cooled will reform the double-stranded molecules, but only with strands with a complementary sequence (or one that is very little different from the complement). Thus the extend of hybrid formation may be used as a measure of the relatedness of two nucleic acid strands (Chapter 12).

Restriction endonucleases

Another indirect measure of the relatedness of DNA sequences is to assess the number, and sometimes the position, of sequences that are cut by sequence-specific enzymes called restriction endonucleases. Each of these enzymes, which are produced by bacteria as part of their nucleic acid defence system, recognizes and cuts a particular sequence of four or six nucleotides. The number and arrangement of the fragments produced can

Nucleotide number	NT 68	HK 71	ENG 72	MEM 72	PC 73	VIC 75	AC 76	VIC 76	TEX 77	BK1 79	Amino acid number	Amino acid change
81	G								A	A	2	asp — asn
83	C		U									
84	C			U	U	C	C	C	C	C	3	leu — phe
89	A									C		
103	A	G	G	G	G	G	G	G	G	G	9	asn — ser
106	C	A									10	thr — lys
113	G						U	U				
128	U									C		
134	G								A			
149	A				G	G	G	G	G	G		
167	A				G	G	G	G	G	G		
168	G		A	A	A	A	A	A	A	A	31	asp — asn
178	U	C									34	ile — thr
188	U						C					
200	G					A	A					
203	A		G	G	G	G	G	G	G	G		
212	C			U	U	U	U	U	U	U		
221	G								A	A		
224	G					U	U	U	U	U		
226	A								G	G	50	lys — arg
234	A						G	G	G	G	53	asn — asp
238	A								G	G	54	asn — ser
245	U									C		
260	A								G	G		
262	U								A	A	62	ile — lys
264	G						A	A	A	A	63	asn — asn
287	A				A	G						
293	G									A		
310	U		G	G	G	G	G	G	G	G	78	val — gly
311	A		C	C	C	A	C	C	C	C		
325	C					A	A	A	A	A	83	thr — lys
338	C							U	U	U		

Only nucleotides that differ from those of isolate NT68 are recorded. Nucleotides are numbered from the 3' terminus of the gene, amino acids from the N-terminal residue of HA1. Most mutations are transitions (i.e. C-U and A-G changes), some are 'silent' (i.e. cause no change in the amino acid) and most of these are in the third nucleotides of triplets. Nucleotide changes in the 'flu haemagglutinin gene are unusual in that so many are 'accepted' by the protein even though result in amino acid changes, mostly to the antigenic sites; in most proteins there are several times (5–10) more silent than expressed changes. Note that some changes have 'reverted' in later isolates.

Fig. 2. Sequential nucleotide changes over the period 1968 to 1979 in the part of the gene of Hong Kong influenza virus coding for the N terminal 100 amino acids of the HA1 protein (modified from Both et. al., 1983).

then be assessed in various ways, and used as a measure of the relatedness of the DNAs (Gibbs and Fenner, 1984). This method may be used directly to compare DNAs from some viruses, or from higher organisms. However restriction endonucleases do not cut RNAs specifically, and they must first be copied into DNA using an enzyme called reverse transcriptase. This method has great potential for characterizing and comparing different isolates of viruses that have RNA genomes; Figure 3 shows the 'restriction patterns' from three isolates of U5 tobamovirus from *Nicotiana glauca*, a common roadside shrub in most of the Mediterranean regions of the world including Australia.

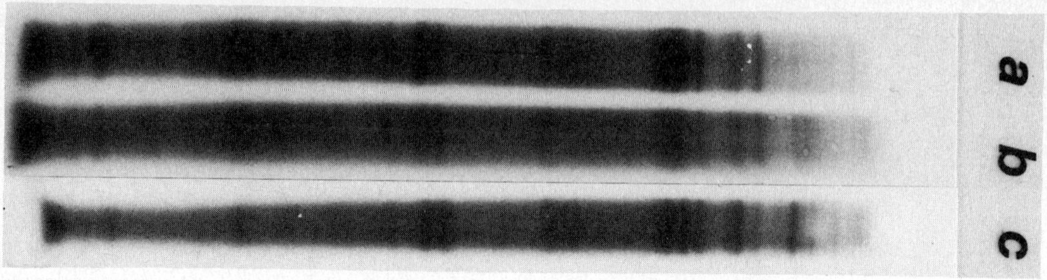

Fig. 3. 'Restriction enzyme patterns' of cDNA copies of the genomes of three tobamoviruses:
  a. isolated in 1981 from a roadside shrub of *Nicotiana glauca* growing at Pooncarrie, N.S.W.
  b. isolated in 1983 from a dried leaf specimen of *Nicotiana glauca* in the N.S.W. Herbarium. The specimen was collected in 1938 at Gilgandra in N.S.W.
  c. the U2 strain of tobacco mosaic virus isolated around 1945 from tobacco.

Note how similar the three patterns are — that of the type strain of tobacco mosaic shows no similarities at all. The close similarity of a. and b. shows how little the virus has changed in over 40 years.

Chromosome banding

The nuclear DNA of higher organisms is arranged in chromosomes, which often small and difficult to study. However certain cells of dipteran insects (i.e. flies and mosquitoes) show the phenomenon called 'polyteny'; the cells replicate the DNA in their chromosomes more than is required for cell division, and the copies, which may number in excess of 1000, remain side by side, so that the chromosomes become huge. These may be specifically stained to show stain absorbing bands. The study of such chromosome banding patterns has been of great importance in evolutionary studies of the vinegar fly, *Drosphila* (Carson, 1970), and the malaria carrying mosquitoes, *Anopheles* (Green, 1982). For example, Figure 4 shows the chromosomes of the mosquito,

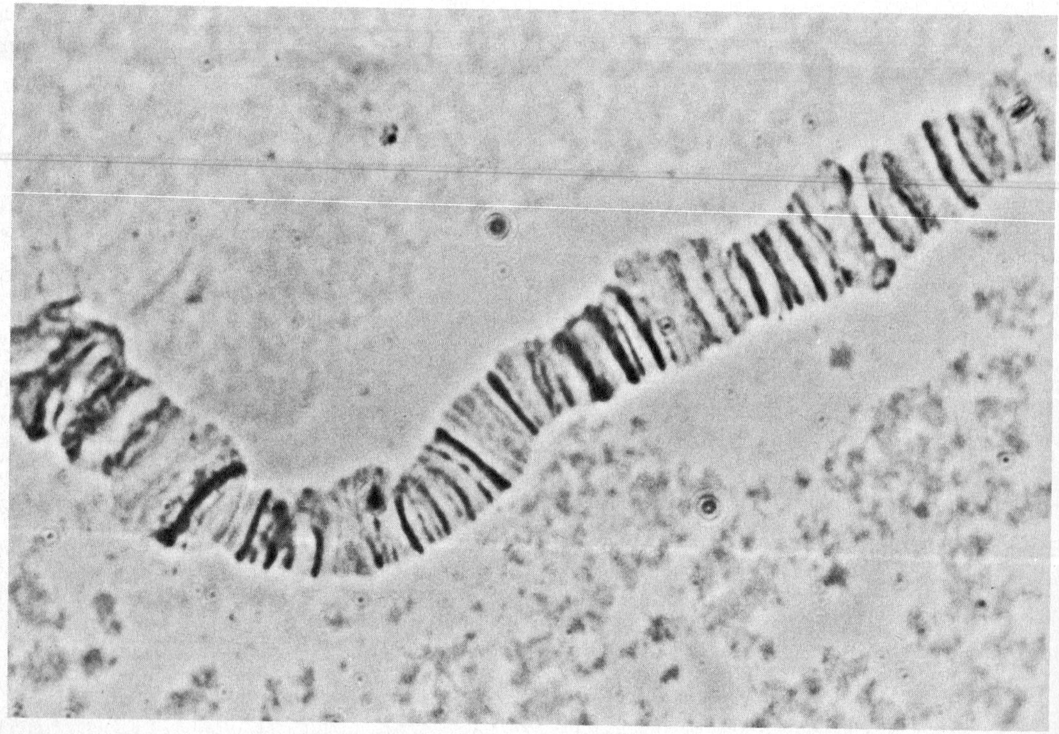

Fig. 4. A polytenized X chromosome from the salivary gland of the mosquito *Anopheles farauti* No. 3. The banding pattern of this chromosome distinguishes this species from four other morphologically indistinguishable species *A. farauti* No. 1, 2 and 4.

Anopheles farauti, a vector of the malaria organism, stained to show chromosome bands. Studies of these banding patterns have shown that populations of some organisms, like *A. farauti*, that are of uniform appearance and appear to be a single species are, in fact, mixtures of more than one species, which differ from one another in the pattern of bands in their chromosomes; it is often found that sections of their chromosomes have bcome inverted or been transfered to other parts of the chromosome. These mixed populations are called 'sibling species' and their component species, even if they can be induced to mate, rarely produce viable offspring. They may also differ subtly in their feeding preferences and ecological behaviour.

One classical study which illustrates the importance of obtaining information on subtle genetic differences is that of the mosquito, *Anopheles gambiae*. This mosquito is found in Africa, south west Arabia and the islands from Cape Verde to Mauritius, and was shown, nearly a century ago (Ross et. al., 1900) to be the principal vector of the malaria parasite and the filariasis nematode in Africa. However, the behaviour of the mosquito was inexplicably variable in such traits as which animals it preferred to bite, and whether it preferred to rest during the day indoors or outdoors. Furthermore the prevalence of malaria was not consistently correlated with numbers of mosquitoes. It was then shown by breeding studies, and later confirmed by studies of chromosome banding patterns, that what was known as *A. gambiae* was in fact six sibling species (White, 1974); they differ in biting habits, resting habits (and hence the ease with which they can be controlled by insecticide sprays) and, most importantly, they differ in their efficiency as vectors of malaria. The five freshwater-breeding species are genetically very similar, and cannot be separated by their appearance. However, as they differ in their ability to transmit malaria, it is important to be able to distinguish between them, and this is done by studying their polytene chromosomes (Coluzzi and Sabatini, 1967; Green, 1972; Davidson and Hunt, 1973), or by electrophoretic analysis of their enzymes (Mahon, Green and Hunt, 1976; Miles, 1976).

A similar situation has been found in the major malaria vector of the south west Pacific region. *Anopheles farauti* s.l. is found in Northern Australia, New Guinea and the nearby Pacific islands. Breeding and chromosome banding studies, like those done with *A. gambiae* s.l., have shown there are at least three sibling species (Mahon, 1984), and there may be a fourth in the Solomon islands (Mahon, unpublished data). So far we have not compared the efficiency of malaria transmission by the members of the *A. farauti* complex, however it is likely that they will differ in this trait, and also other behavioural and ecological traits, and, like the sibling species of the *A. gambiae* complex, they will only be distinguished by cytological examination of their polytene chromosomes, or electrophoretic comparisons of particular enzymes.

Proteins

Just as the nucleic acids may be characterized in various ways, so too can the proteins they encode. Their sequences may be determined and compared either directly, or by using enzymes to fragment them and then studying the resulting pieces (i.e. peptide mapping). However two of the most useful techniques are:

1. comparisons of the rate of movement of proteins in an electrophoretic field; and
2. serological comparisons (Chapter 13).

The differences assessed by these techniques depend on differences in the sequences of the proteins, and hence are often correlated with evolutionary relationships. The relative rates of electrophoretic migration of readily purified proteins can be assessed directly, for example, Figure 5 shows a stained gel in which purified particles of three strains of Kennedya yellow mosaic tymovirus have been directly compared. However the proteins of larger organisms, such as mosquitoes, are not so readily purified and so those compared by electrophoresis are, most frequently, enzymes. This is because after electrophoresis in a suitable gel, such as dilute starch, the position of an enzyme may be detected by its enzymatic properties; the gel is soaked in a solution of a suitable substrate of the enzyme to produce a coloured product, so that the position in the gel of extremely small amounts of enzymes can be detected without having to purify them.

GENES FOR DISEASE RESISTANCE AND PATHOGEN VIRULENCE

There are many clear examples that show that the susceptibility or resistance of host organisms to the attack of parasites is genetically controlled (Chapter 15; Russell, 1978), and there is similar evidence that the virulence of parasites is also genetically controlled. However there is little direct evidence about how the genes control these characteristics. One of the most clearcut examples is that of sickle cell anaemia and malaria (Allison, 1964).

Sickle cell anaemia is a disease common among some human populations. It is caused by a 'missense' mutation in one of the genes that encode the two proteins of the haemaglobin molecule. The mutant gene (Hb-S) produces a haemaglobin with the amino acid valine in place of the 'correct' amino acid, glutamic acid, in the sixth position in its chain of amino acids. Individuals that are homozygous for this gene (i.e. both chromosomes carrying the same form of the gene) have unusual sickle shaped red cells in their blood, because the haemaglobin produces long narrow crystals in these cells. These individuals have a greatly reduced life expectancy because of the inability of their red cells to carry oxygen. Thus we might expect Hb-S genes to be eliminated from the human population by natural selection. However, in regions of the world where the malaria parasite, *Plasmodium falciparum*, is prevalent, Hb-S genes persist because people who are heterozygous for

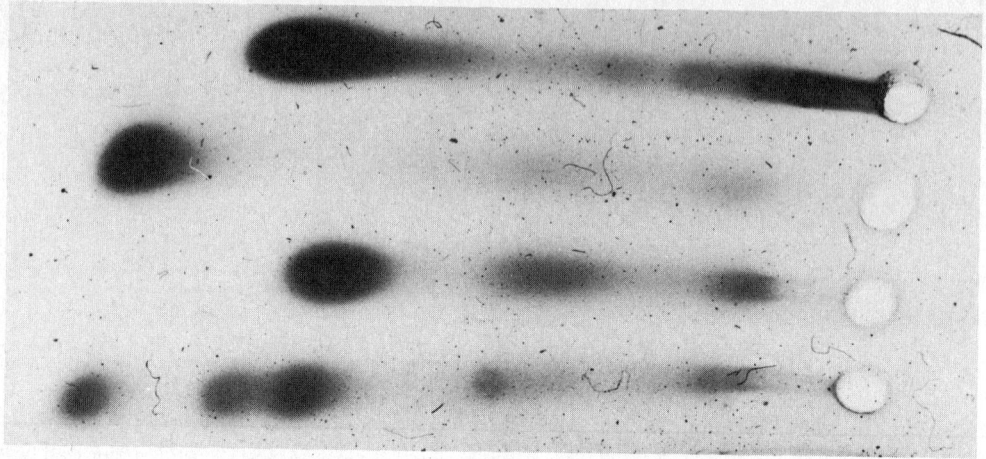

Fig. 5. Electrophoretic pattern of the particles of three isolates of Kennedya yellow mosaic virus. Preparations of particles of the isolates were placed in wells cut in a slab of 1% agraose gel (Tris/EDTA buffer pH 8) before the current was passed. In the top well, particles of an isolate from Port Douglas, Queensland; next, one from Jervis Bay, N.S.W., then one from Wapengo, N.S.W. and finally a mixture of all three.

the genes (i.e. have one copy of the Hb-S gene and one normal one in each nucleus) resist malaria, it is likely that the parasite cannot replicate normally in red cells containing the mutant haemaglobin.

George Bruening and his colleagues have made perhaps one of the only studies to establish the possible molecular basis of the resistance of a plant to a pathogen. They tested more than one thousand lines of cowpea, *Vigna unguiculata*, by inoculating them with the SB strain of cowpea mosaic virus. Sixty five lines resisted infection. Protoplasts from the primary leaves of fifty five of the resistant lines were tested, all but one, Arlington, were as susceptible to infection as the susceptible lines, indicating that the plants were resistant because of some restriction on cell-to-cell spread. Arlington protoplasts specifically resisted infection by the SB strain of cowpea mosaic virus, but not the DG strain of the related cowpea severe mosaic virus, the resistance is probably produced by a specific inhibitor of the proteinase encoded by the virus (Kiefer *et. al.*, 1984).

DISCUSSION

The examples given in this brief review illustrate the importance of studies of subtle genetic characters as tools for understanding the complex battle between pests, parasites and their hosts. In many instances the characters are merely convenient markers, but in others they are important to study as they are the molecular weapons used in the battle. Interactions between resistance and virulence genes may be the balancing selection that enable particular host/parasite combinations to survive; with no balance the parasite or both host and parasite will become extinct. It has been suggested (Clarke, 1976; Haldane, 1949) that there is an 'incessant evolutionary dance' between the virulence genes of parasites and the resistance genes of their hosts, and that this is the driving force that maintains polymorphisms in a wide range of genes in most populations of organisms.

ACKNOWLEDGEMENTS

We are indebted to Anne Mackenzie for the restriction fragment 'map' in Figure 3.

REFERENCES

Allison, A.C. (1964). *Cold Spring Harb. Symp. Quant. Biol.* 29, 137.
Both, G.W., Sleugh, M.J., Cox, N.J. and Kendal, A.P. (1983). *J. Virol.* 48, 52.
Carson, H.L. (1970). *Science* 168, 1414.
Clarke, B. (1976). *Symp. Brit. Soc. Parasitol.* 14, 87.
Coluzzi, M. and Sabatini, A. (1967). *Parasitol.* 9, 73.
Davidson, D. and Hunt, R.H. (1973). *Parasitol.* 15, 121.
Gibbs and Fenner (1984). *J. Virol. Methods.*
Green, C.A. (1972). *Ann. Trop. Med. Parasit.* 66, 143.
Green, C.A. (1982). *J. Heredity* 73, 2.
Haldane, J.B.S. (1949). *Ric. Sci. Suppl.* 19, 68.
Kiefer, M.C., Bruening, G. and Russell, M.L. (1984). *Virology 137.* 371.

Mahon, R.J. (1984). In: *Malaria* (J. Bryan and P. Moodie, eds) Aust. Govt. Publishing Service, Canberra.
Mahon, R.J., Green, A.C. and Hunt, R.H. (1976). *Bull. Ent. Res.* 66, 25.
Miles, S.J. (1976). *Bull. Ent. Res.* 68, 85.
Ross, R., Annette, H.E. and Austen, E.E. (1900). *Mem. Liverpool Sch. Trop. Med. 2*.
Russell, G.E. (1978). *Plant breeding for pest and disease resistance*. Butterworth, London.
White, G.B. (1974). *Trans. Roy. Soc. Trop. Med. Hyg.* 68, 278.

Chapter Fifteen

HOST-PATHOGEN RELATIONSHIPS: THE STRUGGLE OF GENES

J.J. Burdon and D.R. Marshall

THE GENETIC BASIS OF RESISTANCE AND PATHOGENICITY

Any consideration of co-evolutionary interactions between populations of hosts and parasites inevitably relies on a knowledge of the interactions which occur between resistance and virulence factors at the level of the individual. At this level, the genetic basis for resistance in the host and virulence in the pathogen is well understood. Flor's hypothesis (Flor, 1955) of a gene-for-gene relationship between host and pathogen has been extended from the system which he originally investigated, flax and flax rust (*Linum usitatissimum* and *Melampsora lini*), to a wide range of other host-pathogen systems; over 30 according to Sidhu (1980). In most studies, resistance has been found to be dominant; virulence to be recessive; and both are inherited in a simple Mendelian fashion. As a consequence, any single gene-for-gene relationship involves two phenotypes (resistant and susceptible) in the host and interactions of these with two phenotypes (virulent and avirulent) in the pathogen. These four phenotypes when brought together in all combinations produce the 'quadratic check' (Figure 1) which reveals that resistance is expressed only when dominant genes for resistance and avirulence interact. In all other circumstances, the plant is susceptible.

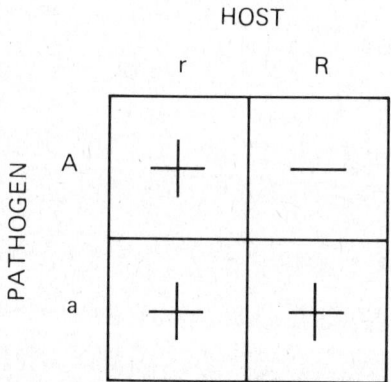

+ HOST SUSCEPTIBILITY/PATHOGEN VIRULENCE

− HOST RESISTANCE/PATHOGEN AVIRULENCE

Fig. 1. The reaction pattern of the 'quadratic check', where the 'resistant response' results from the interaction between a dominant resistance gene (R) in the host plant and a dominant avirulence gene (A) in the pathogen.

Virtually all formal studies of the gene-for-gene hypothesis have involved host-pathogen combinations where single resistance genes have been responsible for large phenotypic effects (e.g. hypersensitive fleck resistance versus large pustule susceptibility). Recently studies of individual host-pathogen interactions have turned more to those involving various other forms of resistance (e.g. so-called 'durable' resistance, slow rusting and tolerance).

HOST-PATHOGEN INTERACTIONS AT THE POPULATION LEVEL

Our knowledge of the effect of individual interactions on the genetic structure of host and pathogen populations is scanty. This knowledge comes from three sources, and is discussed below.

Agricultural experience

Agricultural crops are commonly protected from damaging pathogens by incorporating one or two resistance genes into all members of the cultivar. This ensures that at any given time the entire population is either resistant or susceptible to the prevalent pathogen races. However, even if it is resistant, the uniformity of response of the host population places strong selection pressure on the pathogen population, favouring the appearance of new virulence combinations. The frequency and intensity of economically damaging disease epidemics in agricultural crops illustrates dangers inherent in growing high density, genetically uniform crops. Where man is prepared to continually intervene between host and pathogen populations (usually by producing and growing new cultivars with new resistance combinations) such unstable interactions can continue to operate. In wild plant communities, however, man does not assist beleaguered plant populations and so a quite different outcome is to be expected.

Theoretical expectations

A pathogen attacking a dense plant population may have several effects on the size and structure of that population. In the short term, plant density will fall as individuals succumb to disease; in the longer term, other resistant plant species may invade or changes may occur to the genetic composition of the original population. In virtually all interactions between plants and pathogens at the population level, the final outcome is likely to incorporate both of these effects (Burdon, 1982). Here we discuss possible changes in the genetic structure of host and pathogen populations.

When a genetically uniform plant population with a single gene for resistance is confronted by a genetically uniform pathogen population possessing the complementary virulence gene, any plant mutation producing a novel resistance allele will be favoured. While there are small numbers of plants with this novel resistant gene in the host population, the selective pressure on the pathogen population favouring the emergence of the complementary virulence genes will also remain small. As the frequency of the new genotype increases, however, the selective pressure on the pathogen population also increases, until chance mutation produces a new race with the appropriate virulence gene. When this happens, the selective advantage previously enjoyed by the mutant host genotype is lost and the interaction between host and pathogen comes full circle — except now both host and pathogen are slightly more diverse. So long as the pathogen remains a significant part of the host plant's environment this co-evolutionary interaction will continue, resulting ultimately in genetically diverse host and pathogen populations. Such a scenario was first proposed in detail by Person (1966) who argued that, in a sufficiently complex system, a balance would ultimately be achieved through a series of cyclical polymorphisms between resistance genes in the host and complementary virulence genes in the pathogen. The initial stages of such interactions (the single resistance and virulence gene level) have been modelled mathematically and shown to be logical (Gillespie, 1975). However, further mathematical testing of more complex levels of interaction involving two or more genes in both host and pathogen, frequency and density dependent selection and/or selection at different stages of the life cycle is still in its infancy (Lewis, 1981).

An alternative way of looking at the interaction between host and pathogen populations is that which has developed around the concept of deliberately growing mixtures of cultivars or isogenic lines, differing solely in the resistance genes they possess (mixtures: Wolf et. al., 1981; multilines: Browning and Frey, 1969; Frey, 1982). This strategy aims to thwart the evolutionary tracking skills of the pathogen population. The idea behind this approach is that the closest neighbours of any given individual in the genetically mixed host population are, simply by chance alone, likely to possess different sets of resistance genes. As a result, when the pathogen spreads (by spores or by vector) it is unlikely to be able to infect other plants in the immediate vicinity, as these will be of the resistant cultivar. In this way, it has been suggested, the pathogen population would finally consist of a number of simple races each adapted to only one component of a multi-component ($n > 6$) mixture. In turn, this reduces the effective host density for each individual host-pathogen genotype interaction to $1/n$ that of the total stand and hence reduces disease to acceptable levels.

Unfortunately, however, while this is an attractive solution, theoretical studies indicate that this result will probably only occur if unnecessary genes for virulence make the pathogen less fit (Leonard, 1977; Barrett, 1978; Marshall and Pryor, 1978). Attempts to test these assumptions experimentally have so far proved unsuccessful, and it is still not known whether real host-pathogen systems reflect the results predicted by these 'fitness cost' models.

Evidence from natural systems

The limited evidence we now have concerning host-pathogen interactions in wild plant communities usually confirms the predictions of theory. In environments unsuitable for the growth and development of particular fungal pathogens, plant populations tend to possess relatively little resistance, whereas in more favourable sites the level of resistance may be quite high (Wahl, 1970; Dinoor, 1977; Burdon et al., 1983). For example, in Australia, the average level of resistance possessed by wild oat populations (*Avena* species) in pathogen-favourable environments is much greater than that in unfavourable ones. In addition, however, populations in 'high risk' areas are also more diverse in their infection type response to individual pathogen races, with some individuals being fully susceptible, others fully resistant, while still others showed a range of intermediate responses (Figure 2). In 'low risk' sites on the other hand, populations tend to be much more uniform in their response (Figure 2; Burdon et al., 1983).

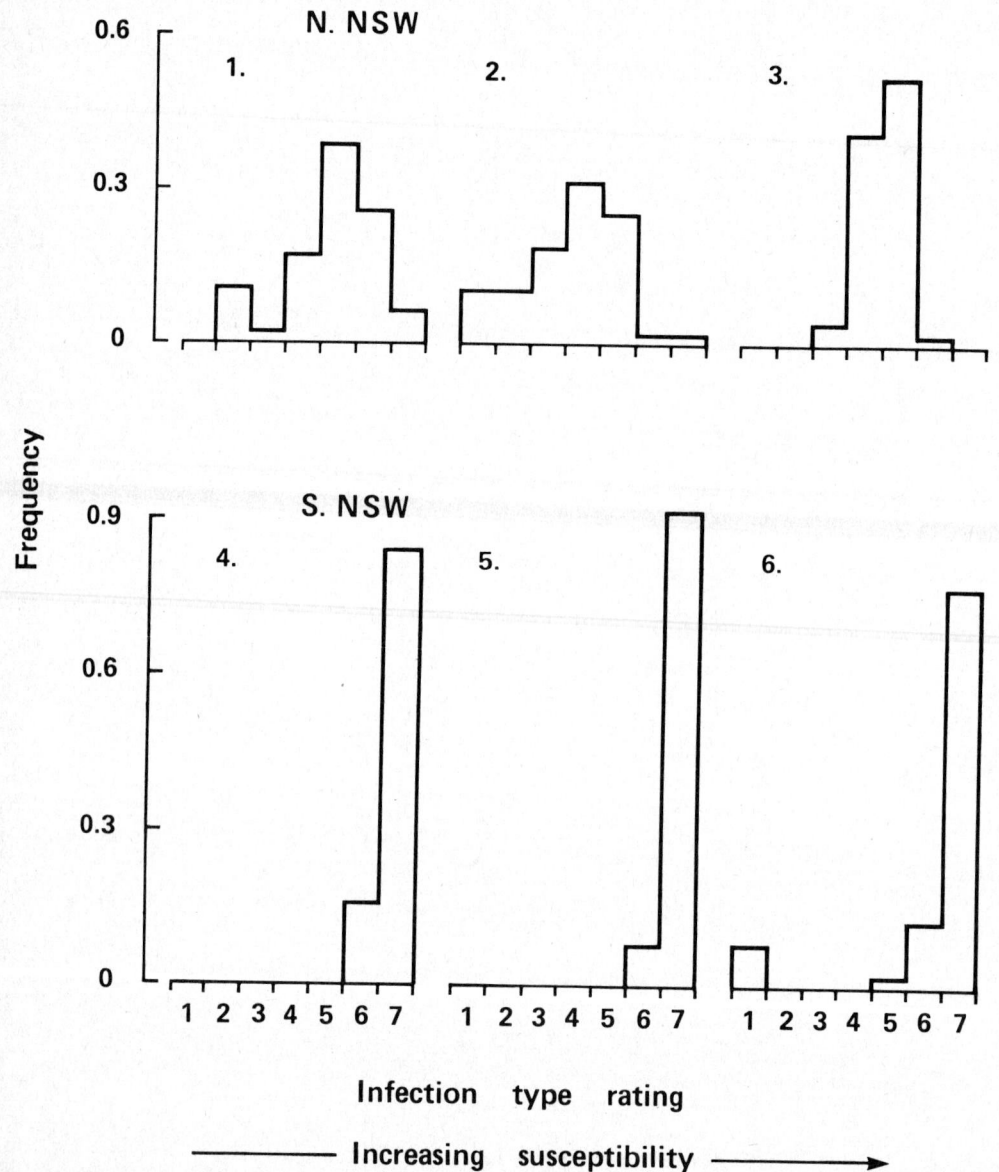

Fig. 1. Frequency distribution of the type of infection responses of three *Avena fatua* populations from environments favourable for (1 to 3) or unfavourable for (4 to 6) development of the pathogen *Puccinia coronata*. (Adapted from Burdon et al., 1983).

As an added level of complexity, the response of individual host lines has also been found to differ according to the pathogen race involved. In many populations a few individuals were fully resistant, but the majority showed responses varying from full susceptibility to full resistance depending on the pathogen race involved (Burdon et al., 1983). A similar result has also been obtained in a study of the interaction between the rust pathogen *Phakopsora pachyrhizi* and its host *Glycine canescens*. There, in one population under study, three of fifteen host individuals selected at random were resistant to six different races of *P. pachyrhizi*, twelve showed various combinations of resistant and susceptible reactions while none were susceptible to all races (Burdon, unpublished data).

Information about the genetic structure of non-agricultural pathogen populations is even more limited. However, the *Avena-Puccinia* system mentioned above, it was found that the pathogen population occurring in areas where the plant populations had considerable diversity in resistance, was itself both more diverse and more virulent than that found in areas where host populations were susceptible (Oates et al., 1983). Similarly, populations of the *Glycine* rust, *Phakopsora pachyrhizi*, contain a number of different races that show a considerable range of pathogenicity; twelve isolates have been examined in detail and eight different races have been identified to date (Burdon, unpublished data).

CONCLUSIONS

The results of studies like those discussed above indicate that in environments which favour the growth and development of pathogens, co-evolutionary interactions will produce multiple disease resistance in the host and racial diversity of the pathogen.

However, a great deal remains to be learnt about the complexities of host-pathogen interactions in populations. Thus, although we know several different resistance genes occur in some host populations, we do not know how many are there, whether only one occurs in each individual plant, or whether they are combined randomly as a truly diverse resistance spectrum. Furthermore, though theoretical studies suggest that there must be a 'fitness cost' for those carrying unused genes, there are no measurements of these costs and hence no estimates of the likely rates of spread of such genes in co-evolving systems.

REFERENCES

Barrett, J.A. (1978). In: *'Plant Disease Epidemiology'* (P.R. Scott and A.W. Bainbridge, eds) p. 128, Blackwells, Oxford.
Browning, J.A. and Frey, K.J. (1969). *Ann. Rev. Phytopathol.* 7, 35.
Burdon, J.J. (1982). In: *'The Plant Community as a Working Mechanism'* (E.I. Newman, ed.) p. 66, Blackwells, Oxford.
Burdon, J.J., Oates, J.D. and Marshall, D.R. (1983). *J. Appl. Ecol.* 20, 571.
Dinoor, A. (1977). *Ann. N.Y. Acad. Sci.* 287, 357.
Flor, H.H. (1955). *Phytopathology* 45, 680.
Frey, K.J. (1982). In: *'Plant Improvement and Somatic Cell Genetics'* (I.K. Vasil, W.R. Scowcroft and K.J. Frey, eds) p. 41, Academic Press, New York.
Gillespie, J.H. (1975). *Ecology* 56, 493.
Leonard, K.J. (1977). *Ann. N.Y. Acad. Sci.* 287, 207.
Lewis, J.W. (1981). *J. Theor. Biol.* 93, 927.
Marshall, D.R. and Pryor, A.J. (1978). *Theor. Appl. Genetics* 51, 177.
Oates, J.D., Burdon, J.J. and Brouwer, J.B. (1983). *J. Appl. Ecol.* 20, 585.
Person, C. (1966). *Nature* 212, 266.
Sidhu, G.S. (1980). *Proc. 14th Intern. Cong. Genetics,* Moscow 1978. p. 391.
Wahl, I. (1970). *Phytopathology* 60, 746.
Wolfe, M.S., Barrett, J.A. and Jenkins, J.E.E. (1981). In: *'Strategies for the Control of Cereal Disease'* (J.F. Kenkyn and R.T. Plumb, eds) P. 73, Blackwells, Oxford.

Chapter Sixteen

THE BOVINE MAJOR HISTOCOMPATIBILITY SYSTEM AND DISEASE

M.J. Stear, M.L. Bath, J.T. Mackie, C.K. Dimmock, S.C. Brown, F.W. Nicholas and B. Morris

Breeding for disease resistance is not fundamentally different from breeding for any other character. Many breeding schemes in plants and animals, including livestock, have reduced the incidence or severity of specific diseases (Hutt, 1958; Russell, 1978). The important questions now are whether breeding for disease resistance is worthwhile and, if it is, what is the best way of doing it.

It is difficult to generalise about the desirability of breeding for disease resistance. Much will depend upon the nature of the disease, its incidence, the protection offered by breeding, other methods of disease control, the cost of not breeding for other characters, the genetic correlations between resistance and other characters and the likelihood of the pathogen adopting to circumvent the hosts's genetic resistance. The breeder has to consider whether he wants hosts which offer resistance to pathogen establishment, growth or dispersal, or whether he wants hosts which tolerate the effects of the pathogen (although tolerant, infected hosts may serve as reservoirs of infection). Also, the genes which confer resistance to one disease may confer susceptibility to another disease. Diseases which are rare and economically unimportant now may devastate a population selected for resistance to another disease.

Generally speaking, total and long-lasting resistance to any specific disease is difficult to obtain, at least in animals, and selective breeding is unlikely to play a role in preventing the establishment of an exotic disease, and for that other measures will be more valuable. If a disease does become established, selective breeding may be able to limit the damage caused by the unwelcome arrival; especially if used in conjunction with other measures. For example, vaccination, selective breeding and careful choice of stock may provide protection against disease where a single measure would be ineffective.

Breeding for general disease resistance could in principle be achieved by selecting animals with a greater ability to tolerate the effects of pathogens or by selecting animals with more efficient non-specific mechanisms for fighting disease or both. In practice, so little is known about genetic variation in non-specific mechanisms of disease resistance that we cannot determine which organisms have the best general resistance to disease unless we expose organisms to many different diseases, which is prohibitively expensive.

Selecting for resistance to a specific disease or group of diseases is the best way to improve genetic resistance to disease. In Australia, selective breeding of cattle for tick resistance is practised on some commercial farms. Plant strains with resistance to specific diseases are commonplace.

PREDICTING DISEASE RESISTANCE IN THE ABSENCE OF DISEASE

The simplest way to discover which animals have superior disease resistance is to infect all stock with the pests or parasites whose incidence we wish to reduce, but this can be very expensive. Testing the breeding population may interfere with measurements of production parameters, yet creating separate populations simply to measure disease resistance is often too extravagant. With enzootic diseases, such as tick or worm infestations of livestock, natural infection may suffice, although natural exposure may not identify resistant animals or plants as accurately as a challenge infection when the dose is standardised. With epizootic or enzootic diseases, it may be impractical to test for disease resistance directly.

In laboratory animals, many genes which are associated with disease resistance have been identified (Rosenstreich et al., 1982). Unfortunately, much of this work has not been repeated in species of economic importance and there are very few genes or indirect measures of disease resistance which are being used by commercial animal breeders to improve disease resistance, although plant breeders have often used resistance genes in their breeding programmes.

The gene coding for the *Escherichia coli* K88 antigen receptor in pigs is one example of a gene being examined by breeders. Neonatal diarrhea is a severe problem for some piglets. It is often caused by *E. coli* bacteria which have the K88 antigen on their cell surface. The bacteria must attach to a receptor for the K88 antigen which is coded for by a dominant allele. Piglets which lack the dominant allele do not have the receptor for the K88 antigen on *E. coli* and the bacteria are unable to cause disease in these piglets (Gibbons et al., 1977).

THE MAJOR HISTOCOMPATIBILITY SYSTEM

The major histocompatibility system (MHS) is a cluster of genes found in higher animals. These genes code for proteins found on the surface of the animal's cells. Those proteins which are recognized by antibodies are called 'antigens' and the antigens are responsible for many effects including the ability of the immune system to recognize the difference between 'self' and 'non-self', thus the MHS is involved in the acceptance or rejection of tissue grafts between animals.

The MHS has been implicated in resistance to diseases in humans (Braun, 1979) mice (Lilley, 1966), chickens (Longenecker and Mosman, 1981), guinea pigs (Geczy and Rothwell, 1981), rats (Gasser et al., 1973; Williams and Moore, 1973) and cattle (Stear et al., 1984). One of the genes of the chicken MHS the B^{21} allele, is associated with resistance to Marek's disease. Although chicken breeders could use the B^{21} allele to select for resistance to Marek's disease, many breeders prefer to vaccinate against this disease. This situation may be changing. Gavora and Spencer (1979) have argued that optimum resistance to Marek's disease can only be achieved by a combination of vaccination and selective breeding, and there is a resurgence of interest among chicken breeders in the United States and in Australia, in the B^{21} allele.

We are studying the role of the MHS in diseases of cattle. Our strategy has been to establish whether there is any association between the economically important diseases and those antigens of the MHS system for which the animals can be tested. Table 1 lists the five diseases where the role of the bovine MHS has been studied.

Tick infestation

Ticks present a severe disease problem to the world's domestic livestock. In Australia, the most important tick species is the single-host cattle tick *Boophilus microplus*. Different cattle have different levels of resistance and the level of resistance achieved is influenced by the immune response to ticks (Willadsen, 1980). The tick resistance of 199 cattle from a ¾ Brahman ¼ Shorthorn synthetic herd was assessed by placing a collar containing 20,000 living tick larvae for four hours around the neck of each animal (Uteck, Seifert and Wharton, 1978). The number of surviving female ticks, 4.5 to 8.0 mm in length, on one side of each animal was counted 20 and 21 days after infestation by two people independently. The tick numbers on cattle with a particular MHS antigen were compared with tick numbers on cattle lacking that particular MHS antigen.

Worm infestation

Cattle from two locations in Central Queensland were examined (Stear, 1985). The first group consisted of 204 eleven month old heifers from the experimental herd owned by the Brigalow Research Station, Queensland Department of Primary Industries, Theodore. Three main breed types were represented. They were Hereford, predominantly Simmental and predominantly Africander. The 204 heifers were the offspring of 19 sires. The number of daughters per bull ranged from seven to seventeen. Worm eggs were counted in faeces on four occasions over four months in 1980. The modified McMaster technique was used to count the eggs (Whitlock, 1948). Worm species were differentiated using the methods described by Roberts, Elek and Keith (1962) and Keith (1953). Four samples, each of four grams were used. Worm egg counts ranged from zero to 2,600 eggs per gram, with a mean of 221.6 and a median of 160. The breed of the heifer was not associated with differences in worm egg counts, so the breed of the heifer was ignored in subsequent analyses. The repeatability of worm egg counts was 0.33 ± 0.10 (estimate ± standard error). The heritability was 0.31 ± 0.11 (estimate ± standard error).

Cattle from the second location consisted of 91 bulls which were the progeny of multiple sire joinings at Mt. Eugene, Jambin, a commercial beef property. There were two major breed groups represented. One group was about seven-sixteenths Africander, three sixteenths Brahman and six-sixteenths Hereford. The other group was more than three quarters Africander. Worm eggs were counted on seven occasions over ten months in 1981–82, and at the start of sampling the bulls were eight to twelve months of age. Worm egg counts ranged from zero to 8,375, with a mean of 291.7 and a median of 175. the repeatability of egg counts was 0.39 ± 0.05, which is not significantly different from the value obtained for the Brigalow data. The worm eggs counts were not significantly different among the breed types.

Counts on animals of different breed types at Brigalow and at Mt Eugene showed that most worm eggs came from *Cooperia* species. However, eggs from *Haemonchus placei* dominated the counts in some cattle.

In both studies, mean worm egg counts were compared between cattle possessing an MHS antigen and cattle lacking that antigen.

Ocular squamous cell carcinoma

Ocular squamous cell carcinoma has been reviewed by Spradbrow and Hoffman (1980). Briefly, it is a spontaneous, superficial carcinoma which is particularly common in cattle. The disease is most common in Herefords but it does occur in other breeds. Tumours develop on the epithelial surfaces of the various structures of the eye. Tumours do not develop on fully pigmented eyelids or on the pigmented portion of eyelids that are partially pigmented, although tumours which start on unpigmented areas may encroach on pigmented areas as they grow. Estimates of the heritability of the disease vary from 0.01 to 0.66 but the late expression of susceptibility makes it difficult to breed for resistance to carcinoma. Chronic exposure to bright light is a factor in the development of disease. Spontaneous regression is rare but carcinomas of most, but not all, cattle regress after intramuscular injections of phenol saline extract of allogeneic or autologous tumour. The casual agent of ocular squamous cell carcinoma is not known but one or other of several viruses are though to be involved.

The frequency of MHS antigens present in 19 Hereford cattle with ocular squamous cell carcinoma was compared with the frequency of MHS antigens in 18 healthy cattle which came from the same three farms as the 19 diseased cattle. The healthy cattle were matched for age, sex, breed and the absence of pigment around the eye.

Enzootic bovine leucosis

Bovine leucosis virus is a retrovirus responsible for lymphosarcoma. Less than 5% of infected cattle develop the disease although many more cattle develop persistent lymphocytosis, which often but not always, precedes the development of lymphosarcoma. Most cases of lymphosarcoma occur in cattle 4 to 6 years of age. The disease is usually fatal (Dimmock, Waugh and Rogers, 1979).

We selected 66 Australian Illawarra Shorthorn cows from five herds. All 66 cattle had antibodies to bovine leucosis virus, which indicates that all animals had been infected; 33 of the cattle had persistent lymphocytosis while the other 33 cattle had normal white blood cell counts. There were significant differences in antigen frequency between the normal cattle and those with persistent lymphocytosis.

TABLE 1. *Disease associations with the bovine major histocompatibility system*

Country	Disease	Number of cattle studied	Antigens possibly associated with resistance (e) or susceptibility (s)	References
Australia	Tick infestation	332	W16 (R), W6 (S)	Stear et al. (1984) Stear (1985)
Australia	Worm infestation	295	W16 (R)	Stear (1985)
Australia	Ocular squamous cell carcinoma	19	W6 (S)	Stear (1985)
Australia	Enzootic bovine leucosis (persistent lymphocytosis)	33	CA12 (S)	Stear (1985)
Norway	Mastitis	12	W2 (R), W16 (S)	Solbu et al. (1982)

Mastitis

Mastitis is an inflammation of the mammary glands and it can be caused by any of several species of bacteria including staphyloccocci, streptococci, *Escherichia coli* and *Corynebacterium pyogenes*. Mastitis is possibly the most economically-important disease of dairy cattle. Solbu et al. (1982) inferred the MHS types of 12 Norwegian Red bulls by testing their male progeny. They then examined the mastitis records of the female progeny of the 12 bulls and concluded that there were significant associations between the antigens present in the bulls and the susceptibility of their progeny to mastitis.

It is necessary to be cautious in interpreting these results as the studies are not yet completed, although it is clear that the MHS antigens do not correlate perfectly with susceptibility to the diseases we studied. This may be because resistance involves the activity of several genes, or because the antigens that we studied are not products of the resistance genes but are only associated with the products of resistance genes.

ACKNOWLEDGEMENTS

We thank Dr R.G. Holroyd for supplying blood samples from cattle of known tick resistance, Mr T. Rudder and Mr T. Tierney for supplying blood samples from cattle with known faecal worm egg counts and Dr P. Spradbrow for supplying blood samples for the ocular squamous cell carcinoma study.

REFERENCES

Braun, W.E. (1979). *HLA and disease — a comprehensive review.* CRC Press, Boca Raton, Fla.
Dimmock, C.K., Waugh, P.D. and Rogers, R.J. (1979). *Aust. Vet. J. 55*, 278.
Gasser, D.L., Newlin, C.M., Palm, J. and Gonatas, N.K. (1973). *Science 181*, 872.
Gavora, J.S. and Spencer, J.L. (1983). *Anim. Blood Grps. Biochem. Genet. 14*, 159.
Geczy, A.F. and Rothwell, T.L.W. (1981). *Parasitol. 82*, 281.
Gibbons, R.A., Sellwood, R., Burrows, M. and Hunter, P.A. (1977). *Theoret. Appl. Gen. 51*, 65.
Hutt, F.B. (1958). *Genetic resistance to disease in domesticated animals.* Publishing Associates, Ithaca, New York, U.S.A.
Keith, R.K. (1953). *Aust. J. Zool. 1*, 223.
Lilley, F. (1966). *Genetics 53*, 529.
Longenecker, B.M. and Mosman, T.R. (1981). *Immunogenet. 13*, 1.
Roberts, F.H.S., Elek, P. and Keith, R.K. (1962). *Aust. J. Agric. Res. 13*, 551.
Rosenstreich, D.L., Weinblatt, A.C. and O'Brien (1982). CRC *Crit. Rev. Immunol. 3*, 263.
Russell, G.E. (1978). *Plant breeding for pest and disease resistance.* Butterworths, London.
Solbu, H., Spooner, R.L. and Lie, O. (1982). *2nd World Congress Applied Animal Production.* October, 1982.
Spradbrow, P.B. and Hoffman, D. (1980). *Vet. Bull. 50*, 449.
Stear, M.J., Newman, M.J., Nicholas, F.W., Brown, S.C. and Holroyd, R.G. (1984). *Austr. J. Exp. Biol. Med. Sci. 62*, 47.
Stear, M.J. (1985). *The mechanisms of disease resistance in cattle* (Ed. W.C. Davis) University of Washington Press (in press).
Utech, K.B.W., Seifert, G.W. and Wharton, R.H. (1978). *Aust. J. Agric. Res. 29*, 411.
Whitlock, H.V. (1948). *J. Counc. Sci. Indust. Res. Aust. 21*, 177.
Willadsen, P. (1980). *Adv. Parasitol. 18*, 293.
Williams, R.M. and Moore, M.J. (1973). *J. Exp. Med. 138*, 775.

Chapter Seventeen

NOVEL VACCINES

F. Brown

Vaccination has been important in the control of many virus diseases. Some which were major scourges even as recently as 30 years ago have been controlled or, in the case of smallpox, eradicated. Inevitably it is the industrialised world which has benefited most from the advances in vaccine technology but the underdeveloped countries have also profited considerably.

The concepts which form the basis of this technology have not changed in the last 50 years. There are two kinds of vaccine currently in use, 'live attenuated' and 'inactivated'. The former contain infective particles of attenuated or avirulent strains of the virus. They have been developed by the empirical procedure of growing the virus in a 'foreign' host until its virulence for the 'natural' host or hosts has diminished so that it replicates in them without causing disease. The molecular differences between the viruent and attenuated strains is not known except for some isolates of poliovirus. By contrast, inactivated vaccines are simply produced by treating particles of the virulent virus with a chemical agent such as formaldehyde or an imine so that they lose their infectivity but retain their immunogenic activity.

There are several very successful vaccines of both types. For example, attenuated strains of polio, measles, rubella and yellow fever viruses, together with naturally occurring avirulent isolates of Marek's disease and vaccinia viruses have been successful, as have inactivated vaccines produced from the particles of virulent isolates of polio, Newcastle disease, and foot-and-mouth disease viruses. One might ask, then, as these vaccines are so successful, why we should attempt to produce new ones. There are several reasons:

1. Even the best attenuated vaccines cause a certain, but thankfully very small, number of clinical cases of the disease. There is, for example, a sufficient number of cases of poliomyelitis each year, probably caused by the vaccines, to cause some misgiving.

2. It is impossible to be certain that an inactivated vaccine is entirely free from infectious virus. However, the record of innocuity is improving. For example, the Wellcome Foundation Laboratories have produced many millions of doses of foot-and mouth disease vaccine by inactivating the virus with acetylethyleneimine without detecting any infective batches. Nevertheless there are still incidents of disease which can be linked to vaccines that have been incompletely inactivated.

3. Some vaccines cause side reactions because they contain lipids, pyrogenic polysaccharides or other substances. Probably the most notorious is the rabies vaccine produced from virus grown in the brains of sheep and goats — unfortunately still the only vaccine available to people in many developing countries.

4. With both types of vaccine there is the problem of storage at temperatures which allow the retention of infectivity and/or immunogenicity. This is particularly important in tropical countries where it is often difficult and expensive to maintain appropriate storage conditions.

5. Work with large quantities of virus requires special containment facilities to prevent infection of the staff handling it (e.g. rabies virus) or its spread to the environment.

6. There are no vaccines against some diseases either because the virus (a) will not grow well enough (i.e. produce a sufficient number of particles) to allow a killed vaccine to be produced; or (b) will not produce variants that are sufficiently attenuated. Hepatitis A and respiratory sycytial disease viruses provide examples of this category.

RECOMBINANT DNA

Several of the problems outlined above can be overcome by recombinant DNA technology. It has been shown over the last 30 years in studies on the structure and properties of the particles of a wide range of viruses, that the immune response to each is usually directed towards a very small part of one of its proteins (Brown, 1984). This protein is usually part of the surface of the virus particle and, in those virus particles that have a lipid containing envelope, this protein projects from the lipid layer. These proteins can often be obtained completely free from the remainder of the virus particles, yet retain their immunological activity.

During the last decade, the genes coding for the individual proteins of many viruses have been identified by 'mapping' experiments, and, over an even shorter period, some of these genes have been isolated and cloned into suitable prokaryotic and eukaryotic cells, which translate them into proteins thereby providing a potential source of immunizing proteins (Emtage, Tacon, Catlin, Jenkins and Porter, 1980).

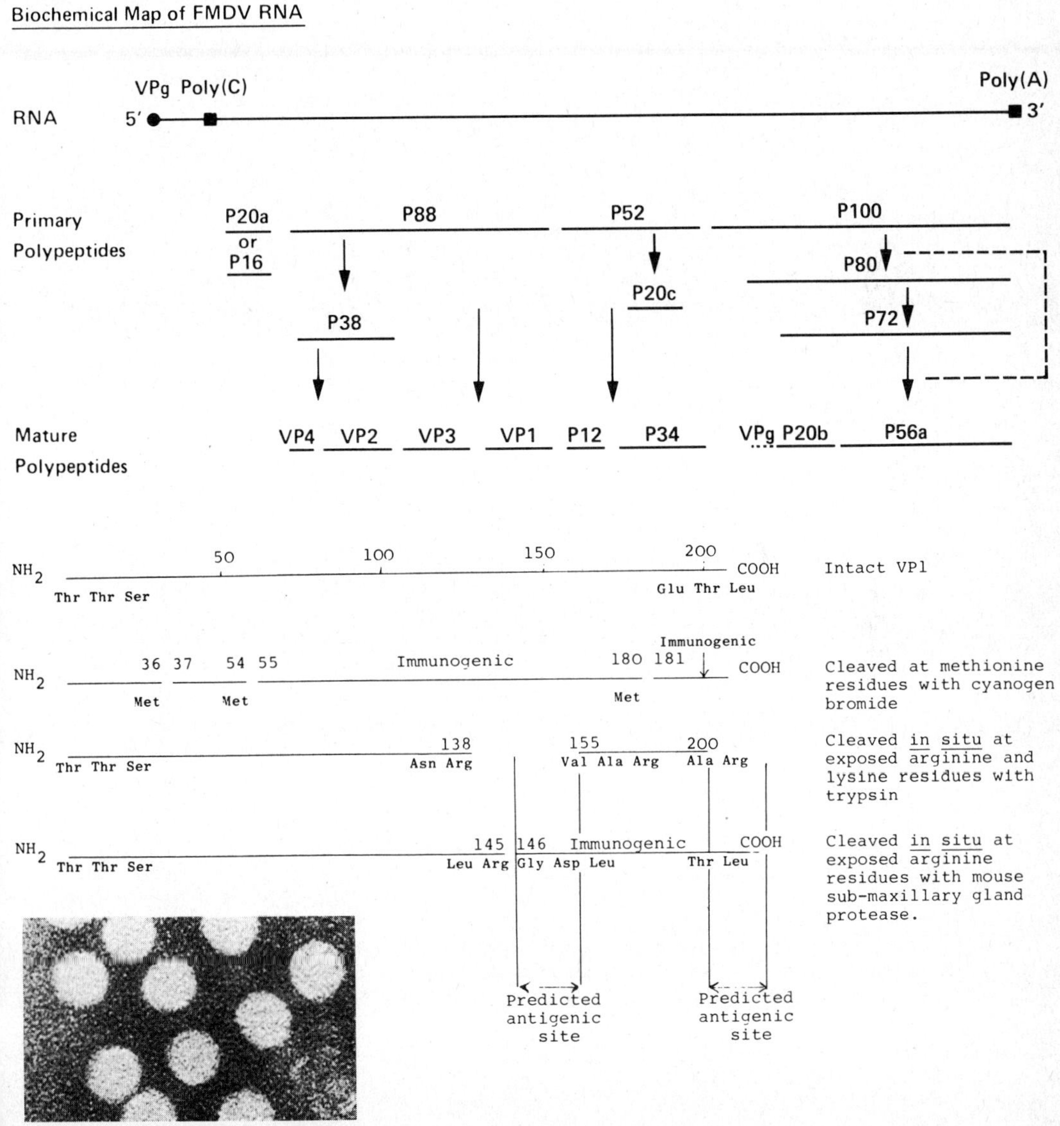

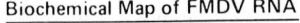

Fig. 1. Identification and Localisation of the immunogenic site of foot-and-mouth disease virus (reference: Strohmaier, K., Franze, R. and Adam, K.H. (1982). *J. General Virology* 59, 295). Inset: micrograph of FMD virus, 2 nm diameter.

However, there are problems. For example, the protein, freed from the steric restraints imposed on it by the other macromolecules in the virus particle (i.e. proteins, nucleic acid, carbonhydrate and lipid) may not fold in the natural manner. In some instances (e.g. vesicular stomatitis virus and some strains of influenza virus) this does not appear to affect its immunogenicity but, with the majority, the all-important antigenic site becomes hidden and does not stimulate the immune system. Such loss of immunogenicity limits the use of sub-unit and synthetic vaccines.

However, the application of recombinant DNA technology to the hepatitis B surface antigen has been very successful. The immune response to hepatitis B virus is directed against a protein at the surface of the intact particles. This protein also assembles, in the blood of asymptomatic carriers of the disease, to form spherical particles 22 nm in diameter. The DNA coding for this surface protein has been cloned and inserted into the genome of yeast cells, and these then produce significant numbers of the 22 nm particles, which have the same antigenic activity as the natural product found in human blood (McAleer, Buynak, Maigetter, Wampler, Miller and Hilleman, 1984).

The genetically engineered proteins of other viruses have disappointingly little immunizing activity, presumably because of the conformational changes mentioned above. For example, the immunogenic protein of foot-and-mouth disease virus can be obtained in large amounts from genetically engineered *E. coli* cells (about $1-2 \times 10^6$ molecules per bacterial cell), yet this protein has a very much lower immunizing activity than a similar amount of protein in intact virus particles (Kleid, Yansura, Small, Dowbenko and Moore, 1981). Whereas a single injection of 10 μg of inactivated virus particles is sufficient to protect cattle against experimental infection, two injections of 500 μg of the protein produced by *E. coli* are required to protect the cattle.

OTHER STRATEGIES

An exciting recent development has been the demonstration that the genes coding for the immunizing proteins of several different viruses can be inserted into the DNA of vaccinia virus without materially affecting its growth (Mackett, Smith and Moss, 1982). Animals infected with the chimaeric virus produced immunologically active proteins from the inserted gene or genes and stimulated the animals to produce protective antibody and cell mediated immunity. The most dramatic experimental result obtained using this system was that in which chimpanzees were protected against infection by hepatitis B virus by vaccinia containing the gene coding for the surface antigen of hepatitis B virus. Other viruses which can act as gene 'vectors' and which do not carry the slight but significant risk associated with vaccinia virus infection will doubtless be tested. One of the many adenoviruses which cause only mild respiratory illness may prove to be more suitable, or genetically disarmed variants of vaccinia virus might be produced by manipulation.

Another recent advance has been the demonstration for several viruses that immunization with short peptides corresponding to their immunogenic sites will afford protection against infection. This work resulted from experiments made by Anderer with tobacco mosaic virus 20 years ago (Anderer, 1963). The peptides used in these experiments have been produced either by recombinant DNA technology or by organic synthesis, but require the immunogenic site to be identified. Although no rules, which allow this prediction to be made with certainty, have been discovered, the results of early experiments are encouraging. Thus with foot-and-mouth disease virus, it has been shown recently that a single injection containing 25 μg of the appropriate peptide will protect against infection (Callis, 1984). This very encouraging result makes me believe that chemically defined antigen peptides of a relatively few amino acids will be the vaccines of the future. They may be completely defined chemically, and relatively easy to adjust to combat the antigenic variation shown by some viruses, such as foot-and-mouth disease virus, and there is of course the possibility that the rules for producing wide spectrum vaccines will be discovered.

REFERENCES
Anderer, F.A. (1963). *Z. Naturforsch Teil B. 18*, 1010.
Brown, F. (1984). *Ann. Review of Microbiology, 3*, 221.
Callis, J.J. (1984). *Personal communication*.
Emtage, J.S., Tacon, W.C.A., Catlin, G.H., Jenkins, B., Porter, A.G. et al. (1980). *Nature 283*, 171.
Kleid, D.G., Yansura, D., Small, B., Dowbenko, D., Moore, D.M. et al. (1981). *Science 214*, 1125.
Mackett, M.A., Smith, G.L. and Moss, B. (1982). *Proceedings of the Natonal Academy of Sciences 79*, 7415.
McAleer, W.J., Buynak, E.B., Maigetter, R.Z., Wampler, D.E., Miller, W.J. and Hilleman, M.R. (1984). *Nature 307*, 178.

Chapter Eighteen

PATHOGEN IMPORTATION FOR BIOLOGICAL CONTROL — RISKS AND BENEFITS

R.J. Milner

Insect pathology has a long and distinguished history. Diseases of bees were studied by Aristotle (384–322 B.C.), and an Italian, Agostini Bassi in 1833 was the first to demonstrate that an infectious organism caused disease. He found that the fungus *Beauveria bassiana* caused disease in silkworm larvae (Steinhaus, 1956). The subsequent decades saw many insect pathogens described, but it was not until the last years of the century that the first attempts to use diseases for insect pest control were made. In Australia, McAlpine (1899) sought to import the fungus *Entomophaga grylli* from South Africa for control of grasshoppers. It was distributed to farmers in New South Wales and Victoria from 1900 to 1904, but was found to be ineffective, presumably because the fungus actually distributed was *Mucor* sp. and not the highly fastidious *E. grylli* (Wilson, 1960). Other early attempts were mostly concerned with another fungus *Metarhizium anisopliae* for control of cane grubs and these were also unsuccessful. Microbial control thus became discredited and only in the past 20 years has research been redirected into this field in Australia.

STRATEGIES FOR USE OF PATHOGENS IN PEST CONTROL

Most insect pathogens, in contrast to many plant and animal pathogens, cause death of the host, and are only transmitted after the hosts death. In addition, insects have no acquired immunity system. The infectious propagules of most insect pathogens form within the tissues of the living host but are released by disruption of the cadaver. In contrast, most entomogenous fungi form infectious spores externally to the host during a secondary saprophytic growth phase.

Two more or less distinct strategies for using pathogens in pest control are recognised — 'microbial insecticides' and 'classical biological control'. A 'microbial insecticide' may be defined as 'a stable formulated insecticidal product with a pathogen as the active ingredient, which results in temporary control of the target insect in the treated area'. While the initial research on potential biological insecticides is usually done by Government or University scientists, subsequent development and registration is dependant on commercial interests. The most successful microbial insecticides to date are those based on the bacterium *Bacillus thuringiensis* (B.t.). The taxonomy and pathogencity of B.t. is complex with some 22 serotypes or varieties currently recognised (de Barjac, 1981) with variable spectra of host activity (Dulmage, 1981). The 3 main commercial varieties are:

1. *kurstaki* (trade-names Dipel, Thuricide, etc.) effective against various lepidopteran larvae notably on crucifers;
2. *galleriae* (trade-name Certan) effective against waxmoth (a pest of bee hives), and;
3. *israelensis* (trade-names Bactimos, Tecknar, Skeetal, etc.) effective against mosquito and blackfly larvae (see below).

Currently the only microbial insecticides registered for use in Australia are the B.t. based products and the *Heliothis* nuclear polyhedrosis virus (trade-name Elcar) for control of *Heliothis* caterpillars on cotton and sorghum. The nematode *Neoaplectana bibionis*, is used commercially for control of currant borers in Tasmania, is associated with a bacterium but is regarded as a 'parasite' exempt from registration. Many other pathogens, both naturally occurring in Australia and exotic, are being studied as potential microbial control agents (Table 1).

Major advantages of microbial insecticides over chemical insecticides are the specificity, environmental safety, compatibility with other biological control agents, and reduced likelihood that resistance will develop. Some disadvantages are inactivation by the UV component of sunlight, that they may be too specific, may be effective only against young larvae, and may be effective only at certain temperatures and humidities. Though some pathogens provide on-going control by means of insect to insect transmission, other pathogens, notably B.t., do not.

The other strategy, 'classical biological control', may be defined as 'the release of an exotic pathogen into the environment to provide long term population reduction of the target insect by means of natural spread and development of epizootics'. This approach has received much less attention because of the lack of commercial incentive, and because of the apparently small number of opportunities for this type of pest control. Many of the most effective insect pathogens occur world-wide, either naturally or because they have been accidentally introduced with their hosts.

A major problem of classical biological control programmes has been to evaluate their effectiveness. This is well illustrated by two 'successful' Australian examples (Table 2): *Zoophtora radicans* for control of the spotted alfalfa aphid (SAA), *Therioaphis trifolii* f. *maculata*, and *Deladenus siricidicola* for control of the wood wasp, *Sirex noctilio*. In the case of the wood wasp the nematode parasizes up to 90% of the larvae and renders the resulting female wasps sterile. However the dominant role of tree physiology in *Sirex* ecology and the presence of introduced parasitoid wasps make assessment of the effect of the nematode extremely difficult (Taylor, 1981). The nematode disperses poorly and it is still necessary to distribute it to new areas of *Sirex* infestation.

Other classical biological control projects in progress in Australia are summarised in Table 2.

TABLE 1. *'Microbial Insecticides'* — *Status in Australia*

Pathogen	Type	Pest	Crop	status[1]
Heliothis NPV	virus	*Heliothis* spp.	cotton	comm.
Bacillus thuringiensis				
var. *kurstaki*	bacterium	various	various	comm.
var. *israelensis*	bacterium	mosquitoes	—	comm.
var. *wuhanensis*	bacterium	sheep blowfly	sheep	lab.
Bacillus moritai	bacterium	sheep blowfly	sheep	lab.
Neoaplectana bibionis	nematode	currant borer	blackcurrants	comm.
Neoaplectana spp.	nematode	weevils	banana, ornamentals	field
Culicinomyces	fungus	mosquitoes	—	field
Verticillium lecanii	fungus	aphids	glasshouse	field
Metarhizium anisopliae	fungus	crickets	pasture	field
M. anisoplae	fungus	*Aphodius tasmaniae*	pasture	field
M. anisopliae	fungus	termites	—	field
Nosema locustae	protozoa	grasshoppers	pasture	field

1. comm. = in commercial use.
   lab. = undergoing tests in the laboratory.
   field = undergoing field testing.

TABLE 2. *'Classical Biological Control' by Pathogens* — *Status in Australia*

Pathogen	Type	Origin	Pest	Crop	Status[1]
Deladenus	nematode	Europe	*Sirex noctilio*	pine	op.
Zoophthora	fungus	Israel	*Therioaphis*	lucerne	op.
Beauveria	fungus	France	*Sitona*	pasture	field
Entomophaga	fungus	U.S.A.	*Phaulacridium*	pasture	field
Octosporea	protozoa	U.S.A.	*Lucilia*	sheep	lab.

1. op. = operational.
   field = undergoing field testing.
   lab. = undergoing tests in the laboratory.

CASE HISTORIES

A microbial insecticide

Bacillus thuringiensis var. israelensis (B.t.i.). Prior to 1977, many tests had been done with various strains of B.t. against mosquito larvae. Some strains were found to be highly active due usually to the B-exotoxin (Rogoff et al., 1969). This exotoxin, formed by some strains of B.t., is thought to be potentially hazardous to mammals (Faust and Bulla, 1982). Consequently the discovery of an isolate of B.t. with high activity towards mosquito larvae, yet which produces very little B-exotoxin, was very exciting (Goldberg and Margalit, 1977). Originally found in the soil of mosquito breeding sites in Israel, it was found to have an LC_{50} of 6×10^3 spores/ml against *Culex pipiens*, and to be harmless to lepidopteran larvae. Biochemical and serological tests showed this isolate to be distinct and it was named *B. thuringiensis* var. *israelensis* (B.t.i.) (de Barjac, 1978).

Subsequent tests showed that mosquito larvae from all 3 major genera (*Aedes*, *Anopheles* and *Culex*) were susceptible. In addition, blackfly and chironomid larvae were susceptible, but other members of the aquatic ecosystem, notably fish, crustaceans, beetles and mayflies were not. The irregularly shaped crystals associated with the spores of B.t.i. were shown to be the main toxic entity, and ultrastructural studies have shown this crystal to be distinct from those in other varieties of B.t. (Huber and Luthy, 1981).

B.t.i. is easily grown on normal bacteriological media (Goldberg and Margalit, 1977). Commercial production in deep tank fermenters required some modification to the system used for lepidopteran strains, but this was readily achieved (Couch, 1981).

Formulations of B.t.i. have been found to be stable. Wettable powders generally retain full activity for 2 to 3 years if kept dry and below 40°C. Survival is much reduced at higher temperatures and low temperatures prolong activity. Liquid formulations are less stable but still have a shelf life of over 1 year. In water activity is quickly lost due to both inactivation and sedimentation. For example, Mulligan et al. (1980) obtained 99% control of *Culex tarsalis* from an aerial application of 1.2 kg/ha but could detect no residual activity 24 h later. At least one product is available as a briquette which provides slow release for up to 4 weeks.

B.t. based products effective against lepidopteran caterpillars have been extensively used for many years in most countries of the world. The safety of these products, as demonstrated in testing prior to registration, has been thus resoundingly confirmed since no harmful effects on non-target species have been detected. Nor has any species developed resistance, despite some 20 or so years of continuous use on vegetable pests in California. Thus the registration of new isolates demands only minimal safety testing. Nonetheless maximum challenge tests have been undertaken for B.t.i. using oral, ocular, intracerebral, and intraperitoneal routes against mice, rabbits, guinea pigs and rats. No toxicity, pathogenicity or replication was detected (unpublished reports WHO).

Commercial formulations of B.t.i. have been extensively tested in many countries for control of mosquito larvae and blackfly larvae. These tests have shown that, provided an adequate dose of the crystal is available for ingestion, control is certain. The non-feeding pupal stage is not affected. Older instars are only up to 10 fold less susceptible than the newly hatched larvae (Mulla and Federici, 1982). Higher rates of application are needed for control in polluted water, but salinity has no effect (Davidson, 1982). B.t.i. is effective over a wide range of temperatures, optimally between 19 and 33°C.

Tests with 'Skeetal' (trade-name) in Australia have shown it to be effective over a range of field sites in Queensland against 6 species of mosquito; *Aedes aegypti*, *A. vigilax*, *Anopheles annulipes*, *Culex annulirostris*, *C. sitiens*, and *C. quinquefasciatus*, (S. Sexton, pers. comm.). It is currently being used by several local councils in Queensland and as its advantages become better known its use is expected to spread.

A Classical Biological Control Agent

Zoophthora radicans. The spotted alfalfa aphid, *T. trifolii*, and the blue-green aphid (BGA), *Acyrthosiphon kondoi*, two serious pests of lucerne, were both first recorded in Australia during 1977. A large cooperative programme of research was established, aimed at developing integrated control strategies for these pests. As part of this programme, I was asked to investigate the feasibility of using exotic fungi for biological control. Surveys of aphid diseases in Australia during the next two years detected 6 species of entomophthoran fungi (Milner et al., 1980). Many epizootics of fungi were found, especially in BGA populations. However disease in SAA populations was rare, never exceeding 10%. Entomophthoran fungi were attributed as assisting control of SAA in California (Hall and Dunn, 1957). Unfortunately no authentic isolates of the main pathogen involved '*Entomophthora exitialis*' could be obtained. Reports from Israel, a country within the region of origin of SAA, indicated that a similar pathogen, correctly identified as *Zoophthora radicans*, was responsible for frequent epizootics in SAA (R.G. Kenneth, pers. comm.). It now seems likely that these two fungi are identical.

Potentially useful isolates of entomophthoran fungi were imported into Australia and screened for activity against SAA. Some 52 tests involving 24 isolates were carried out (Milner and Soper, 1981). On the basis of virulence and ability to form new conidia on SAA, an isolate of *Z. radicans* from Israel was selected for field release. The first releases were made in February 1979 at Armidale, New South Wales (Milner et al., 1982).

Z. radicans grows and sporulates well *in vitro* but the conidia are fragile and short lived. Consequently it was not possible to use it by spraying onto crops, it is best instead to produce conidia *in situ*. Therefore, for field release, sporulating culture dishes are placed directly over aphids feeding on the lucerne crop with a plastic garbage bin or tote box as a high humidity cage (infection only occurs in a very moist atmosphere). The cages are left in place overnight and as a result up to 50% of the treated SAA die of disease 2 to 5 days later. Spread from these foci of infection occurs naturally, either by wind for short distances (up to 1 m) or by means of winged adults incubating the disease for longer distances (up to at least several hundred metres). Control of the pest is not normally achieved within the year of introduction, but in subsequent seasons, given favourable weather conditions, epizootics will occur. It has been shown that the duration of dew period is critical: a minimum of 9 hr being required for effective transmission (Milner et al., 1982). *Z. radicans* is well adapted to the temperatures normally encountered in the field in southeastern Australia and this factor is rarely limiting (Milner and Lutton, 1983). Towards the end of an epizootic, in response to various factors including temperature, *Z. radicans* forms resting spores internally in the aphid which persist from year to year in the surface layers of the soil.

The disease is becoming widely distributed over NSW and southern Queensland and causes epizootics in late summer/autumn. SAA is seasonal in its occurrence and only the first outbreaks in the summer are likely to escape infection by *Z. radicans*. The disease does not affect predators or parasitoids, but other pest aphids are sometimes infected. It is not possible to evaluate the success of *Z. radicans* at present. The disease is still becoming established and is being spread by natural and artificial means. Also many other control measures are now firmly in place, notably the use of resistant cultivars and the parasitoid, *Trioxys complanatus* (Wilson et al., 1982).

SAFETY TESTING AND HOST SPECIFICITY

Safety testing is the quantitative evaluation of the risk to mammals and other warm blooded animals of insect pathogens prior to registration as microbial insecticides. While some general pathogens also kill insects, e.g. *Serratia marcescens*, no true insect pathogen being considered for biological control is known to cause disease in higher animals. A cardinal factor is that most insect pathogens will not replicate at 37°C. Nevertheless concern over safety delayed the registration of *Baculovirus* spp. in the USA during the late 1960's. A major problem was that insect pathogens had to satisfy tests designed for chemical insecticides where adverse reactions could be detected and measured: usually with insect pathogens no adverse effects can be detected.

Many authorities now have established guidelines specifically for insect pathogens which also distinguish between different types of pathogen. These are based on a three tier system (Table 3) (Burges, 1981). All pathogens are subject to tier 1 testing which is mainly short term testing of a maximum challenge dose using a variety of entry routes. Only if potential problems are identified does the pathogen have to be subject to tier 2 testing in order to quantify such hazards. If problems still exist then long term testing in tier 3 is required. Most pathogens are expected to be registered solely on the basis of tier 1 testing.

The safety of insect pathogens registered to date is demonstrated by the exemption from withholding time. Thus pathogens can be used right up to the day of harvesting, a distinct advantage for control of some vegetable pests.

There are a few recognised risks with particular pathogens or groups. The possible hazards to mammals of strains of B.t. containing the B-exotoxin has already been mentioned, though such strains are used in Finland and the USSR without any reported problems. Other groups thought to pose some risk are Rickettsia and non-occluded viruses. This concern inhibits research in these areas. *M. anisopliae* is pathogenic for reptiles such as

TABLE 3. *Safety Testing of Microbial Insecticides*

TIER 1	acute oral
	acute inhalation
	acute intra-peritoneal
	dermal/ocular irritation
	allergen test
	mutagen screens
TIER 2	ED_{50} evaluations of infectivity
	LD_{50} toxicity single dose and 90 day multiple dose
	quantification of persistence
	for irritation — restrictions on label
TIER 3	teratogenicity tests
	long term mutagenicity tests
	long term feeding tests

crocodiles and turtles though such animals are unlikely to be in the target area, so the significance of this finding is unknown (Austwick, 1980). Finally small fungal spores such as those of *Beauveria bassiana* can be allergenic (York, 1958).

GENETIC FACTORS

Of concern in all biological control programmes is the question of whether the host preference of the controlling organism will remain unchanged when released into what is often a new environment. Genetic variation in virulence towards a particular host is well known in most pathogen groups (for example see Milner, 1982). However the actual spectrum of activity is a stable character. The 'multiple embedded' NPV of *Autographa californica* has a relatively wide host range within the Lepidoptera, yet passage of the virus through several alternate hosts failed to change its virulence for the original host (Vail *et al*., 1973). An even wider host range is shown by the fungus *M. anisopliae*, involving Diptera, Coleoptera, Hemiptera, Hymenoptera and Lepidoptera. Yet individual strains are often quite host specific. For example, Fargues and Robert (1983) found that passage of strains through susceptible hosts increased virulence for that host, but it was not possible to change the basic host preferences of the individual strains. Mutants usually have reduced virulence, but on occasions, increased virulence has been demonstrated (Al-Aidroos and Roberts, 1978). Again though, there is no evidence for a change in host specificity.

The techniques of molecular biology are now being applied to insect pathogens, especially *Baculovirus* spp. and B.t. Studies with insect viruses are mainly concerned with basic aspects of identification and the genetic basis of pathogenicity (Miller and Dawes, 1979). Those with B.t. are more commercially orientated and aimed at more efficient production methods and production of compound strains with enhanced host ranges (Brown *et al*., 1984; Klausner, 1984; Martin and Dean, 1981). Requests for approval for field release of genetically manipulated organisms can therefore be expected and any unique hazards of such strains evaluated.

It is much more difficult to generalise concerning the host specificity testing of pathogens to be used for classical biological control. The most obvious beneficial insect at risk is the honey bee. Fortuitously most bee pathogens are not related to the pathogens suitable for biological control. An exception is the Nosema disease caused by *Nosema apis*. While *N. apis* seems to be quite specific, other *Nosema* spp. imported for biological control may need testing. Tests with *N. locustae*, a biological control agent for grasshoppers, showed that bees were not susceptible (Menapace *et al*., 1978). Exotic insects, imported for biological control of weeds and other pests, are more often related to the target host. Tests may than be required. For example it was necessary to test *V. lecanii* against the Harrisia cactus scale prior to release for aphid control. In some cases native beneficial insects may be at risk. The proposed release of *Octosporea musca-domestica* for control of the sheep blowfly may adversely affect the populations of other blowflies which are important as detritus feeders. These risks, as well as the risks to economically unimportant native insects, are extremely difficult to assess. Laboratory tests can indicate an absolute level of susceptibility, but under field conditions contact between such potential hosts and the pathogen may be minimal. Since insect pathogens are density dependent in effect, and most non-pest insect populations are sparse, this factor would ameliorate any adverse effect. Currently this type of risk is usually not evaluated by experimentation but the advent of 'protected species' lists for insects may lead to some tests being required.

Most of the information regarding host specificity of a given pathogen can be obtained from the literature, though this may be difficult to evaluate because of taxonomic problems. Pathogens may be erroneously regarded as host specific and therefore named as new species based on host records. Certainly more research is needed on the host specificity of candidate pathogens and on its genetic basis.

CONCLUSIONS

Pathogens, both exotic and native, are now an accepted part of pest control in Australia. They offer the great advantage of being target specific and free of adverse side effects: in addition insect to insect transmission can sometimes provide ongoing control after a single application. While more research is needed on the genetic basis of host specificity, there are established procedures for assessing the risks involved. In practice few hazards have been recognised. However Australia needs to adopt a set of guidelines for registration of microbial insecticides to replace existing *ad hoc* arrangements. It is likely that approval for field release of genetically manipulated pathogens will be sought in the near future. These will have to be carefully assessed to determine if they have any additional hazards. It is likely that pathogens will become increasingly important in pest control in the future, but no spectacular increase is foreseen.

REFERENCES

Austwick, P.K.C. (1980). The pathogenic aspects of the use of fungi: the need for risk analysis and registration of fungi. *Environmental Protection and Biological Forms of Control of Pest Organisms*. B. Lundholm and M. Stuckerud, eds) *Ecological Bulletin 31*, 91–102, Stockholm.

Al-Aidroos, K. and Roberts, D.W. (1978). Mutants of *Metarhizium anisopliae* with increased virulence towards mosquito larvae. *Canadian Journal of Genetics and Cytology 20*, 211–219.

de Barjac, H. (1978). Un nouveau candidat a la lutte bilogique contre les moustiques: *Bacillus thuringiensis* var. *israelensis*. *Entomophaga 23*, 309–319.
de Barjac, H. (1981). Identification of H-serotypes of *Bacillus thuringiensis*. In: *Microbial Control of Pests and Plant Diseases 1970–1980* (H.D. Burges, ed.) pp. 35–43, Academica Press, London.
Brown, R.W., Smith, D.R.J. and Forrage, R.G. (1984). Cloning of a *Bacillus thuringiensis* crystal protein gene. *Australian Microbiologist* 5, 188.
Burges, H.D. (1981). Safety, safety testing and quality control of microbial pesticides. In: *Microbial Control of Pests and Plant Diseases 1970–1980* (H.D. Burges, ed.) pp. 737–768, Academic Press, London.
Couch, T.L. (1981). Mosquito pathogenicity of *Bacillus thuringiensis* var. *israelensis*. *Developments in Industrial Microbiology 22*, 61–67.
Davidson, E.W. (1982). Bacteria for the control of arthropod vectors of human and animal diseases. In: *Microbial and Viral Pesticides*, (E. Kurstak, ed.) pp. 289–316, Marcel Dekker, New York.
Fargues, J.F. and Robert, P.H. (1983). Effects of passaging through scarabaeid hosts on virulence and host specificity of two strains of the entomopathogenic hypthomycete *Metarhizium anisopliae*. *Canadian Journal of Microbiology 29*, 576–583.
Faust, R.M. and Bulla, L.A. (1982). Bacteria and their toxins as insecticides. In: *Microbial and Viral Insecticides*, (E. Kurstak, ed.) pp. 75–208, Marcel Dekker, New York.
Goldberg, L.J. and Margarlit, J. (1977). A bacterial spore demonstrating rapid larvicidal activity against *Anopheles sergentii, Uranotaenia unculata, Aedes aegypti* and *Culex pipiens*. *Mosquito News 37*, 355–358.
Hall, I.M. and Dunn, P.H. (1957). Entomogenous fungi on the spotted alfalfa aphid. *Hilgardia 27*, 159–181.
Huber, H.E. and Luthy, P. (1981). *Bacillus thuringiensis* delta endotoxin: composition and activation. In: *Pathogenesis of Invertebrate Microbial Diseases*, (E.W. Davidson, ed.) pp. 209–234, Allanheld Osmum, New Jersey.
Klausner, K. (1984). Microbial insect controls: bugs to kill bugs. *Biotechnology 2*, 408–419.
McAlpine, D. (1899). Brief report on locust fungus from the Cape. *Agricultural Gazette of N.S.W. 10*, 1213.
Martin, P.A. and Dean, D.H. (1981). Genetics and genetic manipulation of *Bacillus thuringtiensis*. In: *Microbial Control of Pests and Plant Diseases 1970–1980*, (H.D. Burges, ed.) pp. 299–311, Academic Press, London.
Menapace, D.M., Sackett, R.R. and Wilson, W.T. (1978). Adult honey bees are not susceptible to infection by *Nosema locustae*. *Journal of Economic Entomology 71*, 304–306.
Miller, L.A. and Dawes, K.P. (1979). Physical map of the DNA genome of *Autographa californica* nuclear polyhedrosis virus. *Journal of Virology 29*, 1044–1055.
Milner, R.J. (1982). On the occurrence of pea aphids, *Acyrthosiphon pisum* resistant to isolates of the fungal pathogen, *Erynia neoaphidis*. *Entomologia Experimentalis et Applicata 32*, 23–37.
Milner, R.J. and Lutton, G.G. (1983). Effect of temperature on *Zoophthora radicans* (Brefeld) Batko: an introduced microbial control agent of the spotted alfalfa aphid, *Therioaphis trifolii* (Monell) f. *maculata*. *Journal of Australian Entomological Society 22*, 167–173.
Milner, R.J. and Soper, R.S. (1981). Bioassay of *Entomophthora* against the spotted alfalfa aphid, *Therioaphis trifolii* f. *maculata*. *Journal of Invertebrate Pathology 37*, 168–173.
Milner, R.J., Soper, R.S. and Lutton, G.G. (1982). Field release of an Israeli strain of the fungus *Zoophthora radicans* (Brefeld) Batko for biological control of *Therioaphis trifolii* f. *maculata*. *Journal of the Australian Entomological Society 21*, 113–118.
Milner, R.J., Teakle, R.E., Lutton, G.G. and Dare, F.M. (1980). Pathogens (Phycomycetes: Entomophthoraceae) of the blue-green aphid *Acyrthosiphon kondoi* Shinji and other aphids in Australia. *Australian Journal of Botany 28*, 601–619.
Mulla, M.S. and Federici, B.A. (1982). Evaluation of *Bacillus thuringiensis* (H-14) against mosquitoes with emphasis on field trials. *Proceedings IIIrd International Colloquium of Invertebrate Pathology, Brighton*, 1982. pp. 466–472.
Mulligan, F.S., Schaefer, C.H. and Wilder, W.H. (1980). Efficacy and persistence of *Bacillus sphaericus* and *B. thuringiensis* h.14 against mosquitoes under laboratory and field conditions. *Journal of Economic Entomology 73*, 684–688.
Rogoff, M.H., Ignoffo, C.M., Singer, S., Giard, I. and Prieto, A.P. (1969). Insecticidal activity of 31 strains of *Bacillus* against 5 insect species. *Journal of Invertebrate Pathology 14*, 122–129.
Steinhaus, E.A. (1956). Microbial control — the emergence of an idea. *Hilgardia 26*, 107–157.
Taylor, K.L. (1981). The Sirex woodwasp: ecology and control of an introduced forest insect. In: *The Ecology of Pests*, (R.L. Kitching and E. Jones, eds) pp. 231–248, CSIRO, Melbourne.
Vail, P., Jay, D. and Hunter, D. (1973). Infectivity of a nuclear polyhedrosis virus from the alfalfa looper, *Autographa californica,* after passage through several alternative hosts. *Journal of Invertebrate Pathology 21*, 16–20.

Wilson, F. (1960). A review of the biological control of insects and weeds in Australia and Australian New Guinea. *Technical Communication, Commonwealth Institute of Biological Control* No. 1, pp. 102.

Wilson, C.G., Swincer, D.E. and Walden, K.J. (1982). The introduction of *Trioxys complanatus* Quilis (Hym.: Aphididae), an internal parasite of the spotted alfalfa aphid, into South Australia. *Journal of the Australian Entomological Society 21*, 13–27.

York, G.T. (1958). Field tests with the fungus *Beauveria* sp. for control of European corn borer. *Iowa State College Journal of Science 33*, 123–129.

Section D

Cautionary tales

This is the most varied part of the book, and contains tales of past, present and possible future outbreaks of pests and parasites; both in Australia and overseas. Some accounts are historical, some point to future problems and possible ways to avoid them. The tales differ in style and content, and illustrate the great variety of reasons why, and methods by which, exotic organisms may be studied, from the pragmatic search for biological control agents to the mathematical comparison of pest life cycles.

The tales presented here are by no means exhaustive, several sets of other quite different 'cautionary tales' could have been assembled in this book, such is the scale of the problem that confronts Australians seeking to decide with which other organisms they wish to share their continent.

Chapter Nineteen

ACQUIRED IMMUNODEFICIENCY SYNDROME (AIDS)

Walter R. Dowdle

In June 1981, the Centers for Disease Control (1981, June 5) reported the unusual occurrence of pneumocystosis, later classified as a manifestation of what became known as AIDS, in young homosexual men in Los Angeles, California. This was the first of many AIDS-related events that have tested the scientific, social, sexual, and ethical values of our times. Fear of AIDS has diminished, presumably because of increasing public knowledge and decreasing news media interest in the sensational aspects of AIDS. In the United States, homosexual and bisexual men and intravenous drug users have accounted for nearly 90% of the cases, but AIDS also occurs in Haitians who have recently arrived in the United States, hemophilia patients, heterosexual sexual partners of AIDS patients, and infants born of high-risk mothers, and occasionally in people who have received blood transfusions. AIDS is now recognized in many other countries, notably in central Africa. Elucidation of the natural history of the disease and the spectrum of illness, and opportunities for control and prevention await documentation of the etiologic agent.

The occurrence of *Pneumocystis carinii* at three different hospitals in Los Angeles was unusual in that the five patients had no known underlying disease or condition and that all five were homosexuals. Within weeks, 26 cases of Kaposi's sarcoma and other opportunistic infections, also in young homosexual men, were reported from New York and Los Angeles to the CDC (1981, July 3). One year later, the number of cases of Kaposi's sarcoma and/or opportunistic infections reported to CDC (1982, June 11) had risen to more than 300. By April 1984, more than 4,000 cases meeting the case definition of acquired immunodeficiency syndrome, as the disease was now called, had been reported. Of the first 4,000 cases in the United States, homosexual or bisexual men and intravenous drug users accounted for nearly 90% of the total. Other patient characteristics included being an emigrant from Haiti, 4.0%, a hemophilia patient, 1.0%, a transfusion recipient, 1%, a heterosexual sexual partner of an AIDS patient, 1.0%; and those with no apparent risk. There were also approximately 55 reported cases in children, most of whom, with the exception of the transfusion recipients, were reported as born into families at high risk of acquiring AIDS. Over 80% of the U.S. AIDS patients have been between 20 and 49 years of age. Most cases were reported from the major urban areas, with the largest number in New York City, followed by San Francisco, Los Angeles, Miami, and Newark, New Jersey. Most of the States have reported at least one case. The case-fatality ratio remains about 40%, although it is over 80% for those who have had the disease more than 2 years. Many other countries have reported AIDS, but the highest incidence outside the United States is recognized to be in Haiti and Central Africa, primarily Zaire.

The geographic origins of AIDS are unclear. Epidemiologic studies suggest evidence of the disease as early as 1979 in Haiti, Central Africa, and the United States. Early theories as to the possible cause of AIDS included the illicit use of drugs, multiple infections, hepatitis B, different cytomegaloviruses, and multiple exposures to semen. Evidence that AIDS is caused by a transmissible agent consisted of the identification of clusters of cases (CDC, 1982 June 18), the risk factors among homosexual and intravenous drug abusers and the geographic 'spread' of disease, reports of cases among persons with hemophilia (CDC, 1982, July 16), and reports of transfusion-associated cases and cases in infants in late 1982 (CDC, 1982 December 17).

The social, economic, and political effects of AIDS in the United States as elsewhere have been profound. The widespread fear of acquiring the disease and reports of discrimination against the affected groups reached their zenith in late 1982 and early 1983 amid criticism that the Government was slow in responding to the challenge of AIDS because it affected chiefly homosexuals, drug users, and the economically disadvantaged. Such criticism was unfounded. It failed to take into account the natural evolution of the disease, gradual acceptance of the fact that it was a new disease by the medical community, and recognition by the general public. An early task of the Public Health Service was to provide the data necessary to convince the medical community, public health authorities, and the affected groups that AIDS was an important public health problem. Later our task was to calm the fears of the general public and provide assurance that the disease was not transmitted by casual, non-sexual contact, and that it was unlikely that AIDS would spread to a large segment of the population.

Prevention recommendations were issued by the Public Health Service in late 1982 for clinical and laboratory staffs (CDC, 1982 November 5) and in early 1983 for the high-risk groups (CDC, 1983 March 4). The prevention recommendations were simple, emphasizing opportunities for decreasing transmission among the high-risk groups, hemophiliacs, and transfusion recipients. With the better understanding of the disease and less emphasis in the news media, the public's fear of AIDS has subsided.

Because epidemiologic evidence was consistent with AIDS being caused by a virus, a retrovirus was strongly suspected for several reasons. Retroviruses can cause diseases in animals such as feline leukemia bovine leukemia, and equine infectious anemia, all of whch share immunologic and pathologic characteristics with AIDS in man. In addition, certain animal retroviruses exhibit an affinity for lymphocytes, a characteristic of the etiologic agent of AIDS.

In May 1983, workers in the United States presented evidence of an association of human T-cell leukemia virus-I (HTLV-I) with AIDS (Marx, 1983; Essex et al., 1983; Gallo et al., 1983). HTLV-I had been previously identified as the cause of adult T-cell leukemia, a neoplastic disease common in southern Japan but rare in the United States. Positive antibody findings with HTLV-I were thought to reflect cross-reactions with the possible causative agent. In the same publication (Barre-Sinoussi et al., 1983), workers in France reported a new retrovirus, which they call lymphadenopathy-associated virus (LAV), as being associated with AIDS. They described additional isolations of similar viruses in April 1984 (Vilmer et al., 1984).

In May 1984, a team of U.S. researchers reported multiple isolations of a retrovirus, designated as human T-lymphotropic virus-III (HTLV-III), from AIDS patients (Popovic et al., 1984; Gallo et al., 1984; Schupback et al., 1984; Sarngadharan et al., 1984). Also, importantly, a newly developed culture system was described that facilitated virus isolation and permitted high-level production of virus.

The morphologic characteristics and serologic patterns of HTLV-III and LAV suggest that they are the same virus, but they have not yet been directly compared in biochemical and immunological tests. Ultimate proof of causality awaits further study, but preliminary evidence has shown that viruses of this group or antibody to these viruses may be found in serum from a high percentage of AIDS and lymphadenopathy patients and in temporal association with the development of AIDS or pre-AIDS. Isolation of a similar virus from a blood donor-recipient pair both of whom have AIDS has also been described (Feorino et al., 1984). Transmission of the human virus and reproduction of the disease in experimental animals has not been achieved, but Koch's postulates could be considered to be fulfilled by documentation of virus transmission through blood transfusion with subsequent development of disease. These present findings are promising and suggest the imminent development of better diagnostic and screening tests and, eventually, more focused prevention and control measures.

Many questions remain. What is the significance of a positive or negative test for antibody? What is the significance of isolating the virus in the absence of AIDS? Is a retrovirus vaccine feasible? How would such a vaccine be used? How can the target population be identified?

Clearly, technical achievements alone will not provide all the answers. The successful strategy for the control of AIDS among homosexual and bisexual men and intravenous drug users, who are at highest risk in industrialized countries, must include a strong information/education component. Already the simple lesson of AIDS, that is, that the spread of an infectious disease increases with increased opportunity for transmission, has had a demonstrable affect. The incidence of gonorrhea among homosexuals is down by as much as 50% in such major cities as Denver, New York, and Seattle. We do not yet have accurate figures on the prevalence of infection in Central African countries or know the risk factors for these populations. A still different control strategy may be required there.

ADDED IN PRESS

In the year since this presentation was given, the number of cases in the U.S. have more than doubled. The reported incidence of disease is also continuing to increase elsewhere in the world, although relatively few cases of AIDS have been reported from Eastern Europe, Asia, and the Western Pacific, except Australia. AIDS has become increasingly recognized as a serious public health problem in Central Africa. In the industrialized countries the population affected remains remarkably similar to that reported in 1984. However, heterosexual contact appears to be a major risk factor for transmission in Haiti and several Central African countries. Serosurveys for antibodies to the retrovirus considered to be the etiologic agent of AIDS, LAV/HTLV-III, have demonstrated that 30 to 100 times more persons have been infected with the virus than have acquired AIDS. Other studies suggest that infected persons may carry the virus for a lifetime. Because of the long incubation period and multiple opportunities for transmission during that time, the AIDS virus is anticipated to continue to spread. In several countries, including the United States and Australia, laboratory tests for detecting infection with the AIDS virus are now in place for screening blood intended for transfusion or fractionation. Attempts to develop drugs and antiviral vaccines are underway, but years will be required before either can be realistically assessed as tools for reducing the transmission of virus or the fatal consequences of the disease.

REFERENCES

Centers for Disease Control (1981 June 5). *Morbidity and Mortality Weekly Report 30*, 250.
Centers for Disease Control (1981 July 3). *Morbidity and Mortality Weekly Report 30*, 305.
Centers for Disease Control (1982 June 11). *Morbidity and Mortality Weekly Report 31*, 294.
Centres for Disease Control (1982 June 18). *Morbidity and Mortality Weekly Report 31*, 365.
Centres for Disease Control (1982 November 5). *Morbidity and Mortality Weekly Report 31*, 577.
Centres for Disease Control (1982 December 17). *Morbidity and Mortality Weekly Report 31*, 665.
Centres for Disease Control (1983 March 4). *Morbidity and Mortality Weekly Report 32*, 101.
Barre-Sinoussi, F., Chermann, J., Rey, F., Nugeyre, M., Chamaret, S., Gruest, J., Dauguet, Ci, Axier-Blin, C., Vezinet-Brun, F., Rouzioux, C., Rozenbaum, W. and Montagnier, L. (1983). *Science 220*, 868.
Essex, M., McLane, M., Lee, T., Falk, L., Howe, C., Mullins, J., Cabradilla, C. and Francis, D. (1983). *Science 220*, 859.
Feorino, P., Kalyanaraman, V., Haverkos, H., Cabradilla, C., Warfiled, D., Jaffe, H., Harrison, A., Goldfinger, D., Gottlieb, M., Chermann, J., Barre-Sinoussi, F., Spira, To, McDougal, J., Curran, J., Montagnier, L., Murphy, F. and Francis, D. (1984). *Science,* in press.
Gallo, R., Sarin, P., Gelmann, E., Robert-Guroff, M., Richardson, E., Kalyanaraman, V., Mann, D., Sidhu, G., Stahl, R., Zolla-Pazner, S., Leibowitch, J. and Popovic, M. (1983). *Science 220*, 865.
Gallo, R., Salahuddin, S., Popovic, M., Shearer, G., Kaplan, M., Haynes, B., Palker, T., Redfield, R., Oleske, J., Safai, B., White, G., Foster, P. and Markham, P. (1984). *Science 224*, 500.
Marx, J.L. (1983). *Science 220*, 806.
Popovic, M., Sarngadharan, M., Read, E. and Gallo, R. (1984). *Science 224*, 497.
Sarngadharan, M., Popovic, M., Bruch, L., Schupbach, J. and Gallo, R. (1984). *Science 224*, 506.
Schupbach, J., Popovic, M., Gilden, R., Gond, M., Sarngadharan, M. and Gallo, R. (1984). *Science 224*, 503.
Vilmer, E., Rouzioux, C., Brun, F., Fischer, A., Chermann, J., Barre-Sinoussi, F., Gazengel, C., Dauguet, C., Manigne, P. and Griscelli, C. (1984). *Lancet,* 753.

Chapter Twenty

BLUETONGUE

B.M. Gorman

Before 1977 Australia was considered to be an area free of bluetongue disease. Strict quarantine regulations applied to livestock and animal products were designed to exclude exotic pathogens including bluetongue viruses. Fears were often expressed of the serious economic losses which would be incurred if the viruses were introduced into Australia. A typical comment (Bowne, 1971) suggested that bluetongue had "explosive potential especially in countries like Australia where the sheep industry is of great economic importance and BT (bluetongue) could raise havoc with an extremely susceptible sheep population".

In a symposium on Bluetongue held in Adelaide in May 1974 many speakers referred to the potential dangers posed by bluetongue viruses. Geering (1975) outlined the plans for control of bluetongue in an epizootic situation. Those plans were based on three lines of attack. These were "slaughter of ruminants in the infected areas, disinfection" (the killing of insects) "of the infected area . . ." (since bluetongue is transmitted by biting midges) ". . . maintenance of a larger quarantine zone in which the standstill of ruminants . . . was to be . . . enforced". Geering suggested that "a virulent strain of bluetongue could cause a mortality rate of up to 70% in the highly susceptible sheep population" and he proposed that it might "be necessary to blanket vaccinate a large proportion of the Australian sheep population to lessen the serious effects of the disease". As part of that plan the Commonwealth Serum Laboratory in Melbourne held seed lots of most of the known serotypes of bluetongue which had been isolated in South Africa. If necessary an appropriate vaccine could be made quickly in the event of an outbreak of bluetongue.

In November 1977 it was announced that workers at the CSIRO Division of Animal Health had isolated a virus, from insects collected in the Northern Territory, which was indistinguishable from bluetongue (St George et al., 1978). Despite the fact that the insects had been collected and the virus isolated more than two years before that announcement, and that there was no evidence of disease, 32 countries imposed bans on our livestock exports and control measures were introduced in northern Australia (Lehane, 1981).

BLUETONGUE VIRUSES

Bluetongue viruses are classified as Orbiviruses and are part of a genus of the Family Reoviridae. The reoviruses are unusual in that their genetic material consists of many pieces of double-stranded ribonucleic acid. Some reoviruses are pathogens of plants, of insects and of vertebrates but the majority are not known to produce disease in the hosts that they infect (Joklik, 1983).

Most of the known orbiviruses have not been associated with disease in animals but some are important pathogens of livestock (Table 1). The orbiviruses are transmitted by insects or by ticks. Bluetongue and horse sickness were desribed in the scientific literature in the 19th century and were among the diseases which threatened the establishment of pastoral industry in southern Africa (Howell, 1963a, b).

Orbiviruses as agents of disease

Bluetongue viruses can cause a degenerative and sometimes fatal disease in sheep. It is characterized by hyperaemia of the mucous membranes of the mouth, nose and intestines. It is often associated with inflammation of the coronary tissue of the hooves and by stiffness due to muscle damage (Henning, 1956). The name is derived from the Afrikaans description of haemorrhage and cyanosis of the tongue. Signs of the disease can range from subclinical to severe depending on the susceptibility of the infected animal and on the virulence of the virus strain. It is often difficult to reproduce clinical signs of bluetongue under experimental conditions. The degree of susceptibility of goats is reported as "markedly less than of sheep" (Erasmus, 1975). Infection in cattle is often mild and inapparent but outbreaks of clinical bluetongue have occurred in cattle. The lesions produced can resemble those of foot-and-mouth disease (Goltz, 1978).

TABLE 1. *Orbivirus Serological Groups*

Serogroup	Number of Serotypes	Presence in Australia
Bluetongue	21	+
Eubenangee	3	+
Epizootic Haemorrhagic Disease	11	+
Palyam	9	+
Corriparta	4	+
Warrego	2	+
Wallal	2	+
African horsesickness	9	−
Equine encephalosis	5	−
Changuinola	12	−
Umatilla	2	−
Ungrouped viruses:		
Paroo River	1	+
Ieri	1	−
Ife	1	−
Japanaut	1	−
Lebombo	1	−
Orungo	1	−

Epizootic haemorrhagic disease of deer (EHD) is similar in many clinical and pathogenic signs to experimental bluetongue in deer (Karstad and Trainer, 1967). A serotype of EHD (Ibaraki virus) causes an acute febrile illness in cattle but not in sheep (Inaba, 1975).

The viruses of African horsesickness and equine encephalosis are important causes of disease in equines in Africa (Howell, 1963a; Erasmus et al., 1970).

A serotype of the Palyam serogroup (Nyabira virus) was isolated from the aborted foetus of a cow in Zimbabwe (Swanepoel and Blackburn, 1976).

THE ORBIVIRUSES

The bluetongue viruses are part of a large group of structurally similar viruses. The name orbiviruses is derived from the ring-like arrangement of protein units which make up the nucleocapsid of the virus particle (Figure 1A). The virus is composed of seven proteins, two of which are at the surface of each particle. The extra-capsid coat gives the virion an amorphous appearance (Figure 1B).

After attachment of virus particles to susceptible cells the surface proteins are removed and an enzyme associated with the nucleocapsid (a transcriptase) copies the 10 genes into single-stranded messenger RNA. At least 10 proteins are produced in virus-infected cells. In the assembly of progeny virions one or more of these proteins attaches to each of the 10 single-stranded RNA molecules, converts them to double-stranded RNA (polymerase) and packages the 10 genes into a virus particle consisting of seven proteins. Two major virus-induced proteins are not incorporated into assembled virus particles and probably have an important role in virion assembly.

Detailed descriptions of the structural features and morphogenesis of orbiviruses can be found in review articles (Verwoerd et al., 1979; Gorman et al., 1983) but there are two important features which will be referred to later. In infected animals, antibodies are produced against all of the virus-induced proteins but specific antibodies which neutralize virus infectivity are directed mainly to one of the surface proteins. In serological tests antibodies to the nucleocapsid proteins of bluetongue viruses are used to define the group: antibodies to the surface protein defines the type.

The other feature of the replication cycle of bluetongue viruses is that in cells simultaneously infected with two different viruses, the genomes may reassort so that progeny virus particles may derive genes from each of the parent viruses. The reassorted viruses may have a range of biological properties distinct from either of the parental viruses. the potential for generating diversity among genetically interacting viruses portends problems in the design of vaccines to protect against the viruses and difficulties in their classification.

The classification of the orbiviruses

The methods used to classify orbiviruses and the problems associated with the classification can be illustrated by reviewing the isolation and identification of orbiviruses in Australia.

The first orbiviruses were isolated as by-products of a programme designed to isolate, from mosquitoes, the viruses that cause epidemic polyarthritis and Murray Valley encephalitis in man (Doherty, 1974). Mosquitoes

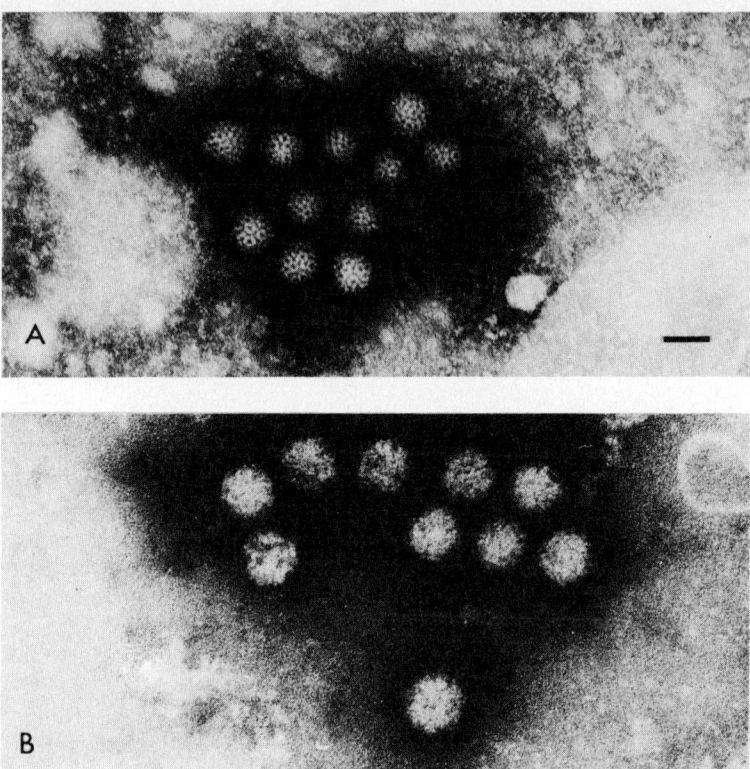

Fig. 1. Electron micrographs of bluetongue virus serotype 20. A. Nucleocapsids. B. Virions. Virions have an outer shell which obscures the ring-like capsomeres of the nucleocapsids.

were collected, ground in suitable saline slutions, and the suspensions inoculated into baby mice which are highly susceptible to these viruses. Paralysis and death of the mice indicates a possible virus isolate and extracts of mouse brain are used as antigens in serological tests. Initially the tests used are broadly reactive for groups of viruses and the isolate is tested against a range of reference antisera to viruses known to occur in Australia. If the virus cannot be related to these it is sent to an International Reference Centre for Arthropod-borne Viruses at Yale University. Conventionally, distinct serotypes are named from the locality in which they were isolated or from the disease with which they are associated. Ibaraki virus was isolated in the prefecture of Ibaraki in Japan and after the name became established was found to be a serotype of the EHD group. Bluetongue viruses were designated types 1 to 21 as the viruses were often isolated by workers concerned with the problem of bluetongue and they recognised distinct serotypes of the infecting viruses. The different approaches in designating serotypes can be confusing. The serotypes of the Corriparta viruses are distinguished by names: the serotypes of bluetongue viruses by numbers.

From 1960 to 1965, isolates of Corriparta and Eubenangee viruses were shown not to be previously recorded in Australia and were subsequently related to viruses isolated in Africa. In the electron microscope they were structurally similar to bluetongue viruses (Carley and Standfast, 1969; Schnagl et al., 1969). In complement fixation tests Eubenangee virus was related to Pata virus from Africa which in turn was related to a virus of the EHD serogroup (Borden et al., 1971). The EHD viruses were known to be distantly related to bluetongue viruses so Eubenangee viruses were recognised as having a distant relationship to bluetongue viruses (Doherty, 1974).

Following an epizootic of ephemeral fever in cattle in 1967–68 Doherty and his colleagues at the Queensland Institute of Medical Research extended the range of insects collected and processed for virus isolation. The collection included species of *Culicoides* (biting midges), potential vectors of bluetongue (Murray, 1975). Orbiviruses of the Wallal, Warrego and Palyam serological groups were isolated as part of that programme. The Palyam viruses were known to occur in India and Africa.

In November 1974 the CSIRO Division of Animal Health began a major programme of isolation and identification of viruses from insects collected in the Northern Territory. The emphasis was on collection of *Culicoides* as potential vectors of bovine ephemeral fever. The viruses were compared with the known arthropod-borne viruses at the Queensland Institute of Medical Research and those which could not be identified there were sent to Yale. The isolate CSIRO19 was not identified and when tested against the international reference collection in 1977 was found to be indistinguishable in a group-reactive test from bluetongue viruses. The complement fixation test used measures primarily the relatedness of the nucleocapsid proteins of the viruses. No identification could have been made at the Queensland Institute of Medical Research since the viruses and antisera prepared against them were prohibited imports to Australia.

The isolate CSIRO19 was sent to the Veterinary Research Institute, Onderstepoort, South Africa and tested in virus neutralization tests against the 19 known serotypes of bluetongue. None of the reference antisera inhibited the infectivity of the Australian virus and it was designated as a previously unknown serotype 20. A known serotype had not been introduced into Australia. The imported South African serotypes of bluetongue could not have been used to develop a vaccine.

The isolation of bluetongue and related orbiviruses in Australia

Bluetongue type 20 was isolated from a mixed pool of *Culicoides* collected at Beatrice Hill in the Northern Territory (St George et al., 1978). Surveys for antibodies in stored animal sera revealed neutralizing antibodies to the virus in sera collected in 1973. The group-reactive test for bluetongue antibodies (the agar gel precipitin test) provided evidence of infection with bluetongue viruses by the CSIRO Division of Animal Health was expanded in an attempt to find these viruses (Snowdon, 1979).

Between 1978 and 1983, 24 strains of five serotypes of bluetongue viruses were isolated by various workers from the blood of healthy cattle (St George, 1984). Isolate CSIRO154 was a new serotype (type 21) but isolate CSIRO156 was serologically indistinguishable from the South African serotype 1. Seven isolates of type 1 bluetongue virus were made between 27 March and 8 May 1979 at Beatrice Hill (St George et al., 1980). Two serotypes DPP90 and DPP192 have yet to be compared with exotic serotypes (G.P. Gard, personal communication). Two of the serotypes have been isolated from the collections of *Culicoides*. Apart from the original isolation of type 20 from a mixed pool, type 1 has been isolated once from *Culicoides fulvus* collected at Beatrice Hill (Standfast, 1984) and twice from *culicoides brevitaris* collected in south-east Queensland (St George and Muller, 1984).

The list of orbiviruses isolated in Australia has expanded. Five serotypes of epizootic haemorrhagic disease of deer have been isolated. In Japan (St George et al., 1983); the other viruses will be registered as new serotypes (St George, personal communication). Three new serotypes of the Palyam serogroup have been registered in the Catalogue of Arthropod-borne Viruses. Numerous isolates of the Eubenangee, Corriparta, Wallal and Warrego serogroups have been made but have not been compared directly with previously known serotypes (reviewed in Gorman et al., 1983). There will almost certainly be a number of new serotypes in each group. Viruses of seven of the 11 serological groups of insect-transmitted orbiviruses have been isolated in Australia.

Of the viruses which were thought to be exotic to Australia and were considered threats to our livestock, only horsesickness and equine encephalosis have yet to be recognised in Australia. None of the orbiviruses isolated in Australia have been shown to cause disease under natural conditions. Antibodies to bluetongue viruses have been found in cattle, buffalo, deer, goats and sheep but indigenous mammals are apparently not infected (St George, 1984). The prevalence of antibodies in cattle is greater in the northern part of the continent, and the southern part seems free of infection. The sheep population which is in the drier and more temperate areas of the continent is virtually free of infection.

The genetic relationships between bluetongue viruses

Bluetongue viruses synthesize single-stranded RNA (ssRNA) molecules from the dsRNA genome segments using a virion-associated transcriptase. Each of the 10 ssRNA molecules is an exact replica of one of the template genome segments. By convention of the ssRNA molecules are designated + to denote their capacity to associate with cellular ribosomes as messenger RNA, and be translated into protein. The genome segments are complementary (±) molecules.

The genome dsRNA molecules can be dissociated into + and − single strands by gentle heating. If the mixtures are allowed to cool the strands reanneal to form the original ± segments. If radiolabelled ssRNA (+) is mixed with unlabelled genome RNA the mixture heated and then cooled, some of the radiolabelled + strands reassociate with unlabelled − strands. The reaction mixture now contains some radiolabelled double-stranded RNA molecules. If the mixtures are electrophoresed in a polyacrylamide gel the radiolabelled dsRNA molecules migrate at the same rate as the original dsRNA genome segments.

A measure of the genetic relationships between bluetongue viruses can be obtained by mixing ssRNA molecules derived from one virus with dsRNA molecules from another and melting and reannealing as described above for the homologous system. If the two virus RNA molecules are identical, 10 radiolabelled dsRNA molecules similar to the homologous genome segments would be obtained. If there is significant variation in the molecules, the hybrid molecules will be mis-matched and their electrophoretic migrations will differ from perfectly matched hybrids. The degree of mis-matching can influence the patterns of electrophoretic separation obtained.

If radiolabelled ssRNA derived from cells infected with the Australian serotype 1 virus is reacted with unlabelled dsRNA of types 20 and 21, eight hybrid molecules are formed. Some electrophoretic migration changes are detectable but the viruses are closely related (Gorman et al., 1983). In similar tests Huismans and Howell (1973) found from six to eight duplex molecules when comparing South African serotypes. They also found that RNA of bluetongue virus type 16 isolated in Pakistan hybridized to eight of the genome segments of type 3 in South Africa.

Gorman et al. (1981) found no significant reassociation of ssRNA derived from cells infected with bluetongue type 20 with strands of opposite plarity of types 1, 4 and 10 isolated in South Africa or with type 17 isolated in the United States of America. Huismans and Bremer (1981) compared individual genome segments of type 4 with RNA of type 20 and found regions of homology on two genome segments but the alterations in electrophoretic mobilities of the hybrids compared with the original genome segments suggested extensive regions of mis-matched base pairs.

If one assumes that the bluetongue viruses have common ancestry, and that bluetongue virus populations evolving in isolation accumulate differences in their gene pools, then the differences between Australian and South African bluetongue viruses suggest that the Australian isolates are not recent introductions from South Africa.

Genetic reassortment between bluetongue viruses

In cells simultaneously infected with two or more bluetongue virus serotypes, genetic reassortment occurs and recombinant viruses deriving genome segments from each can be isolated from cell cultures (Gorman et al., 1983; Kahlon et al., 1983). The observations of Sugiyama et al. (1981) and Collisson and Roy (1983) suggest that reassortant bluetongue viruses exist in nature.

Cells simultaneously infected with three Australian bluetongue virus serotypes yielded reassorted viruses (Gorman et al., 1982). Despite the RNA sequence divergence between the Australian and South African bluetongue virus strains, recombinant viruses have been obtained from cells infected with the South African serotype 1 and the Australian serotype 20 (Gorman et al., 1982).

CAUTIONARY TALES

Although the announcement in November 1977 of the isolation of a bluetongue virus in Australia led to significant economic loss, in some respects we were fortunate that the isolated virus was a previously unknown serotype. If the first serotype isolated had been a type 1 virus there may have been a more serious outcome.

Reliance on serological tests alone to identify bluetongue viruses would have suggested that a type 1 virus had been introduced into Australia. Australia had a policy of eradication of affected animals and of vaccination. A Working Party on Bluetongue Seed Virus met at the National Biological Standards Laboratory in Melbourne on August 2, 1977 to consider the safety of working on South African serotypes of bluetongue in the new high security facility at the Commonwealth Serum Laboratory. It was intended that we would be prepared for a possible outbreak of bluetongue. If the first bluetongue virus isolated had been type 1 there might have been pressure to vaccinate. Since live attenuated viruses are used as bluetongue vaccines, a decision could have been made to introduce the South African virus into Australia. Bluetongue virus type 1 from South Africa reassorts genetic material with the Australian blutongue virus type 20. We can only speculate on the consequences of introducing novel genes into the population of Australian bluetongue viruses. For example the molecular basis of virulence is not known; it may be multigenic and is better studied prospectively in the laboratory than retrospectively in nature.

It may appear obvious to import potential pathogens ahead of an outbreak of disease. The argument that vaccine may be needed quickly is used to justify the introduction of exotic pathogens into secure laboratories. The major risk may be in the use of the vaccines and not in the holding of inocula in laboratories. It is likely that fewer risks are involved in using the outbreak strain. Little is known about the structure of virus populations but it is becoming clear that geographical clones of microorganisms exist. Our work on the apparent genetic divergence of populations of South African and Australian viruses is consistent with the observations of Rentier-Delrue and Young (1980) on genetic divergence of the alphavirus Sindbis in different geographic areas. The application of restriction endonuclease mapping of the genomes of some DNA viruses suggests clonal isolation rather than continuous global circulation of the viruses. Studies of genotypic variation in natural populations of *E. coli* suggests that the genetic structure of the populations are clonal (Ochman and Selander, 1984). As we have demonstrated for bluetongue viruses, the serotypes of *E. coli* are unreliable indices of clonal identity.

The bluetongue viruses isolated in Australia produce mild clinical signs of bluetongue when inoculated into sheep (St George et al., 1980; Uren and Squire, 1982). It has been suggested that failure to recognise bluetongue disease may be due to complex considerations of the distributions of sheep, vectors and viruses. Workers at the CSIRO have shown that the distribution of viruses and their vectors coincides with the distribution of cattle in Australia (St George, 1984; Standfast et al., 1984). The main vectors of bluetongue are biting midges which feed on cattle and breed in dung. Infection of cattle is inapparent, so that transmission of viruses in midges and cattle is the primary cycle; there is less important involvement of other insects that bite cattle, and of other animals on which they feed.

Failure to detect antibodies to bluetongue viruses in significant numbers of sheep suggests that bluetongue viruses may have a restricted distribution in Australia because they require a specific ecosystem (i.e. combination of host, vector and climate). Disturbance of that ecosystem could lead to demonstration of disease in susceptible sheep. However failure to detect antibodies in sheep may be because the serological tests used to monitor infection are inadequate, because detectable antibodies do not persist in sheep, or because there is an inherent resistance of Australian sheep to bluetongue viruses. Inadequately supported published statements on bluetongue suggest that British breeds of sheep vary in susceptibility, whereas Merino sheep in South Africa are susceptible to bluetongue disease.

The recognition of bluetongue viruses in Australia should influence our policy on exotic viruses. Some of our ideas on introduction and eradication of pathogens should be modified to take account of the experiences with bluetongue. We should discourage the view that the bluetongue viruses present no risks to our livestock industry. We know too little of the genetics of the viruses, their insect vectors and of the susceptibility of populations of ruminants at risk, to adopt a complacent attitude to bluetongue.

ACKNOWLEDGEMENTS

Orbivirus research at the Queensland Institute of Medical Research is supported by grants from the Australian Meat Research Committee and the National Health and Medical Research Council. I thank Mr H.A. Standfast for providing me with a copy of a manuscript (Standfast et al., 1984) before publication and Miss R. McDowell and Miss C. Hawkins for typing the manuscript.

REFERENCES

Borden, E.C., Shope, R.E. and Murphy, F.A. (1971). *J. gen. Virol. 13*, 261.
Bowne, J.G. (1971). *Adv. vet. Sci. Comp. Med. 15*, 1.
Carley, J.G. and Standfast, H.A. (1969). *Am. J. Epidemiol. 89*, 583.
Collisson, E. and Roy, P. (1983). *Amer. J. vet. Res. 44*, 235.
Doherty, R.L. (1974). *Prog. med. Virol. 17*, 136.
Erasmus, B.J. (1975). *Aust. vet. J. 51*, 165.
Erasmus, B.J., Adelaar, T.F., Smith, J.D., Lecatsas, G. and Toms, T. (1970). *Bull. off. Int. Epizootol. 74*, 781.
Geering, W.A. (1975). *Aust. vet. J. 51*, 220.
Goltz, J. (1978). *Can. vet. J. 19*, 95.
Gorman, B.M., Taylor, J., Walker, P.J., Davidson, W.L. and Brown, F. (1981). *J. gen. Virol. 56*, 251.
Gorman, B.M., Taylor, J., Finnimore, P.M., Bryant, J.A., Sangar, D.V. and Brown, F. (1982). In: *Arbovirus Research in Australia.* Proc. 3rd Symp. (T.D. St George and B.H. Kay, eds) p. 101, CSIRO-QIMR, Brisbane.
Gorman, B.M., Taylor, J. and Walker, P.J. (1983). In: *The Reoviridae* (W.K. Joklik, ed.) p. 287, Plenum Press, New York.
Henning, M.W. (1956). *Animal Diseases in South Africa.* 3rd ed., Central News Agency, South Africa.
Howell, P.G. (1963a). In: *Emerging Diseases of Animals.* FAO Agricultural Studies No. 61, p. 73, Rome.
Howell, P.G. (1963b). *ibid.*, p. 111.
Huismans, H. and Bremer, C.W. (1981). *Onderstepoort J. vet. Res. 48*, 59.
Huismans, H. and Howell, P.G. (1973). *Onderstepoort J. vet. Res. 40*, 93.
Inaba, Y. (1975). *Aust. vet. J. 51*, 178.
Joklik, W.K. (1983). In: *The Reoviridae* (W.K. Joklik, ed.) p. 1, Plenum Press, New York.
Kahlon, J., Sugiyama, K. and Roy, P. (1983). *J. Virol. 48*, 627.
Karstad, L. and Trainer, D.O. (1967). *Can. vet. J. 8*, 247.
Lehane, L. (1981). *Rur. Res. 112*, 4.
Murray, M.D. (1975). *Aust. vet. J. 51*, 216.
Ochman, H. and Selander, R.K. (1984). *Proc. Natl. Acad. Sci. USA, 81*, 198.
Rentier-Delrue, F. and Young, N.A. (1980). *Virology 106*, 59.
Schnagl, R.D., Holmes, I.H. and Doherty, R.L. (1969). *Virology 38*, 347.
Snowdon, W.A. (1979). In: *Arbovirus Research in Australia.* Proc. 2nd Symp. (T.D. St George and E.L. French, eds), p. 16, CSIRO-QIMR, Brisbane.
Standfast, H.A., Dyce, A.L. and Muller, M.J. (1984). *Proc. Int. Symp. on Bluetongue and Related Orbiviruses.* In press, Alan R. Liss.
St George, T.D. (1984). *Proc. Int. Symp. on Bluetongue and Related Orbiviruses.* In press, Alan R. Liss.
St George, T.D. and McCaughan, C.I. (1979). *Aust. vet. J. 55*, 198.
St George, T.D. and Muller, M.J. (1984). *Aust. vet. J. 61*, 95.

St George, T.D., Standfast, H.A., Cybinski, D.H., Dyce, A.L., Muller, M.J., Doherty, R.L., Carley, J.G., Filippich, C. and Frazier, C.L. (1978). *Aust. vet. J. 54*, 153.

St George, T.D., Cybinski, D.H., Della-Porta, A.J., McPhee, D.A., Wark, M.C. and Bainbridge, H.M. (1980). The isolation of two bluetongue viruses from healthy cattle in Australia. *Aust. vet. J. 56*, 562.

St George, T.D., Cybinski, D.H., Standfast, H.A., Gard, G.P. and Della-Porta, A.J. (1983). The isolation of five different viruses of the epizootic haemorrhagic disease of deer serogroup. *Aust. vet. J. 60*, 216.

Sugiyama, K., Bishop, D.H.L. and Roy, P. (1981). Analyses of the genomes of bluetongue viruses recovered in the United States. I. Oligonucleotide studies that indicate the existence of naturally occuring reassortant BTV isolates. *Virology 114*, 210.

Swanepoel, R., and Blackburn, N.K. (1976). A new member of the Palyam serogroup of orbiviruses. *Vet. Rec. 99*, 360.

Uren, M.F., and Squire, K.R.E. (1982). The clinico-pathological effect of bluetongue virus serotype 20 in sheep. *Aust. vet. J. 58*, 11.

Verwoerd, D.W., Huismans, H., and Erasmus, B.J. (1979). Orbiviruses. In *Comprehensive Virology, Vol. 14* (H. Fraenkel-Conrat and R.R. Wagner, eds.), pp 285–345, Plenum Press, New York.

Chapter Twenty-one

FOOT AND MOUTH DISEASE

Ulrich Kihm

Foot-and-mouth disease (FMD) is one of the most feared of animal diseases because it is very contagious, and infects and causes severe disease in many important livestock animals. It is a disease of great international importance, and the few countries, including Australia, that are free of the disease are constantly on guard against it. They restrict trade of most animal products and only allow livestock to be imported, if at all, after rigorous laboratory tests and quarantine procedures. Countries where the disease is established sometimes object to such limits on trade, and this has provoked much discussion between governments on how best to control the disease.

There are many excellent descriptions of FMD and the virus which causes it (Brooksby, 1982; Joubert and Mackowiak, 1968; Rohrer, 1967; Gillespie and Timoney, 1981) so I will discuss some aspects of the disease that will be of particular interest to those living in a country where there is no FMD.

HISTORY

The first record of a disease in cattle resembling FMD was made in 1514 in northern Italy. During the 17th and 18th centuries the disease occurred frequently in France, Germany and Italy, and was reported during the 19th century in other European countries, such as Poland, Switzerland and Britain. This spread was probably caused by increasing trade in livestock and the poor sanitary measures operating at the time.

FMD was also recorded in Asia and Africa in the early 19th century, in Argentina from 1860 to 1870, in the U.S.A. in 1870, and in Australia in 1871. There are only a few privileged places in the world where FMD has never been recorded, New Zealand is one of them, and some countries have been able to eradicate the disease; the northern part of Europe, North America, Japan and Australia.

THE VIRUS

What is the agent that causes so much trouble? Loeffler and Frosch in 1897 reported that the agent which caused FMD was not a bacterium and could pass through bacteria-proof filters. This was the first study of a virus of mammals, and FMD virus has become one of the most thoroughly investigated animal viruses.

Several factors make FMDV different from other viruses and complicate its control:

1. it can infect many different animal species, and some carry the virus without showing symptoms;
2. it is extremely contagious, and the infective particles are very stable in saliva, milk and meat;
3. it is antigenically complex and variable.

ITS HOSTS AND EFFECTS

All domesticated and wild cloven-hooved animals, such as cattle, pigs, sheep and goats, are naturally susceptible to infection. In addition a number of other animals including rats, rabbits and suckling mice can be infected experimentally. Interestingly, horses cannot be infected and man has only rarely been infected.

The clinical signs are the result of damage caused by the virus to cells in particular tissues. Vesicles appear in the lining of the mouth, between the hooves and in the coronary regions of the feet, so animals with the disease drool saliva and are very lame. Other parts of the skin, teats, rumen and myocardium may also show signs, and the animal develop a fever of 40–41°C.

In pigs, lesions are found on the snout and, in particular, the feet.

In sheep and goats the lesions in the mouth and on the feet are sometimes small and cause little inconvenience. So the first affected animals in a herd may not be noticed before all are infected.

Usually fewer than 5% of adult animals are killed by the virus, but up to one half of young animals may die. Direct losses are usually less costly than the indirect consequence of infection, such as wasting, the premature end of lactation, a long period of infertility and, above all, widespread restrictions on trade.

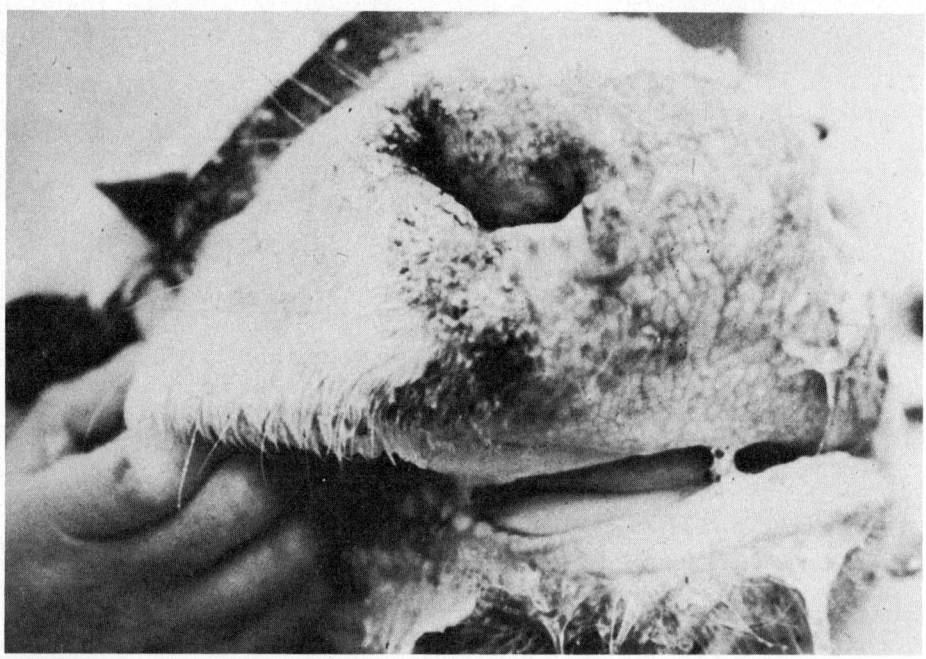

Fig. 1. Foot and Mouth Disease in a cow — lesions on the mouth and muzzle.

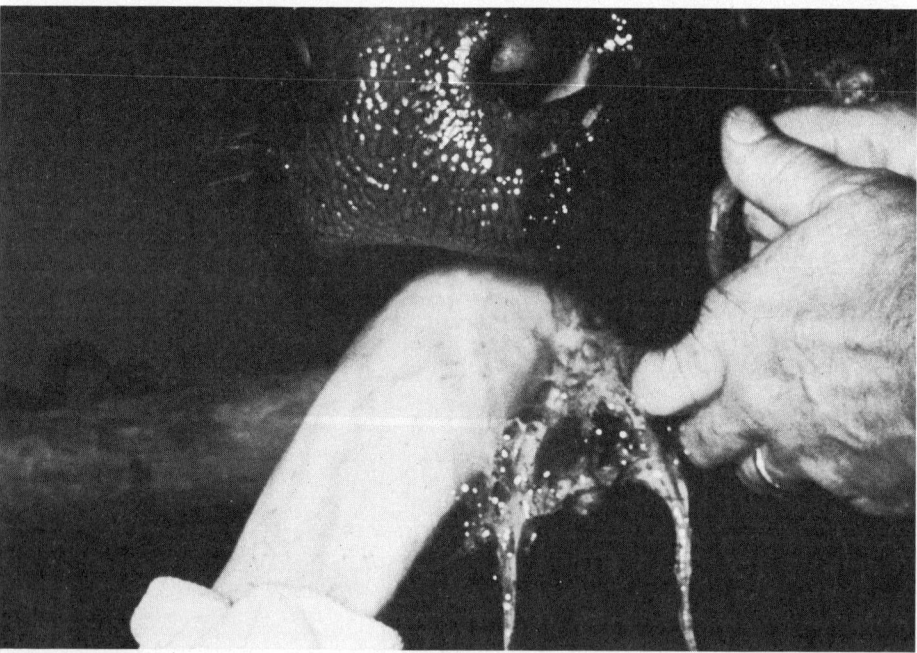

Fig. 2. Foot and Mouth Disease — primary, non ruptured vesicles of the tongue epithelium.

CONTAGIOUSNESS AND TENACITY

Usually FMD infects the nasopharynx first (McVicar et al., 1971), but although no lesions appear in this area, virus may be excreted within a few hours. The virus is then distributed via blood and lymphatic systems throughout the body, but infects and grows particularly well in certain tissues, such as the epithelium lining the mouth. When virus is actively growing and circulating in the body it is found in all secretions and excretions of the animal such as faeces, urine, saliva and milk.

Susceptible animals may be infectd by contact with an infected animal or by contact with contaminated food or bedding, or they may be infected indirectly by droplets (aerosols) (Donaldson, 1979). The virus may be carried by air over long distances.

The virus can persist for long periods on, for example, hay, clothing, hides and skins. However, FMDV particles are very sensitive to acidity, so the virus does not persist in well-matured acid cheeses. However, carcasses of infected cattle may act as virus reservoirs, because although muscle becomes acid after death, lymph nodes and bone marrow do not, hence the importance of trading only boneless beef. Hence meat exported from countries where the virus persists may carry the virus, and many FMD outbreaks have originated from feeding uncooked meat scraps to pigs (Reid, 1968). The chance of spreading the virus throughout the world in meat or meat products, or on the 'hooves' of the growing flood of international tourists is obvious.

CARRIERS AND RESERVOIRS

The existence of 'carriers', animals that after infection recover clinically yet carry the virus, has been demonstrated many times (Bekkum et al., 1959; Bekkum et al., 1966; Sutmoller et al., 1968). In carrier animals, the virus seems to be confined to the nasopharynx and is fully virulent when inoculated to susceptible animals. However there is no direct evidence that it passes naturally from carriers to susceptible contact cattle, so the significance of carriers in the spread of the disease is uncertain.

The importance of susceptible wild animals as reservoirs of FMDV is also uncertain. In regions of Africa where FMDV is widespread, native breeds of stock and wild ruminants develop much milder FMD symptoms than exotic breeds and improved native breeds; the virus and its 'natural' hosts seem to have established an equilibrium (Brooksby, 1982).

ANTIGENIC VARIABILITY, IMMUNOLOGICAL COMPLEXITY

FMD virus is a picornavirus. Its particles are spherical and about 25 nm in diameter. Each particle contains, in its centre, a single long molecule of RNA, which encodes the genetic information of the virus. Around this is a protective shell of protein sub-units, and it is these to which an animal produces antibodies when it is infected, or when it is vaccinated. These particles have a sedimentation coefficient of 146 S. Similar empty particles, lacking the RNA, have a sedimentation coefficient 75 S. Additional virus specific antigens have been identified such as 12 S particles, which are probably the individual sub-units of the protein shell, and the virus-infection-associated antigen (VIA-Antigen). The 146 S particles contain four major structural proteins, called VP1, VP2, VP3 and VP4. VP3 is located mostly on the surface of the virus particle, and hence is the one against which the animal produces antibodies.

There are several antigenically distinct types of FMDV, the most important are the seven serotypes: A, O, C, SAT 1, SAT 2, SAT 3 and ASIA 1. These serotypes mostly occur in different regions of the world and are so distinct that an animal, which has recovered from infection by one serotype remains fully susceptible to infection by any of the other types. The clinical disease caused by different serotypes is the same. In countries where more than one serotype is found, vaccines must contain antigens effective against each serotype. Within each type, there are many subtypes that have smaller antigenic differences. However, some subtypes are immunologically so distinct from others of the same type, that specific subtype vaccines are needed to combat them.

The duration of immunity in recovered or vaccinated animals is, at best, quite short; a few years in cattle and only a few months in pigs. Not all the antibodies produced by recovered or vaccinated animals are able to neutralize the infectivity of FMD virus and protect against infection. Other immunological mechanisms may also be responsible for protection.

DIAGNOSIS

The extreme contagiousness of FMD, and its ability to spread widely even before animals show signs of infection, make it absolutely essential to establish a diagnosis of the disease as rapidly as possible. One should always suspect FMD when a number of animals of a susceptible species develop sore mouths and/or lameness at about the same time. The mouths and feet should be examined carefully for vesicles, and, if any are found, a veterinary officer should be called immediately as confirmatory laboratory tests are essential.

Samples of epithelial tissue, particularly from small and developing lesions, are taken for laboratory tests. The traditional quick method for detecting the presence of FMD virus in extracts of tissue samples is a serological test called the complement fixation test. For a complete test, reference antisera against all known

FMDV serotypes are used. Reference antigens are included to show that the test system is operating properly; non-infectious virus preparations (e.g. inactivated particles) should be used as reference antigens, to avoid any risks (O.I.E. 1984).

Clinical signs, similar to those of FMD, may be caused by other diseases including swine vesicular disease, vesicular stomatitis, vesicular exanthema, bovine virus diarrhoea. Any vesicular disease found in domestic livestock should be considered to be FMD until differential tests have shown it to be otherwise. A well equipped laboratory with well trained staff can do such tests rapidly and accurately.

Another problem is to distinguish between animals that have been vaccinated against FMD and those that have recovered from the disease and might therefore be carriers. This is of importance for those seeking to control FMD in countries where it is endemic, or to determine the 'FMD status' of stock being imported to FMD-free countries. Fortunately it has been found that a virus-infection-associated antigen (VIA antigen) elicites specific antibodies after virus infection (Cowan and Graves, 1966). The VIA antigen is probably the RNA polymerase of the virus; those of different FMD types are serologically indistinguishable. VIA antigen is produced in cells of infected animals. Consequently, animals vaccinated with inactivated virus should not produce VIA antigen or antibodies against it. The ability to distinguish antibodies induced by vaccination and antibodies elicited by FMDV infection is essential in eradication or quarantine programmes or in epidemiological studies.

CONTROL

The goal of every FMD control programme is to eliminate the virus. The style of the campaign will depend on whether the virus is permanently established (enzootic) in the country, or whether it is an occasional migrant. However, for success it is best to use a number of measures in combination. These include:

— a well organized and efficient veterinary service
— good diagnostic facilities
— power to stop transport of animals
— effective disinfectant producers
— controls on important meat or meat products
— good quality vaccine

Quarantine

Control of the imports of animals and animal products, is the primary weapon for those countries free from FMD. The aim is to prevent virus being introduced into country. The import regulations are based on research on the persistence of FMDV in such products under different conditions (Cottral, 1969). The absence of FMD from many countries, including USA and Australia for more than 50 and 100 years respectively, is testimony to the efficacy of quarantine regulations. However even the best regulatory system may fail and so appropriate control procedures must be organized.

Eradication

This is the second defence line of countries, like Australia, usually free from FMD. It is of prime importance to have a diagnosis of the disease as quickly as possible. All infected animals and their immediate contacts must be slaughtered immediately and their carcasses safely disposed. Animals in the neighbourhood must be quarantined, and all transportation of cattle stopped.

The major disadvantages of eradication are the cost and the loss of valuable breeding stock. Eradication is only feasible in regions where the disease occurs sporadically, and where the country is properly administered.

Ring vaccination

A strategy that has been used in some outbreaks of FMD, when slaughter was proving to be ineffective. The scheme is to vaccinate all animals in a zone several kilometres wide around the advancing front of infection. Although cattle would be the first to be vaccinated, other susceptible species, especially pigs, would have to be vaccinated for the method to be effective, and hence this strategy would probably be ineffective in a country, like Australia, which has a large population of feral pigs that cannot be shot let alone vaccinated. The importance of strict quarantine measures in the area of the outbreak cannot be understated, even when vaccination is used. Movement of livestock and of the people who are in contact with livestock should be restricted. The feeding of uncooked garbage to pigs should be prohibited and unpasteurized milk should not be fed to calves and pigs. In several widespread outbreaks in Europe, disease control by vaccination was not very successful because the disease spread faster than animals could be vaccinated. Subsequently it was realized that not enough attention had been paid to sanitation and quarantine because it had been thought that these were unnecessary once cattle had been vaccinated.

Systematic vaccination

A strategy which has been used successfully and is still in use in some European countries. To be effective, animals must be compulsorily vaccinated at regular intervals and, once the incidence of FMD has been decreased, it is best to slaughter any remaining infected stock. This policy led to the dramatic decrease in FMD incidence in Europe from over 30,000 cases in 1965 to some hundred cases in 1983 (Boldrini, 1978; F.A.O., 1983; O.I.E., 1983).

Although FMD has been virtually eradicated from Europe, there have been recent outbreaks when virus has been imported in meat from elsewhere, when the virus has escaped from laboratories and when inadequately inactivated vaccines have been used.

For effective control by vaccination, it seems that as near as possible to 100% of the cattle population, must be vaccinated at regular intervals. It seems also that other measures, such as cooking of swill for pig feeding in a country with a significant pig industry, may be as important as vaccination of cattle.

Last but not least the quality of the vaccine is crucial. The vaccines produced in several countries of the world are of indifferent quality, even though the technology to produce potent vaccines does exist. The vaccine used at present in Europe protects about 80% of vaccinated animals. There is no doubt that vaccines are important in the global campaign to control FMD, but only as an adjunct to slaughter programmes, which unequivocally eliminate the virus.

CONCLUSIONS

Whenever FMD appears in countries, like Australia, that depend upon animal production as a major part of their agricultural economy, its immediate effect and the rapidity of its spread will demand strong measures to be taken. A well organized, well supported and efficient veterinary service with good laboratory diagnostic facilities is essential if FMD is to be controlled and eradicated before it becomes established endemically. The policy of choice for countries or areas free of the disease is to immediately eradicate by slaughter. Countries occasionally infected must maintain preventive programmes and must continue research on control of the disease. Countries in which the disease is enzootic must undertake the greatest effort possible to control and finally to eradicate this virus infection.

REFERENCES

Bekkum, J.G., van Frenkel, H.S., Fredericks, M.H.J. and Frenkel, S. (1959). *Dijdschr. Diergeneesk. 84*, 1159.
Bekkum, J.G., van Straver, P.J., Bool, P.H. and Frenkel, S. (1966). *Bull. Off. intern. Epizoot. 65*, 1949.
Brooksby, J.B. (1982). *Intervirology 18* p. 1–2, 1.
Boldrini, G.M. (1978). *Vet. Rec. 102*, 194.
Cottral, G.E. (1969). *Bull. Off. Intern. Epizootiol. 71*, 549.
Cowan, K.M. and Graves, J.H. (1966). *Virology 30*, 528.
Donaldson, A.I. (1979). *Vet. Bull. 49*, 653.
F.A.O. (1983). *Rapport de la 25eme Session de la Commission Europeanne de lutte contre la fievre aphteuse,* p. 19.
Gillespie, J.H., Timoney, J.F. (1981). Hagan and Brunner's *Infectious Diseases of Domestic Animals,* Section III, The Picornaviridae, Cornell University Press.
Joubert, L., Mackowiak, C. (1968). *La Fièvre Aphteuse,* Vol. I, II, III.
Loeffler, F. and Frosch, P. (1897). *Zentralbl. Bakteriol. (Orig. A) 22*, 257.
O.I.E. (1983). *Situation Zoo-Sanitaire dans les pays membres en 1982,* p. 69.
O.I.E. (1984). *Report on the Meeting of the O.I.E. Foot and Mouth Disease Commission,* p. 1.
Reid, J. (1968). *Vet. Rec. 82*, 286.
Rohrer, L. (1967). Maul and Klavenseuche, *Handbuch der Virusinfelctionein bei Tiefen.* Band 2, Spezieller Teil 1 Jena: VEB Gustav, Fischer Verlag.
Sutmoller, P., McVicar, J.W. and Cottral, G.E. (1968). *Arch. Gesamte Virusforsch. 23*, 227.
McVicar, J.W., Graes, J.H. and Sutmoller, P. (1971). *Proc. 74th Ann. Meeting U.S. Anim. Health Assoc.,* p. 230.

Chapter Twenty-two

LETHAL AVIAN INFLUENZA (H5N2) IN THE USA: IS A SIMILAR OUTBREAK IN AUSTRALIA OR NEW ZEALAND POSSIBLE?

Robert G. Webster, Yoshihiro Kawaoka, William J. Bean, Clayton W. Naeve, John M. Wood and William G. Laver

Influenza virus infections of chickens occur relatively infrequently in the United States; outbreaks of disease that killed many birds occurred in 1924–25 and again in 1929 (Beaudette et al., 1929) and were caused by a fowl plague-like virus. The last highly pathogenic virus outbreak in domestic poultry in North America occurred in turkeys and was caused by A/turkey/Ontario/6632/66(H5N9). This virus was so pathogenic that the outbreak of disease was self-limiting (Lang et al., 1968). Other outbreaks of disease causing mild respiratory infection and few deaths have occurred infrequently in chickens. A self-limiting outbreak of disease in chickens in Alabama in 1975 was associated with a H4N8 influenza virus.

In April 1983, an H5N2 influenza A virus appeared in chickens in Pennsylvania, USA and caused few deaths. Subsequently, in October 1983, virulent influenza viruses were isolated from chickens, that caused up to 100% mortality. The virus spread to Virginia and limited outbreaks of disease occurred in New Jersey and Maryland. From November 1983 to October 1984, the virus was responsible for the destruction of 448 flocks of chickens and turkeys, and limited numbers of guinea fowl and chuckars were also affected. Slaughter of infected poultry flocks by a State, Federal and Industry 'Influenza Task Force' has resulted in the death of over 17 million birds with a value of about 61 million. The Task Force has been successful in eradicating the disease from poultry in Pennsylvania and Virginia.

We decided to study isolates of the virus obtained during this epidemic to see whether molecular studies could shed light on its epidemiology, in particular its sudden increase in virulence. We report here some of our findings.

EXPERIMENTAL INFECTION OF CHICKENS AND VIRUS SHEDDING

The influenza viruses used in this study were provided by Dr James Pearson, National Veterinary Service Laboratory, Ames, Iowa:

a. *A/chick/Penn/1/83* was the virus isolated from the index case in April 1983 and designated UP8125 by NVSL. It was non-pathogenic;

b. *A/chick/Penn/3/83* was isolated in May 1983 and designated 83-25929 by NVSL, and was non-pathogenic;

c. *A/chick/Penn/1370/83* was isolated in October 1983, and was highly pathogenic.

A highly pathogenic virus was defined as one that caused not less than 75% mortality within 8 days when used to infect at least 8 healthy, susceptible chickens (4–8 weeks old), inoculated by the intramuscular, intravenous, or caudal air sac route with bacteria-free infections allantoic or cell culture fluids (Easterday et al., 1981).

To test the pathogenicity after intranasal inoculation, and to study virus shedding, groups of chickens were inoculated with the chick/Penn/83 influenza viruses. The viruses isolated in April and May 1983 (low pathogenic strains) were inoculated into the nasal clefts of adult laying birds (approx. 10^7 EID_{50}/bird), and the birds showed no disease signs. In contrast, when large doses of these viruses were injected intravenously into adult birds, 50% of the animals died by the sixth day after inoculation and virus was recovered from all organs tested, including the brain.

Adult birds inoculated nasally with the highly pathogenic chick/Pen/1370/83 virus (10^4 EID_{50}) died within 4–7 days with obvious clinical signs. Virus was isolated from the blood on the second day after inoculation and from all tissues tested (brain, kidney, lung, intestine); the birds showed signs of damage to the central nervous system and were unable to stand and twisted their necks. The birds usually had swollen heads and wattles with hemorrhage of the intestinal tract, ovary and oviduct. They shed large amounts of virus in their feces with titers of up to 10^7 EID_{50}/gram from the second day after infection (Table 1). Laying hens continued to produce eggs until the day of death, and the last eggs laid by some birds contained much virus in both the egg white and yolk (Table 1). The virus spread between birds in the same cage and between birds in different cages in the same isolation cubicle. By contrast, young birds (5 weeks) inoculated in the nasal cleft with 10^4 EID_{50} took longer to die (5–9 days) with hemorrhage of the legs and edema of the hocks. Disease signs were less apparent in young birds and less virus was shed in the feces (Table 1). Transmission between birds in a cage occurred irregularly and the virus did not spread to birds in adjacent cages.

TABLE 1. *Virus shed from chickens infected with A/chicken/Pennsylvania/1370/83 influenza virus*

	Infectivity titer EID_{50}/gm on days					
	1	2	3	4	5	6
Virus in Feces						
Young — 5 weeks	< 1.0	2.5	3.5	3.8	3.8	dead
Adult hens	< 1.0	6.0	7.2	6.5	dead	
Virus in Eggs		Egg White		Yolk		
		5.6		3.6		

Feces   Groups of 6 chickens were infected with 10^4 EID_{50} and pooled fecal samples were collected over a 30′ period each 24 hours and titrated in embryonated eggs. Infectivity estimate log $_{10}$/gm of feces.

Eggs   Mean infectivity titer from 3 eggs. Two infected birds (10^4 EID_{50}) and 4 contact birds layed 37 eggs before death, virus was isolated only from 3 eggs layed the day the birds died.

The above studies show that the influenza viruses isolated in April/May 1983 differ in virulence from virus isolated in October 1983 and that laying birds show more pronounced disease signs and shed more virus than young birds.

ANTIGENIC CHARACTERIZATION OF THE HEMAGGLUTININS AND NEURAMINIDASES OF VIRUSES

The hemagglutinin and neuraminidase of the influenza virus isolates were characterized to determine their antigenic subtype and to see whether there were any antigenic differences between the viruses isolated in April 1983 and October 1983. All three chick/Penn/83 viruses were inhibited to high titers by monospecific antisera to the hemagglutinin of A/tern/South Africa/61 and by other antisera to H5 subtypes (Table 2). The non-pathogenic (chick/Penn/1/83) and highly pathogenic (chick/Penn/1370/83) isolates were indistinguishable when examined with monospecific or postinfection antisera. The neuraminidase on the chick/Penn/83 viruses was inhibited by monospecific antisera to N2 neuraminidase from human influenza isolates in 1957 (Table 2). The neuraminidase of the chick/Penn/83 viruses was related to the prototype 1957 human N2 neuramindase and typical of the N2 neuraminidases of avian influenza viruses, such as A/turkey/Mass/3740/65 (H5N2). There was no evidence for antigenic differences between the viruses.

ANALYSIS OF VIRAL GENOMIC RNAs FROM CHICK/PENN/83 INFLUENZA VIRUSES

Studies, by Sriram et al. (1980) have shown that avian influenza viruses posessing antigenically indistinguishable hemagglutinin and neuraminidase glycoproteins can differ significantly in the migration of the different parts of their segmented RNA genome. To determine if differences could be detected between the RNAs from the virulent and avirulent chick/Penn/83 viruses, their RNAs were examined by electrophoresis in polyacrylamide gels. The results (Figure 1.) show that the RNA segments from the virulent chick/Penn/1370/83 viruses were indistinguishable from the electrophoretic mobilities of the RNAs from the avirulent chick/Penn/1/83 virus. However, the chick/Penn/1/83 showed additional small molecular weight RNAs not found in the chick/Penn/1370/83 virus. The additional small molecular weight RNA segments found in chick/Penn/1/83 may be defective RNA species like those described by Davis et al. (1980). The similar migration patterns of the virion RNAs of the avirulent and virulent chick/Penn/83 influenza viruses suggests that the virulent strain was probably not derived by genetic reassortment in a mixed infection with other influenza A viruses.

Further analysis of total RNA from the avirulent and virulent viruses by oligonucleotide mapping showed a small number of differences between the avirulent (chick/Penn/1/83) and the virulent virus (chick/Penn/1370/83) (Figure 2). These results confirm that the virulent virus probably originated from the avirulent viruses and that reassortment with other influenza A viruses had not occurred.

TABLE 2. *Characterization of the HA and NA of chick/Pennsylvania/83 influenza viruses*

Antigen	HI titer with antisera to:				
	Ck/Penn/83	Tn/SA/61	DK/Alb/57/76	Sh/Tryon 264C/75	Ty/Ont/66
Ck/Penn/1/83 (H5N2)	320	320	80	< 40	80
Ck/Penn/1370/83 (H5N2)	320	320	80	< 40	80
Tn/SA/61 (H5N3)	80	320	80	< 40	< 40
Dk/Alb/57/76 (H5N3)	160	320	160	< 40	< 40
Sh/Tyron/264C/75 (H5N3)	80	320	80	160	80
Ty/Ont/7732/66	40	80	< 40	< 40	640

Antigen	NI titer with antisera to:		
	RI/5+/57	Aichi/2/68	Texas/1/79
Ck/Penn/1/83 (H5N2)	1000	300	< 50
Ck/Penn/1370/83 (H5N2)	1000	300	< 50
Ty/Mass/3740/65 (H6N2)	1000	< 50	< 50
RI/5+/57 (H2N2)	0	230	< 50
Texas/1/79	< 50	64	1500

Hemagglutination and neuraminidase inhibition (HI and NI) assays were done as described by Webster *et al.* (1981).

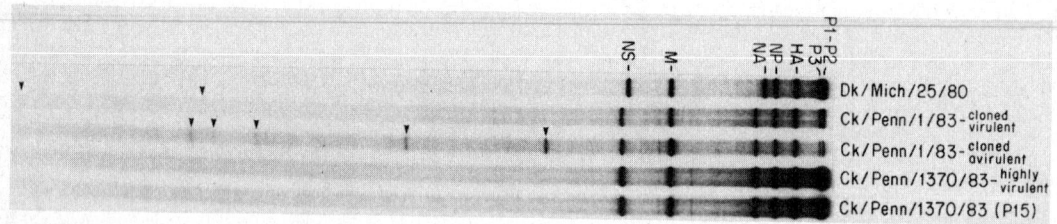

Fig 1. Analysis of the RNAs from virulent and avirulent chick/Penn/83 influenza viruses by polyacrylamide gel electrophoresis. Electrophoresis was done in 3% gels in Tris-borate-EDTA buffer (pH 8.3) and 7 M urea as described by Maxam and Gilbert (1977).

Oligonucleotide mapping of the individual RNA segments showed a small number of changes between the genes of the three viruses analysed. There were as many differences between the oligonucleotide maps of the two avirulent viruses (chick/Penn/1/83) and chick/Penn/3/83) as between the avirulent and virulent viruses. Thus, changes had occurred in more than one gene, and it was not possible to tell which genes were responsible for the acquisition of virulence.

INTERFERENCE *IN VIVO*

The above experiments suggests that defective interfering particles in the avirulent virus might be responsible for its lack of virulence. Therefore studies were done to determine if the avirulent chick/Penn/1/83 (H5N2) influenza virus could modify the severity of disease or mortality caused by the highly virulent chick/Penn/1370/83 influenza virus. Chickens were infected with mixtures of virus containing 10^7 EID$_{50}$ of avirulent virus and 10^4 EID$_{50}$ of virulent virus and the results are shown in Table 3.

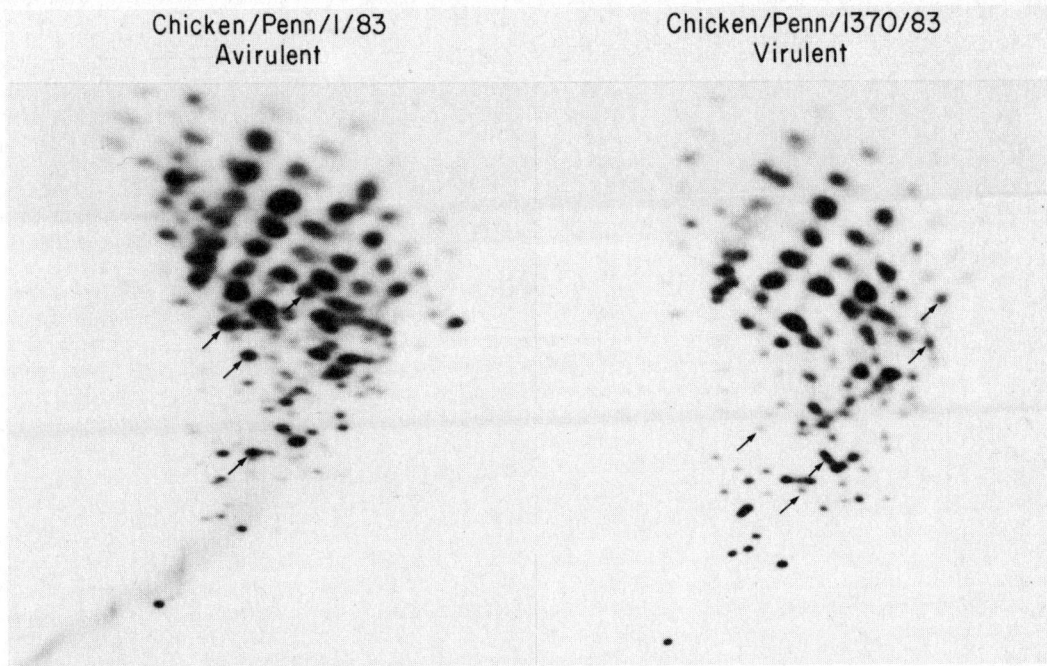

Fig 2. Oligonucleotide mapping of ^{32}P labelled chick/Penn/83 RNA. Viral RNA segments were isolated as described by Bean (1984). T-1 digests of viral genome RNA and isolated genome segments were done by the method of Pedersen and Haseltine (1980) as modified by Lee and Fowlks (1982).

TABLE 3. *Interference between avirulent and virulent chick/Penn/83 influenza viruses in chickens*

H5N2 Influenza Viruses	Dose	Disease Signs No. infected/inoculated (Day)	Mortality dead/inoculated (Day)
Chick/Penn/1/83 + Chick/Penn/1370/83	10^7 + 10^4	4/14 (4–5)	2/14 (5)
Duck/Mich/25/80 + Chick/Penn/1370/83	10^7 + 10^4	6/6 (4)	5/6 (5)
Chick/Penn/1370/83	10^4	10/10 (5–7)	10/10 (6–10)
Duck/Mich/25/80	10^7	0/7	0/7
Chick/Penn/1/83	10^7	0/5	0/5

Groups of White Leghorn chickens (5 weeks) were infected with the above influenza viruses by putting 0.1 ml of inoculum into the nasal cleft. The disease signs included sneezing, rales, hemorrhage of the legs and feet, and swelling of the hocks.

The virulent virus caused 100% mortality, but when mixed with avirulent virus there was a reduction in severity of disease signs and in mortality; 12 of 14 birds infected with the mixture of viruses survived. On the other hand, if virulent chick/Penn/1370/83 influenza virus was mixed with another H5N2 strain (an avirulent strain isolated from wild ducks, A/Michigan/25/80), most (5/6) of the birds died. The genomic RNA of the duck isolate was analysed by electrophoresis in polyacrylamide gels and small molecular weight RNAs were not detected (Figure 1).

These studies indicate that, in mixed infections, avirulent chick/Penn/1/83 influenza virus containing small molecular weight RNAs significantly reduced the mortality caused by the virulent chick/Penn/1370/83 virus, whereas an avirulent virus not containing these RNA species did not cause a significant reduction in mortality.

EXPERIMENTAL INFECTION OF DUCKS

Some of the gene segments of the chick/Penn/83 influenza viruses were genetically closely related to influenza viruses from ducks, and H5N2 viruses have been isolated on several occasions from apparently healthy wild ducks (Hinshaw et al., 1980), therefore we decided to tests whether the chick/Penn/83 viruses would cause disease in ducks. The avirulent (chick/Penn/1/83) and the virulent virus (chick/Penn/1370/83) were inoculated to ducks by the oral-tracheal or rectal routes. After oral-tracheal inoculation, virus was isolated from the trachea of approximately half of the ducks for only one day after infection, but not from the lung, upper or lower intestine, bursa, or rectum. The majority of the ducks were infected, and 2/3 of the animals 'seroconverted' (developed antibodies in their serum against the virus that was inoculated). After rectal inoculation, virus was recovered only from the bursal tissue. Neither the avirulent or virulent chick/Penn/83 viruses caused disease signs in ducks.

SUSCEPTIBILITY OF HUMAN BEINGS, AND THE ISOLATION OF CHICK/PENN/83 INFLUENZA VIRUS FROM FLIES AND PIGS

Attempts were made to isolate influenza viruses from individuals involved in the slaughter of infected chicken flocks, and an H5N2 virus was isolated from two of 40 people immediately after they had left infected chicken houses but not 12 hours later. However none of 109 human beings involved in the slaughter of chickens showed rises in serum antibody titers in paired sera. The results indicate that human beings are not susceptible to infection with H5N2 influenza virus but can act as short term carriers. The H5N2 influenza virus was also isolated from one pig in close contact with infected chickens. This virus will infect pigs experimentally, but there is no evidence for transmission between pigs. The H5N2 virus was isolated from flies caught in the chicken houses of one of the first highly pathogenic virus outbreaks, indicating that they could spread the virus.

DISCUSSION

There are several questions concerning the appearance of chick/Penn/83 influenza viruses and the subsequent disastrous outbreak in domestic poultry in Pennsylvania and Virginia:

1. Where did the virus come from?
2. How did it become virulent?, and
3. What, if any, is the role of 'defective interfering particles?

In additionally, we can ask is it possible that such an outbreak will occur again in the USA or other parts of the world.

Where did the virus come from?

Influenza A viruses, representative of each of the 13 hemagglutinin subtypes and 9 NA subtypes, have been isolated from aquatic birds in North America (Hinshaw et al., 1980), indicating that a large influenza gene pool is maintained in nature. The most likely source of the H5N2 viruses in chickens in Pennsylvania was therefore from wild birds. Competitive RNA-RNA hybridization studies have indicated that the H5N2 viruses we studied are very closely related to, and therefore probably originated from, viruses currently circulating in wild birds. However, so far we have not found a single isolate that is the parent of the virus that appeared in chickens in April, 1983. Another possibility is that the virus arose by recombination and spread to chickens.

Oligonucleotide mapping experiments also indicate that the H5N2 viruses obtained in April and May 1983 were not necessarily the earliest viruses introduced into chickens. There were significant differences between the oligonucleotide maps of these two avirulent viruses, suggesting that it is unlikely that the May isolate was derived by mutation from the April isolate. It is more likely that both viruses were derived from a common ancestor. It is therefore possible that the H5N2 was introduced into chickens before April 1983. An alternative possibility is that the initial virus introduced into chickens was a mixture of viruses and that rapid selection occurred during spread in the chicken population.

How did the virus become highly pathogenic?

There are at least two possible mechanisms by which the low pathogenic strain became a highly pathogenic strain. The first possibility is that the virus present in chickens in April 1983 may have recombined with a second influenza virus and produced the highly pathogenic strain. The second possibility is that mutations occurred in the original virus and viruses possessing these mutations were selected by passage in chickens during the summer of 1983 to give a highly pathogenic form of the virus.

The similarity in RNA migration patterns and inability to distinguish the viruses by competitive hybridization, along with the small number of differences between the oligonucleotide maps of the virus genomes indicates that recombination did not occur in the development of the highly pathogenic strain. The studies suggest that the highly pathogenic strain was probably derived directly from the low pathogenic strain by mutation. Based on the studies to date, we believe that during the summer of 1983, a mutant form of the H5N2 virus, highly pathogenic for chickens, was selected from the avirulent form of the virus.

What if any is the role of 'defective interfering' (DI) particles?

Although there is a considerable body of evidence about the molecular biology of DI particles of influenza viruses, there role in nature is not known (Nayak and Sivasubramanian, 1983). Studies on viruses such as Semliki Forest virus have shown DI particles protect adult mice from the lethal effects of the parental virus (Dimmock and Kennedy, 1978). DI particles have also been shown to suppress the cytopathic effect of standard virus and aid in initiating persistance (Huang and Baltimore, 1970). Oligonucleotide analysis of the small molecular weight RNAs present in the low pathogenic virus, chick/Penn/1/83, indicate that they were probably derived from the polymerase genes, indicating that the small molecular weight bands are from DI particles. The interference between the low pathogenic and high pathogenic viruses demonstrated in these experiments, together with the absence of low molecular weight bands in cloned avirulent virus (Figure 1) supports the notion that DI particles in the original isolate could, in part, be responsible for the reduced mortality caused by this virus in the field.

Virus spread

Despite the introduction of quarantine measures in Lancaster Countary, Pennsylvania in September 1983, the H5N2 influenza virus spread locally and to limited areas of Maryland, New Jersey, and Virginia. There was concern that the virus may have been spread by wildlife or by the wind. Extensive studies, done in collaboration with Dr Frank Hayes, and his staff, of the Southeastern Cooperative Wildlife Disease Study Group, Athens, Georgia, failed to implicate wildlife in transmission and there is no conclusive evidence for aerial spread of virus between chicken houses. The high concentration of virus in fecal material (up to 10^7 EID_{50}/gm) provides a likely source of virus. Mechanical transmission of the virus by human beings and flies was probably the main factor in the spread of this virus. The presence of infectious virus in the albumen and yolk of eggs layed by hens experimentally infected with fowl plague virus has been reported (Moses *et al.*, 1948). The number of eggs infected with H5N2 virus was limited to those layed just before death and could provide another mechanism for transmitting virus to other flocks. Egg shells on smaller farms are frequently given to poultry as a source of calcium.

Potential for similar outbreaks of disease in other parts of the world

The available evidence indicates that the highly pathogenic H5N2 influenza virus in the USA originated from influenza virus circulating in wild aquatic birds. In 1983, a limited outbreak of highly pathogenic avian influenza occurred in Ireland and the causative virus were identified as H5N3 (D. Alexander, personal communication). This outbreak was controlled by slaughter.

Influenza viruses of the H5 subtype have been isolated from shearwaters (*Puffinus* spp.) in Australia (Downie *et al.*, 1977) and the majority of influenza A sybtypes have been isolated from wild ducks or seabirds in New Zealand and Australia (Downie and Laver, 1973; Austin and Hinshaw, 1984; J. McKenzie, personal communications). Thus, the potential exists for the appearance of a highly pathogenic avain influenza viruses in domestic poultry in almost any country. The only approaches to preventing such an incident are to keep wildlife separate from domestic poultry and to practice good animal husbandry.

ACKNOWLEDGEMENTS

We thank the United States Department of Agriculture for cooperation in these studies, both the personnel in the Avian Influenza Task Force and Dr James Pearson, NVSL, Ames, Iowa, for providing the virus strains, and Dr Robert, J. Eckroade from the University of Pennsylvania for providing viruses and for valuable discussions. The authors thank Mary Ann Bigelow, Charlene Carrell, Lisa Newberry, Elizabeth Bordwell, Hunter Fleming and Kenneth Cox for thier excellent technical assistance. This work was supported by U.S. Public Health Research Grant AI 08831 and AI 02649 from the National Institutes of Allergy and Infectious Diseases, Cancer Center Support (CORE) Grant CA 21765, and ALSAC, and was also facilitated by international telephone facilities provided to Dr W.G. Laver by the Australian Overseas Telecommunications Commission.

REFERENCES

Austin, F.J.and Hinshaw, V.S. (1984). *Aust. J. exp. Biol. Med. Sci.* (in press).
Bean, W.J.(1984). *Virology* 133, 438.
Beaudette, F.R., Hudson, C.B. and Saxe, A.H. (1929). *J. Agric. Res.* 49, 83.
Davis, A.R., Hiti, A.L. and Nayak, D.P. (1980). *Proc. Natl. Acad. Sci. USA* 77, 215.
Dimmock, N.J.and Kennedy, D.J.T. (1978). *J. gen. Virol.* 39, 231.

Downie, J.C. and Laver, W.G. (1973). *Virology 51*, 259.
Downie, J.C. Hinshaw, V.S., and Laver, W.G. (1977). *Aust J. exp. Biol. Med. Sci. 55*, 635.
Easterday, B.C., Bachmann, P.A., Bankowski, R.A., Hinshaw, V.S., Land, G., Pearson, J.E., and Rott, R. (1981). In *Proc. First Int. Symp. on Avian Influenza*. p. 7 Beltsville, USA.
Hinshaw, V.S., Webster, R.G., and Turner, B. (1980). *Can. J. Microbiol. 26*, 622.
Huang, A.S. and Baltimore, D. (1970). *Nature 226*, 325.
Lang, G., Narayan, O., rouse, B.I., Ferguson, A.E., and Connell, M.C. (1968). *Canad vet. J. 9*, 151.
Lee, Y.F. and Fowlks, E.R. (1982). *Anal. Biochem. 119*, 224.
Maxam, A.M. and Gelbert, W. (1977). *Proc Natl Acad Sci. USA 74*, 560.
Moses, H.E., Brandly, C.A., Jones, E.E., and Jungherr, E.L. (1948). *Am. J. Vet Res. 9*, 314.
Nayak, D.P. and Sivasubramanian, N. (1983). In *Genetics of Influenza Viruses* (P. Palese and D.W. Kingsbury, eds.) p. 255.
Pedersen, F.S. and Haseltine, W.A. (1980). *Meth. Enzymol. 65*, 680.
Sriram, G., Bean, W.J., Hinshaw, V.S., and Webster, R.G. (1980). *Virology 105*, 592.
Webster, R.G., Hinshaw, V.S., Bean, W.J., van Wyke, K.L., Geraci, J.R., St. Aubin, D.J., and Petursson, G. (1981). *Virology 113*, 712.

Chapter Twenty-three
MYXOMATOSIS
Frank Fenner

Myxomatosis has been included as a case-study in this book because it is caused by a virus exotic to Australia, in an animal pest exotic to Australia, and was deliberately introduced as a method of pest control and worked dramatically well for many years.

I will address four aspects of myxomatosis:

1. The early history of its introduction into Australia (Fenner and Ratcliffe, 1965).
2. Its effects on the population numbers of Australia's major vertebrate pest, the rabbit; and the influence of this on vegetation and pastures (Fenner and Myers, 1978).
3. Briefly, the changes in virus and host during the last 33 years (Fenner, 1983).
4. Its unique nature, in the literal meaning of that word — having no like or equal or parallel.

BACKGROUND

Myxoma virus is a native to the Americas, where it occurs as a benign tumor-producing virus in the tropical forest rabbit in South and Central America (*Sylvilagus brasiliensis*) and the brush rabbit (*Sylvilagus bachmani*) in California. In 1943 Aragao in Brazil showed that it was transmitted mechanically by biting arthropods, notably mosquitoes. This is nothing remarkable; several other viruses of wild animals are transmitted in a similar way. However, myxoma virus had an effect which was unusual, and which had a direct impact on man. As so often happens with infectious diseases of wild animals, their existence is recognized when man or one of his domestic animals is infected. With myxoma the domestic animal was the laboratory rabbit; the place, the Pasteur Institute in Montevideo, Uruguay, and the year 1896 — i.e., at the dawn time of virology. Sanarelli, the Director of the Institute, could hardly overlook what he described as infectious myxomatosis of rabbits, for this 'spontaneous' disease, which spread from one laboratory rabbit to another, was always fatal.

INTRODUCTION INTO AUSTRALIA

Subsequently Dr Henrique de Beaureparie Aragao, of the Oswaldo Cruz Institute in Rio de Janeiro, did a lot of work with the disease in both laboratory rabbits and its natural host. Having read about the severity of the rabbit pest in Australia, Aragao suggested, in 1919, that myxomatosis might be useful for rabbit control. He received what can only be described as a 'wipe-off' from the Australian Institute of Science and Industry:

"... The trade in rabbits, both fresh and frozen, either for local food or for export, has grown to be one of great importance, and popular sentiment here is opposed to the extermination of the rabbit by the use of some virulent organism".

This was a change from 1887, when the Government of New South Wales offered a reward of £25,000 for a method of controlling rabbits, and differed from sentiment after the Second World War, when the Wildlife Survey section of the CSIRO was set up primarily to control rabbits.

About 25 years later, a young Melbourne pediatrician who later became Dame Jean MacNamara, saw myxomatosis in Shope's laboratory at the Rockefeller Institute. Being a country girl and well aware of the depredations of rabbits, she jumped at the idea of rabbit control with myxoma virus. She was also an astute politician, and, through the Australian High Commissioner in London, succeeded in persuading the newly-formed Council for Scientific and Industrial Research (CSIR) that they should try it out. But the Australian Quarantine authorities intervened; in no way would they admit an exotic virus without evidence of its usefulness for rabbit control and its innocuity, as far as other Australian animals, domestic and wild, were

Fig. 1. Rabbit in advanced stage of myxomatosis. The virus strain with which this animal is infected would kill around 70% of affected animals. Note drooping ears (heavy with myxomatous tumours), swollen eyelids and nasal discharge. (Courtesy Professor R. Fenner, ANU.)

concerned. At that time, not much thought seems to have been given to its possible pathogenicity for man. CSIR therefore organized a study of myxomatosis in rabbits in a large enclosure near Cambridge, England, by Sir Charles Martin, who reported that the virus should be suitable for control of a population of rabbits in a restricted area. Meanwhile CSIR scientists, working in England, reported that the virus did not infect native Australian or domesticated animals. Permission was given for importation of the virus into Australia, still under quarantine. A few field trials were done under the direction of Dr Lionel Bull of CSIR, but in 1943 he concluded that:

> "... myxomatosis cannot be used to control rabbit populations under most natural conditions in Australia with any promise of success. Nevertheless, it seems possible that in some parts of Australia under special conditions, including the presence of insect vectors in abundance and the absence of predatory animals, the disease could be used with some promise of temporary control of a rabbit population.".

Further work was rendered impossible by the Second World War. After the war it was found that the rabbit pest had never been worse. CSIRO, the successor to CSIR, established the Wildlife Survey Section in 1949 and in 1950 they began field experiments. In spite of Bull's comments about the value of "an abundance of insect vectors", there was initially a tendency in CSIRO to say that the virus should be used in the arid parts of Australia, where other methods of rabbit control were very difficult. Jean MacNamara entered the scene again, with a series of biting and rather unfair articles castigating the failure of Bull and Ratcliffe, the Officer-in-Charge of the Wildlife Survey Section, to try myxomatosis in well-watered country.

In May 1950, the virus, which was still under quarantine, was liberated at several sites in the Murray Valley. The disease remained localized and then apparently died out. However, in December 1950 myxomatosis flared up at one of the control sites and was simultaneously reported at several points along the river. In the following few weeks it spread with dramatic speed and unparalleled intensity over the Murray-Darling River system. Aided by farmers carrying and releasing diseased rabbits, and by widespread inoculations by officers of the State Departments of Agriculture, myxomatosis was soon established wherever there were rabbits in Australia, and has remained enzootic throughout the continent ever since.

THE EFFECTS OF MYXOMATOSIS

The effects of myxomatosis on Australian wild rabbits were studied at Lake Urana in the Riverina. Regular rabbit counts were made at dusk — about 5,000 rabbits in a standard transect on the sandy lake bank. Several hundred of these were captured, inoculated and released during September and October, and in October a few natural cases were observed. By the end of the summer the population count on the same transect had fallen to 50.

Fig. 2. Before myxomatosis. Rabbit proof fence dividing heavily infested paddock from one in which rabbits had been eliminated. (From Country Life.)

Because of the emergence of less virulent strains of virus and the genetic resistance in the rabbit, these enormous mortalities were not maintained. But there was a sustained reduction in rabbit numbers, which produced major ecological changes. Dr Ken Myers of CSIRO carried out quinquennial surveys of rabbit numbers between 1950 and 1975 of three 100-square mile sample areas in the valuable agricultural region of the eastern Riverina. In 1950-51, just before myxomatosis first swept through the region, nearly one third of the region was rabbit infested (38%, 26%, and 52% of the sample areas). It was common then to see morning swarms of rabbits on many of the properties. In 1975 the same three areas showed infestation rates of 0.26%, 0.04% and 13% respectively. There were equally spectacular changes in the pattern of distribution, with the disappearance of rabbits from open agricultural lands and their restriction to timbered and rocky hills, creek frontages and stands of *Callitris* in sandy soils. The range was restricted to about 5% of what it was before myxomatosis.

In 1953 it was calculated that there was an increase in the wool clip worth £24 million, plus some £10 million representing increases in sheep numbers. To this should be added the increased income from increased beef, dairy and crop production and the greatly reduced costs of rabbit control. These advantages are now built into the system, but nevertheless represent an enormous contribution to the wealth of Australia.

Most important of all, perhaps, is that myxomatosis converted what looked in 1949 like an impossible task, bringing rabbits in agricultural lands under control, into a reality — and has continued to make it possible ever since. However, there is still a problem in semi-arid and arid Australia.

CHANGES IN VIRUS AND HOST SINCE 1950

The initial virulence (summarized in a single measure, the mortality, or as epidemiologists say, the 'case fatality rate') was unbelievable. Something like 2 to 5 out of every *thousand* rabbits infected in the first outbreak survived. Later, mutant strains of virus arose, that we called "attenuated", which had a 90% case fatality rate.

For the last thirty years about 70% of strains of myxoma virus recovered from the field fall into the "90% mortality" group measured in genetically unselected laboratory rabbits. One result of the survival of 10% of rabbits is that there are enough rabbits left for the selection of genetic resistance to operate. Where outbreaks were an annual summer event, selection acted rapidly and in such areas most rabbits infected with the original highly virulent strain now survive. In such places more virulent virus strains are now being recovered, for what is selected is a virus that causes the optimum kind of disease for transmission — not too rapidly lethal, so that live rabbits have lesions for mosquitoes to bite, and not too mild a disease, in which the lesions heal rapidly and so become non-infectious.

Many other factors affect the outcome of a virus-rabbit interaction. Hot weather favours survival. Very young rabbits are highly susceptible. Rabbits with myxomatosis are easy prey for predators. Nowadays, myxomatosis doesn't often cause the dramatic effects of the early years, for in most places rabbits are much less common. Occasional very destructive outbreaks still occur, usually where most of the rabbits are kittens and the weather is cool. Myxomatosis is an inescapable part of the ecology of the rabbit in Australia, which tends to keep populations in most parts of the continent much smaller than they were in 1950.

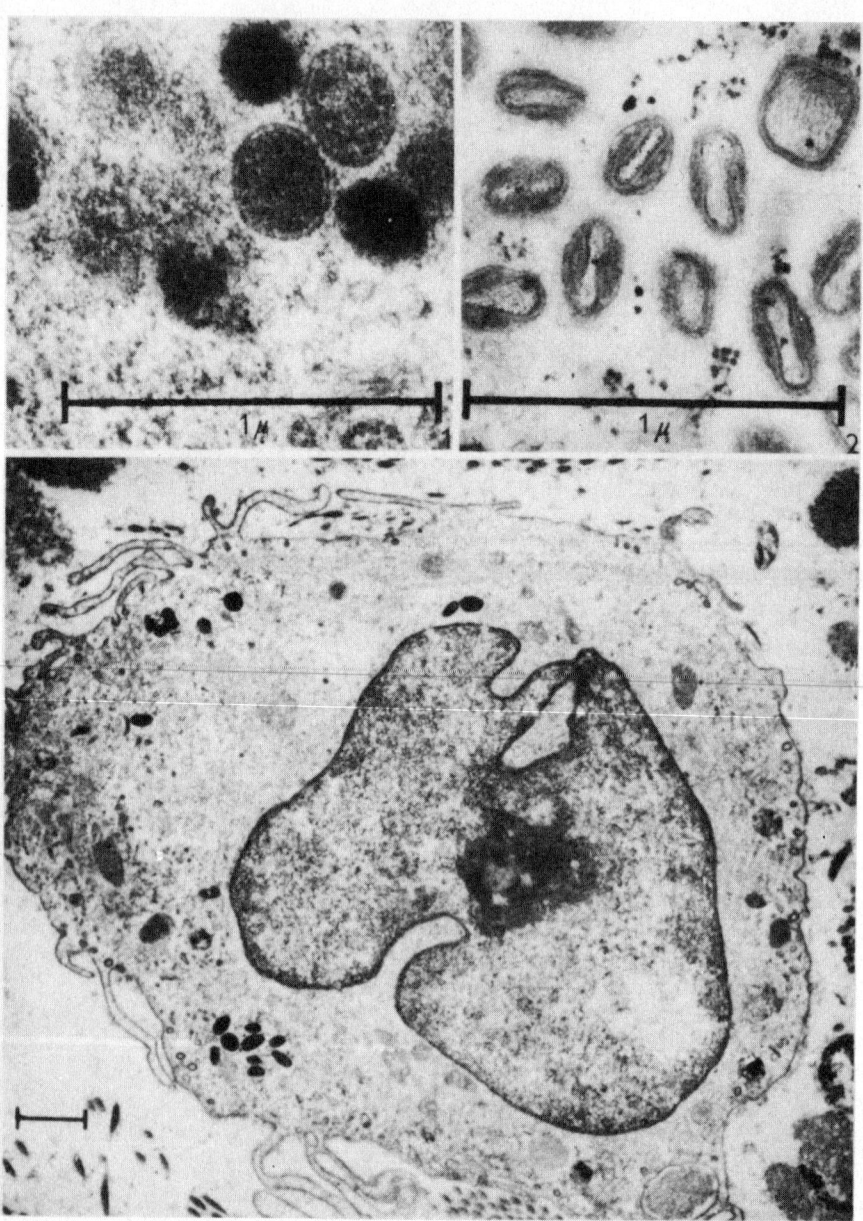

Fig. 3. Thin section electronmicrographs of myxomatous infected cell (main photo) 1. Immature virus particles in cell cytoplasm 2. Mature virus particles in cell cytoplasm. Bars ^{3} 1½. (from Myxomatosis, Fenner, F. and Radcliff, F.N., (1965)).

THE UNIQUE NATURE OF MYXOMATOSIS

Impressed by the spectacular success of myxomatosis, the question arises — what next, for rats or mice or sparrows or starlings or whatever man regards as a serious pest? I have to answer that I don't think that there is anything else that might be expected to repeat the myxomatosis story. Consider the prerequisites:

1. very host-specific, not causing disease in many or any animal that we value, wild or domesticated;
2. very high case fatality rates — over 99% before evolution began to operate;
3. very effective natural transmission, by vectors which could travel for considerable distances.

I know of no other disease with these characteristics, of any kind of animal, let alone one of the few that man regards as a serious pest. Myxomatosis was, and is, unique, a 'one-off'.

REFERENCES

Fenner, F.(1983). *Proc. Roy. Soc. B 218*, 259.
Fenner, F.and Ratcliffe, F.N. (1965). *Myxomatosis*. Cambridge University Press, Cambridge.
Fenner, F.and Myers, K.M. (1978). In *Viruses and Environment*, (E. Kurstak and K. Maramorosch, eds) p. 539 Academic Press, New York.

Chapter Twenty-four

ECHIUM; CURSE OR SALVATION

Linton Briggs

This paper summarizes a debate that has questioned, in a particular case, the notion that an incursion of an exotic weed necessarily detracts from the public benefit. The Australian Beekeeping Industry has contended throughout long discussions that exotic species of the genus *Echium* are of value to Australia and should not be the subject of a National Biological Control Program, unless it could be demonstrated through public scrutiny that control would add to the public benefit.

PATERSON'S CURSE: THE PLANT

Echium plantagineum, formerly *E. lycopsis*, is an annual herb which is also commonly known as Salvation Jane (because of its use during drought as fodder for sheep and cattle), Purple Bugloss, and Lady Campbell Weed. Viper's Bugloss, (*E. vulgare* L.), a closely related species, and Italian Bugloss (*E. italicum* L.) are other *Echium* species found in Australia; *vulgare* occurs in most States although not as widespread as *plantagineum, italicum* is also found in most states, but much more sporadically. The generic name, *Echium*, comes from the Greek word meaning viper, because in ancient times, it was believed the plant's roots, leaves and seeds would, if eaten, protect against and cure snake bite. *Echium* species are native to the Western Mediteranean region of Europe.

E. plantagineum is mostly an annual. However, in some places and in some seasons it may be biennial or even perennial. Seedlings emerge after substantial summer or autumn rains, and develop a rosette of leaves (Fig 1). In the following spring, flower stems up to 0.2 m high emerge fromthe rosettes of mature plants. *E. vulgare* matures later and may, in the N.S.W. highlands, flower well into January. The flowers are commonly purple although blue, pink and occasionally white flowering plants occur. They yield copious amounts of nectar, and a purple pollen which contains up to 40% crude protein.

Research at the Victorian Keith Turnbull Institute during the 1970's showed the plant to be palatable to sheep and cattle, to have a nutritive value similar to tht of subterranean clover, and, when growing among other species, to contribute substantially to pasture yield. Much seed is produced; up to 30,000 seeds per square metre have been recorded as a result of many years seed production. Seeds are dispersed by birds, grazing animals, human beings and surface water. They are usually black, about 2mm across and can remain dormant in the soil for at least five years, germinating erratically. The population can therefore survive despite seedling mortality in several successive years. It is these seeding characteristics, and the deeply penetrating tap root system of *Echium* which make it such a successful plant particularly during and following drought.

Echium successfully competes with other pasture species particularly in moist, fertile soils when undergrazed, and may become dominant unless deliberately controlled. In this way the plant diminishes the yield of other species, and, as it spreads so readily, *Echium plantagineum* has been proclaimed a noxious weed in many parts of Australia. Obviously, prior to research into *Echium* during the 1970's there was little real information or understanding abouut *Echium's* pasture nutrition status.

The presence of pyrrolizidine alkaloids in *Echium plantagineum* is well known. Occasional tests have been made to determine if the alkaloids have any effect on grazing sheep and cattle and other animals. Since the start of the debate on whether the biological control of *Echium plantagineum* should proceed or not, there have been more studies on this subject in New South Wales and Victoria.

THE *ECHIUM* INVASION

Echium species are believed to be natives of Southern Spain, Portugal, and Morocco. They are now widespread throughout the world especially in areas with Mediteranean climate. *Echium vulgare* is well established in New Zealand where the spread of the plant is encouraged, particularly in high country, to provide cattle fodder and prevent gully erosion. It is also valued greatly as a source of food for honeybees in that country.

Echium species are thought to have been introduced to England in 1658 and from there to Australia as a flower garden species in the 1850's. The earliest recorded plantings of *Echium plantagineum* in Australia were by a Mr Paterson in about 1880, near Albury, N.S.W. In 1881, *Echium* seeds were listed in the flower seed catalogue of Mr Heyne, an Adelaide nurseryman, and its deliberate spread was encouraged. Soon after 1900, Lady Campbell is recorded as having grown *Echium* in Western Australia. Seed was also undoubtedly incidentally brought to Australia by ships from ports in England, Madiera, the Canary Islands, South Africa and, after the Suez Canal opened in 1869, from Mediterranean ports. It is also thought that *Echium* seeds were brought to Australia in canary and cumin seed imported from Morocco and France.

Echium is thought to have first naturalised in Victoria and New South Wales in the late 1850's, and in South Australia around 1890. *Echium plantagineum* is now common in all Australian states, and most abundant in New South Wales, South Australia, parts of north east Victoria, and the south west of Western Australia.

In Victoria, *Echium plantagineum* was first proclaimed a noxious weed in the Shire of Towong (Tallangatta) in 1904, for the Shire of Maldon in 1908, and for the whole state in 1911. In 1969 *Echium vulgare* was similarly proclaimed. *Echium* species have been proclaimed noxious in other states, although in some, not for all regions.

Historically, there have been conflicting attitudes towards *Echium* species among farmers and some Government bodies. Some have regarded *Echium* as a weed that should be controlled because it is a conspicuous volunteer, allegedly as a low fodder value, and dominates other highly regarded pasture species. Others have accepted *Echium* species as useful pasture plants, particularly for sheep in lower rainfall country.

Various methods of control have been practiced such as grazing with sheep, cultivation, and spraying with herbicides. However, biological control, if effective, would offer the most efficient from of control, particularly in terms of cost and effort.

ECHIUM, A RESOURCE FOR AUSTRALIAN HONEYBEES

Echium has been worked commercially by Australian beekeepers for about eighty years. As the honey industry has developed and expanded, and *Echium* has spread in south east Australia, beekeepers have made use of this resource. The FCAAA estimates that honey produced from *Echium* constitutes up to 15% of the national crop. Most of this comes from areas that suit the plant, namely the Riverina and Central regions of New South Wales, and the mid north of South Australia. In Victoria, comparatively little *Echium* is worked by beekeepers because the plant is not plentiful enough; most Victorian commercial apiarists work the N.S.W. Riverina region. Nonetheless Victorian land owners and weed controllers have always expressed concern about the possible expansion of *Echium*, and particularly since the 1982 drought, landowners in some of north east Victorian hill districts, where access is difficult, have reported that it is occurring.

Several plants growing in south east Australian pastures are useful to the beekeeping industry, and also support feral honeybees. Such plants include clovers and trefoils, capeweed, onion weed, thistles, lucerne, *Erodium* species, ryegrass, Paterson's Curse, etc, and in recent years, some oil seed crops have shown potential. However, in south east Australia, by far the most important pasture plant for honeybees is *Echium plantagineum*. Its pollen has a high protein content, an important factor in brood rearing. In most springs it provides the vital early food that builds the large honeybee colonies required to harvest efficiently the main nectar flows. In addition, *Echium* provides in most years a useful crop of light premium grade honey that is sought by packers for its outstanding blending properties. Although *Echium* is only one of the honey crops used by most migrating Australian beekeepers, some harvest from it consistently, and in some years, it is the only spring crop available to beekeepers in south east Australia.

Native *Eucalyptus* species provide the main honey for Australian beekeepers. Few eucalypts flower at the same time as *Echium plantagineum*, and then not every year. Furthermore the eucalypt population is gradually diminishing from farmland Australia because of land clearing, insect damage, salinity, and other man induced die back pressures.

To summarize, it seems likely that if a biological programme effectively controlled *Echium* species in south east Australia, sections of the Australian beekeeping industry would be economically and therefore socially disrupted.

THE BIOLOGICAL CONTROL PROGRAM

Until the mid 1970's, the Standing Committee on Agriculture (S.C.A.) Australian Weeds Committee (A.W.C.) listed Paterson's Curse as a major weed requiring control. On several occasions the A.W.C. endorsed research in the Division of Entomology of the Commonwealth Scientific and Industrial Research Organisation (C.S.I.R.O.) to obtain overseas agents that could control *Echium* in Australia. Although some awareness existed within government departments that *Echium* species were valued by some sections of the community, it was not appreciated how this extended until the complex debate erupted in 1978.

By 1974 it had become evident that prospects for effective biological control of *Echium* in Australia were good. CSIRO workers at Montpellier, in France, had identified four insect species whose specificity to *Echium* species was satisfactory, and whose ability to reduce *Echium* had been demonstrated in laboratory and field experiments. In 1974, the CSIRO Division of Entomology approached all State Departments of Agriculture, seeking their views on whether the insects should be released in Australia, and the States unanimously agreed. They believed that the programme would benefit the nation, though the South Australian Department of Agriculture noted at the time that some minor sections of the community would be affected by the proposed program. Even though the beekeeping industry is serviced as an industry by all State Departments of Agriculture, the industry was not consulted about the effect the proposed biological control program might have on it.

In March 1978, the Federal Council of Australian Apiarists Associations heard of the proposed control programme, now well advanced, and with the insects about to be imported. Strong representation to the Australian Agricultural Council by the FCAAA identified that there was a serious conflict of interest, and that the traditional decision making process was clearly inadequate. There followed a complex and controversial debate which at first involved AAC deciding not to proceed with the program, and then a reversal of that position by AAC. The debate is still unresolved. The Australian beekeeping industry has consistently

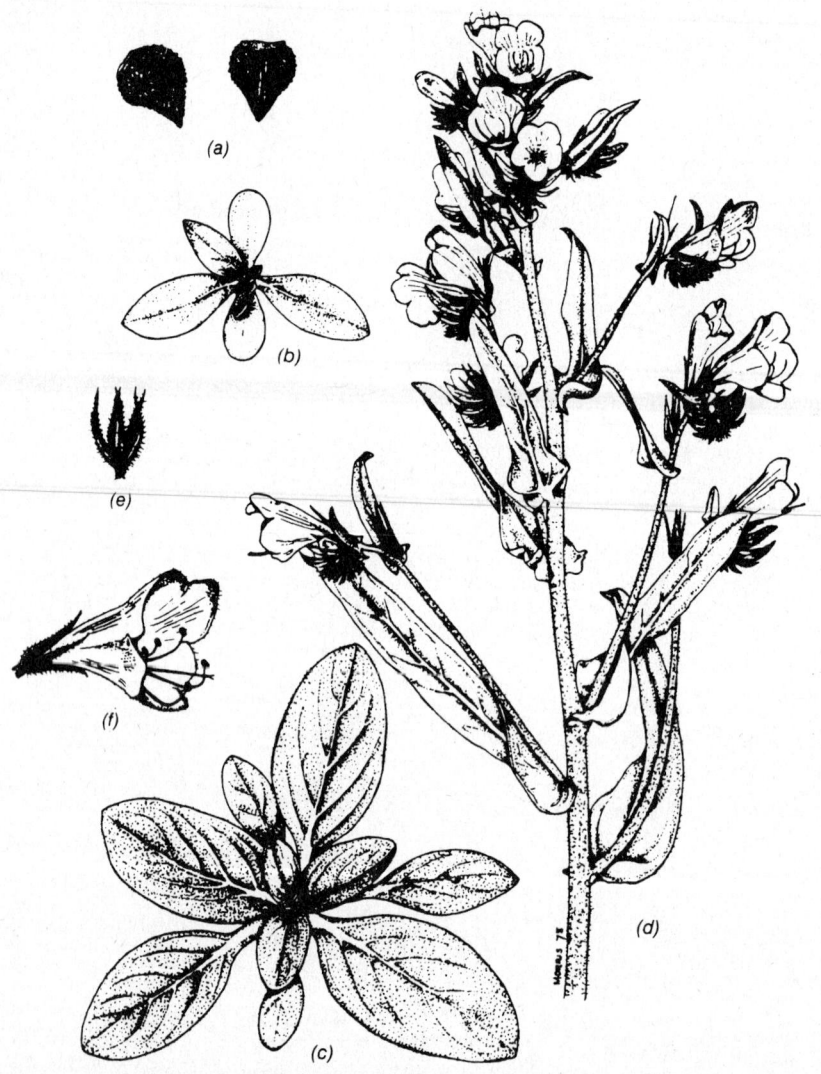

Fig. 1. Detail of *E. plantagineum*. *(a)* seed 7; *(b)* seedling 1; *(c)* rosette 1; *(d)* flowering branch 1; *(e)* calyx 2; *(f)* flower 2. [*(b), (c)* and *(d)* from Hyde-Wyatt and Morris (1980)].

questioned not only the overall value of controlling *Echium* by biological agents, but also the decision making processes which have led to the proposed programme. Other Chapters two chapters highlight the basis for such questioning.

ECHIUM AS A PASTURE PLANT

In the mid 1960's, farmers and shire councils in the north east of Victoria, where Paterson's Curse seemed to be spreading, noted the lack of information on the growth and control of the plant. As a result, a project on its ecology and control was started at the Keith Turnbull Research Institue by Dr C.M. Piggin. The report of this study was published in June 1976.

Dr Piggin reported that the plant has a high growth rate and produced large quantities of herbage, especially when nitrogen and phosphorus fertilizers were applied and/or grazing pressure was low. Consequently, the plant grew well in fertilized, undergrazed pastures and ungrazed roadsides. The total herbage produced by pure Paterson's Curse or mixed Paterson's Curse/subterranean clover pastures was much greater than from pure subterranean clover pastures. In mixtures, there was some competition, and subterranean clover was suppressed especially when there was no grazing (e.g. on roadsides), when fertilizers had been added, or when Paterson's Curse established before subterranean clover. However, competition between the species was not important because stock readily ate Paterson's Curse, and in well managed pastures the species became balanced. For example, in a Rutherglen trial in winter of 1970, continuous grazing by 10 sheep per hectare reduced Paterson's Curse cover in a pasture from 30 to 3% and increased subterranean clover from 20 to 50%.

The nutritive value of Paterson's Curse was similar to that of sub clover throughout the growing season; greatest early in the season and falling as the plants mature. For Paterson's Curse, values ranging from 5–20% dry matter, 55–70% digestibility and 15–40% protein have been recorded, while corresponding values for subterranean clover were 10–15% dry matter, 6–70% digestibility and 25–40% protein. Sheep and cattle readily ate the plant, though as cattle cannot graze the rosettes are closely to the ground as sheep, the plant was often more noticeable in areas grazed by cattle.

Dr Piggin concluded that the plant only became dominant when inadequately grazed, and was a useful pasture species because it produced large quantities of nutritious herbage, especially in *adverse* seasons.

LIKELY EFFECTS OF *ECHIUM* CONTROL ON FARM PRODUCTIVITY

The indirect effect of honeybees on other branches of agriculture grows in significance, year by year. The University of Adelaide, twenty years ago, assessed the agricultural value of incidental pollination by honeybees at $100,000,000 per year. It may be $350,000,000 per year by now. The FCAAA is pressing for more scientific study and assessment of the pollination question. Many crop and pasture species depend on honeybees for pollination; inadequate pollination invariably means reduced seed yields and hence decreased profits, and an increasing number of farmers are seeking the aid of beekeepers. The FCAAA believes that the indispensable role of honeybees for the successful growing of many crops and pasture species will become even more widely realized in the future.

Echium species provide an important link in the chain of factors which maintain a viable Australian beekeeping industry, and a viable pollinating industry. A significant decrease in the *Echium* population in many areas would greatly diminish honeybees populations, both commercial and feral. *Echium* species provide the only regular honey resource sufficiently large to maintain feral honeybee populations of the size useful to farmers, except where large stands of native timber still remain. Furthermore, a decrease of regular visits through annual migration of commercial hives, which swarm and replenish the feral honeybee populatons, combined with an significant increase in the incidence of diseases and use of insecticides, would inevitably lead to a dramatic decline in the feral honeybee population in *Echium* regions.

To summarize, *Echium* helps to maintain in various ways feral and commercial honeybees populations which mainly through incidental pollination maximize production from those agricultural crops which benefit to varying degrees from insect pollination.

THE PRE LITIGATION DEBATE

In August 1978, the CSIRO Division of Entomology, recognizing that a clear dispute had arisen over the status of *Echium*, proposed to the Australian Weeds Committee that the Standing Committee on Agriculture (SCA) should form a working party comprised of all interested parties to resolve the dispute. Thus, CSIRO hoped to be given a path along which it could confidently tread. The Division undertook not to proceed with the release of insects until the proposed working party had concluded its task, nor to proceed even then unless the States were unanimously in favour of any proposed action. In August 1978, at its 114th meeting, the SCA rejected the proposal.

However, no State other than New South Wales, which strongly favoured the programme, supported the release of control agents, and some States doubted whether the proclamation of Paterson's Curse as a noxious weed had been justified. The SCA at its 114th meeting did, however, support a second proposal, put forward by its Plant Production Committee and CSIRO to overhaul the mechanisms for deciding which weed species should be the subject of biological control programmes. They suggested that the discussions should involve the SCA's Plant Production Committee and other technical sub-committees, together with relevant Commonwealth and State research groups. This proposal was endorsed later by the Australian Agricultural Council (AAC).

In August 1978, the 105th meeting of AAC considered the initial representations made by the FCAAA on behalf of the Australian honey industry, and by a majority of 5–1 held that the programme should not proceed. However, the SCA had already decided to determine whether testing and/or release of biological control agents for *Echium* should proceed.

Accordingly, the N.S.W. Department of Agriculture was asked to prepare a paper outlining the pluses and minuses of the proposed programme against *Echium*. A working party involving other States assisted N.S.W. The paper was presented at the 116th meeting of the SCA in January 1979 in Christchurch, New Zealand. N.S.W. argued strongly for control programme to proceed, and on this occasion there was unanimous support within SCA for the programme to proceed. The 106th meeting of AAC in January 1979 endorsed this view, reversing its previous decision. Both the SCA and AAC considered that the evidence presented by the N.S.W. Department of Agriculture indicated that the public interest would be best served if control was to proceed. On the basis of this judgement, AAC requested CSIRO to proceed with releasing the chosen insects.

After the Christchurch meetings, the FCAAA, in conference with the chairman of the AAC, argued that the N.S.W. Department of Agriculture paper was inadequate in some respects, and should not have been the basis for such an important decision. The chairman of AAC agreed that the matter would again be discussed at the next meeting of AAC. At the 107th meeting of AAC in Perth during August 1979, the South Australian member of AAC advised the meeting that his State was preparing a cost benefit study to further investigate the issue, and was given an understanding that the study would be considered by AAC at the meeting in January 1980. However, a change of Government in South Australia occurred before this meeting, and the initiative lost momentum. The AAC subsequently decided that the control programme should proceed.

To summarize. The debate revealed that a quite inadequate amount of properly substantiated evidence had been assembled to assess the value of a biological control programme for Paterson's Curse. Furthermore, the decision making process:

1. Made no provision for the public advertisement of an intention to proceed with biological control for a targeted pest species.
2. Made no provision of equitable ways to examine conflicts that might arise from a proposal to proceed with biological control.

LITIGATION

By April 1980, the FCAAA had exhausted all avenues for reasoned argument. Experimental field releases by CSIRO of *Dialectica scallariella*, a moth whose larvae are leaf miners of *Echium*, were imminent. Although the FCAAA had considered the possibility of litigation for some time, it had consciously decided to avoid such action unless supported by those sections of the grazing community who saw *Echium* as an overall pasture contributor.

In June 1980, graziers, beekeepers and other concerned people met in Adelaide to consider what further steps should be taken in the dispute. Legal representatives attended the meeting as advisors. The meeting heard that the leaf mining moth had been released in N.S.W. near Deniliquin, Jugiong and Braidwood. The meeting was advised that a strong legal case could be prepared on the narrow legal grounds of detriment, and that much evidence had been collected and would provide a basis for broader legal argument. The meeting decided to pursue legal action to stop the biological control programme; CSIRO was named as defendant in action, and two grazier plaintiffs (from S.A. and N.S.W.), and two beekeeper plaintiffs (from S.A. and Victoria) represented the south east Australian grazing and beekeeping communities.

In July 1980, the plaintiffs sought an injunction in the High Court of Australia to restrain CSIRO from importing from Europe further specified insects for the control proramme and from releasing them in Australia, and to have the existing stock of insects destroyed. The High Court of Australia, with Mr Justice Stephen presiding, granted the plaintiffs an interim injunction. Later in July 1980, Mr Justice Stephen continued the order for the interim injunction. The CSIRO agreed to encase or enclose any plants at the sites where the insects had been released, and when all insect life at the sites was extinct, to remove and destroy the material.

In August 1980, legal proceedings were transferred to the South Australian Supreme Court, exercising federal jurisdiction. In September 1980, Mr Justice Zelling in the Supreme Court of South Australia continued Mr Justice Stephen's order for an injuunction and trial. In May 1982, the hearing before Mr Justice Zelling began, but was adjourned by mutual consent of the parties to discuss possible alternative forms of resolution. Both parties were interested in an independent equitable and broadly based inquiry that would hear evdience from all interested parties with the view of resolving the dispute on purely public benefit grounds. CSIRO had found themselves embroiled in a situation over which, in the beginning, they had been unable to exercise much control. The plaintiffs, although in a position of considerable legal strength, were prepared to negotiate towards, in this particular case, a decision making process which provided the fairest consideration of all points of view.

In June 1982, agreement was reached to set up a three person independent tribunal to determine whether or not to proceed with the biological control of *Echium* species in Australia. However, all efforts to establish the inquiry failed, despite exhaustive effort, and some parties began to focus attention on ways to enact legislation which would enable the authorisation of biological control programs. Such proposed legislation would be designed to supercede or override the common law rights of individuals or minority groups opposed to biological control programs, when such programs could be demonstrated to be in the public interest.

In February 1983, the SCA established a working party to suggest a form of legislation whose intent would be to enable biological control programmes to be positively approved. In June 1983, litigation ceased after three years when the South Australian Supreme Court gave consent to an agreement between the parties. CSIRO agreed to use its best endeavours to ensure the plaintiffs views were properly considered during the development of legislation. CSIRO was ordered by the Supreme Court to pay costs of the litigation amounting to $93,000.

LEGISLATION

The Federal Government will probably introduce legislation to the House of Representative during 1984, and the FCAAA has been encouraged by the fact that its views were sought while the legislation was being drafted. Complementary State legislation will be required to make the proposed legislation effective.

Details of the legislation are not known but it is understood to be broad in scope, designed to embrace the biological control of any targeted weed or animal pest, and that it will include such essential elements as the need for public advertisement of an intention to proceed with biological control, and, where a conflict of interest is identified, that such dispute may be referred to a public inquiry for an assessment of where the true public interest may lie.

Such processes will be welcomed by the beekeeping industry. The industry has specific views on the structure of an inquiry to examine the *Echium* question, and looks forward to public discussion of the proposed legislation.

POSTSCRIPT

During the long and emotional debate on *Echium* control, considerable hostility from some sections of the rural community has been directed towards those grazing and beekeeping interests that opposed the proposed programme. This is unfortunate, because the programme has been opposed not only because of likely direct economic losses but also to question whether, in the light of all the available information, the program will be in the overall public interest.

The hostile reaction is understandable. Any landowner who believes that the productivity of his or her farm will be improved by a cost efficient and effective form of control for *Echium*, is likely to be hostile. This unfortunate consequence was well illustrated in 1981 when beekeepers from Victoria and New South Wales were refused their historic access to public land in N.S.W., on which to operate their apiaries. The Albury Pasture Protection Board Reserves refused permission for occupancy during the spring of 1981, on the grounds that while an injunction remained in force restraining the CSIRO from releasing agents to control *Echium*, then permission to occupy reserves would be denied. Reasoned argument by the FCAAA failed to change the Board's attitude. The Board remained hostile to beekeepers and other farmers who supported the injunction. The matter was not resolved until the FCAAA appealed to the N.S.W. State Government, and argued that the Board's decision appealed was emotionally based and against the spirit of the provisions of the Pastures Protection Board Act. Access to the reserves was then restored.

A more positive social interaction has been the interest of the Australian Academy of Science in the complex debate. The Academy in consultation with the Victorian Vermin and Noxious Weeds Destruction Board, the Keith Turnbull Institute, and the FCAAA developed an audio-visual 'kit' based on the *Echium* debate to help secondary students to consider a social issue that involves biology, environmental science and environmental studies courses.

The most important social consequence the debate may ultimately have on the Australian community may be to change, and hopefully improve, an existing decision making process through legislation. The proposed legislation could well serve as a model for resolving other similar disputes. The sensible resolution of community conflicts of interest through community funded public inquiries seems an optimum democratic goal. In striving for such perfection, it must always remain a fundamental aim that the rights of individuals and minority groups to express their views and receive equitable treatment is strictly preserved.

CONCLUSION

The development of biological control systems for Australian animal and plant pest organisms, approved as being environmentally sound and in the overall public interest, has the full support of the Australian beekeeping industry.

The decision to subject *Echium* species in Australia to biological control was not properly based on a factual identification of the true public interest. The AAC Christchurch decision, reversing the previous position taken by the AAC was based largely on a document now widely held to be inadequate as a determinant of the true public benefits or costs of such a program.

The current legislative based, community funded inquiry into whether *Echium* spp. in Australia should be subjected to biological control, achieved after 7 long years of debate, is welcomed by the Australian beekeeping industry. The only real elements of concern within the industry about the inquiry is that so many people in the community, by inference or expression, see the biological control of *Echium* spp. proceeding automatically as a result of the inquiry, and also whether the period of inquiry is long enough to adequately address the question.

The beekeeping industry position throughout the debate, as the record does show, has always been to accept the plus or minus outcome of any equitably conducted inquiry into the costs and benefits of *Echium*. The industry's position also has been to recognize that as a pasture contributor the benefit of *Echium* spp. in Australia extends beyond the beekeeping industry's horizons to a far greater extent than that which accrues to the industry, and it is this element which has been the crux of the industry's opposition to biological control of *Echium* — whether, in other words, the public benefit would really be served by a successful biological control program, and in that event whether it would be wise to push on with the program.

The comprehensive scientific study of *Echium* at the Victorian Keith Turnbull Institute during the 1970's undertaken over several years by Dr. Colin Piggin, represents the main extent of scientifically substantiated knowledge about the pasture contribution aspects of the species. Dr. Piggin concluded the plant, despite some obvious detractions particularly in above average rainfall, high fertility and undergrazed situations, is an overall contributor to pasture and therefore livestock production.

Attempts by some parties to invalidate Dr. Piggin's results are now being made. The FCAAA has submitted that before alteration of the status quo occurs as the result of this public inquiry, firm data as the result of objective experimentation should be brought forward for critical examination by this Commission of inquiry, so there can be a careful weighing of new information against the established record, and decisions then made which can be confidently expected to be in the true public interest.

There are many other grey areas, or 'information gaps' which have a crucial bearing on the central question of this inquiry — whether the public benefit would be best served through a biological control program for *Echium* species in Australia. The 'information gap' areas would all benefit from closer study with the view of providing firmer data on which to base such an important community decision. This submission therefore focuses on the proposition that the duration of this inquiry is not long enough to allow meaningful studies to be mounted and completed, and until they are, it would be difficult for any Commission of inquiry to recommend, with confidence, in any direction of interest and be sure the public benefit question has been adequately addressed.

ADDENDUM IN PRESS

During the 1984 spring session of the Federal Parliament, the Biological Control Bill, 1984, was passed by both Houses, establishing procedures for assessing, and where appropriate, authorising programs for the biological control of animal and plant pest organisms in Australia. Complementary legislation needs to be passed in the States and Northern Territory Parliaments to establish a uniform system throughout Australia. An essential element of the legislation establishes the Biological Control Authority, which, if a serious community conflict of interest is identified for any program, may authorize a community funded public inquiry into the matter, under the provisions of either the Industries Assistance Commission or the Environment Protection Acts. In this way, the true public benefit of any program may be established before proceeding further with the program.

The legislation also provided, for the first time in Australia, legislative authority to release exotic control organisms such as insects and rusts into the Australian environment. The legislation also addresses the question of assistance for people disadvantaged through the implementation of a program.

Although the record shows and continues to show FCAAA support for the nature of the legislation, some elements worried the FCAAA. For example, the rigid, fiscally oriented IAC mechanism did not appear suited to resolving the *Echium* conflict, with its attendant complex environmental and social interactions. Further, the legal rights of the community in the event of a less than adequate inquiry, appeared obscure. During the passage of the Bill through the Senate, the Australian Democrats moved 19 ammendments to the Bill. None were successful.

On the 26 October, 1984, the Biological Control Authority (John Kerin, Minister for Primary Industry), announced the Industries Assistance Commission would enquire into the biological control of *Echium* species in Australia. An associate commissioner, external to the IAC, would be appointed to the inquiry to broaden the expertise of the inquiring tribunal. The media statement and enquiry terms of reference are attached hereto.

Currently, the inquiry is proceeding. The inquiry is scheduled to terminate with final IAC recommendations to the Federal Government being framed by the end of July, 1985. More than 500 submissions have been received by the Commission from interested parties.

"MEDIA RELEASE

Public Inquiry into Biological Control of Echium

The Minister for Primary Industry, Mr John Kerin, announced today that the Industries Assistance Commission would conduct an inquiry concerning biological control of *Echium*, commonly called Paterson's Curse or Salvation Jane. The Commission is to report within nine months.

The Minister said that the public inquiry would consider whether or not there should be biological control of *Echium*. The inquiry will take into account economic, scientific and environmental issues as well as the question of alternative forms of control. It would also consider whether any industries or persons would be significantly disadvantaged if a control program were to proceed and, if so, whether any assistance should be provided.

The Minister said that Mr John Carey, an IAC Commissioner, has been appointed to head the inquiry and he will sit with an Associate Commissioner. The Associate Commissioner, whose name will shortly be announced, will be able to offer substantial scientific expertise to the inquiry.

The inquiry has been timed to coincide with the introduction of the Biological Control Act 1984 and the subsequent introduction of State and Northern Territory complementary legislation.

The Minister said 'the legislation now provides a rational and equitable legal basis for assesing biological control programs and, where it is established that the program should proceed in the public interest, authorizing such programs. The inquiry is a significant step toward the resolution of the question of biological control of *Echium* which has been a controversial and divisive issue for several years.'

The specific terms of reference of the inquiry which have been passed to the IAC by Senator Button, the Minister for Industry and Commerce, are:

1. whether *Echium sp.* (including species commonly known as Paterson's Curse, Salvation Jane or Riverina Bluebell) should be a target for biological control;
2. that in its inquiry and report the Commission shall have regard to and comment on:
   a. the economic advantages and disadvantages of *Echium sp.* to the various sectors of Australian agriculture and the economy in general;
   b. the means by which biological control of *Echium* may be affected;
   c. whether any industries or persons would be significantly disadvantaged as a outcome of biological control of *Echium ap.* and, if so, whether any assistance might be provided;
   d. the availability, effectiveness and efficiency of alternative methods of control such as chemical, mechanical and pasture management; and
   e. conservation of the natural environment, with regard to the above;
3. that the Commission be free to take evidence and, where necessary, to make recommendations on any matters relevant to its inquiry under this reference.

Mr Kerin said, that under the machinery of the Biological Control Act, it would also be necessary for him to invite submissions in his capacity as the Biological Control Authority. Notices inviting submissions concerning the proposal to biologically control *Echium* and the means of effecting control should be published in major weekend newspapers in early December. All submissions to the Authority would be made available for the Commission's perusal.

Mr Kerin said that he hoped all interested parties would make full use of the process of public comment and public inquiry so that the *Echium* issue can be fairly resolved.

Contact: Walter Pearson (062) 72 6661 — Minister's Office
Michael Ryan (062) 72 5046 — Department."

Chapter Twenty-five

WESTERN GALL RUST OF PINES

K.M. Old

During the last 50 years the forestry industries of Australia and New Zealand have become increasingly reliant on exotic conifer plantations. Except for subtropical parts of Australia the plantations are mostly of a single species, *Pinus radiata*. In New Zealand this species represents over 90% of the nation's timber production, whereas in Australia, although softwood plantations occupy less than 5% of the forested area, by the year 2020 they are expected to produce 64% of pulp and sawlogs.

Part of the success of *P. radiata* in the southern hemisphere is probably due to the limited number of pathogens which are present. Excluding wood rotting fungi, Offord (1964) listed 38 pathogens in the California native stands of *P. radiata* compared to 21 recorded in New Zealand. A similar number of pathogens have been found in Australia as in New Zealand.

Of the array of pine pathogens that could breach Australia's quarantine barrier the stem rust fungus, *Endocronartium harknessii*, presents a particular threat to the softwood industry if it were accidentally introduced (Old, 1981).

SYMPTOMS OF DISEASE AND DAMAGE TO THE TREE

So far, the fungus has been recorded only in North America where it is known by a variety of common names including western gall rust. Unlike many pine stem rusts which require an alternate host to complete their life cycles, western gall rust can spread from pine to pine (Peterson and Jewell, 1968). The characteristic symptom is a branch or stem gall (Figs 2, 3) spherical or oblong in shape, at the surface of which the fungus produces masses of spores (Fig. 1), covered at first with a membranous layer. This ruptures to release millions of orange coloured spores which are spread by the wind and infect actively growing pine shoots. These shoots normally produce sporulating galls after two years, but this latent period is of uncertain length during which time infected trees are difficult to detect.

Damage to the plant includes the formation of 'witches brooms' on infected shoots, stimulation of branching which spoils tree form (Fig. 2), and the predisposition of stems to invasion by secondary organisms which often girdle shoots and may kill the tree. Some infections partially girdle the stem, and although the tree may grow to maturity, defective logs are of reduced value and galls act as points of breakage during windstorms.

There is some evidence that in trees older than 10 years new stem infections are unusual, although new galls may be formed on branches. Girdling of branches by secondary infections causes the death of the fungus within the galls and in older native stands the level of infection may fall dramatically (Byler, 1972).

LIKELY CONSEQUENCES OF INTRODUCING WESTERN GALL RUST

Despite the international dispersion of a number of pathogens, western gall rust has remained within North America. Hepting (1964) pointed out that pathogens cannot be assumed to thrive wherever their hosts flourish and it may be argued that the best course of action is to rely on quarantine protection and to delay hazard appraisal or research on the disease until the need arises. However, in view of the importance of *P. radiata* in Australia, the severity of the damage which the fungus can inflict and the fact that rusts have in the past demonstrated a capacity for transoceanic spread, this disease seems worthy of special attention.

Firstly it is reasonable to assume that much of the *P. radiata* grown here is highly susceptible to the disease. The genetic base of the crop in Australia is probably narrower than that of the native stands (Brown and Moran, 1980). Even in the native stands which could be expected to benefit from their genetic diversity with regard to disease susceptibility (Gibson and Jones, 1977), two of the three mainland California stands of *P. radiata* are heavily infected. In addition to *P. radiata*, *P. caribaea*, *P. elliottii*, *P. pinaster* and *P. taeda*, which are grown in various parts of Australia, are listed by Gibson (1979) as being susceptible to *E. harknessii*.

Fig. 1. Spores of *Endocronartium harknessii* showing wall ornamentation as revealed by scanning electron microscopy.
Fig. 2. Pine heavily infected by **E. harknessii** showing massive stem damage and induction of branching which reduces the value of the tree. (About one third life-size.)

Fig. 3. Branch gall producing millions of spores from discontinuous fissures in the gall surface. The spores infect growing pine shoots in the spring. (Approximately life-size.)

The existence of environmental conditions favourable for epidemics to occur cannot be readily predicted. However, both the work of Nelson (1973) and the evidence of the wide geographical distribution of western gall rust across North America indicate that suitable combinations of temperature and moisture could readily occur in the main *P. radiata* growing areas of Australia.

The disease could have a number of consequences once established in Australia. The impact of the rust is most severe on young trees and presents a threat to plantation establishment. Nursery stock may also be infected and seedlings with latent infections can spread the disease to new locatons when transported to planting sites. The relative resistance of older trees to western gall rust indicated by field observations and controlled propagation studies (Zagory and Libby, 1984) suggests that the standing crop would be little damaged, except for the youngest plantations. In the long term however, the disease would be likely to exert a profound effect on the health of all age classes of *P. radiata* and reduce the potential of the species for timber production.

MANAGEMENT OPTIONS IN THE EVENT OF INTRODUCTION OF WESTERN GALL RUST

In the event of the discovery of diseased trees in Australia or New Zealand, it would have to be immediately decided whether to attempt eradication. The New Zealand Forest Service has the advantage of a nationwide responsibility for forests, a permanent forest biology survey staff who visit all exotic plantations on a regular basis, and a central research and diagnostic laboratory. These services increase the likelihood of early diagnosis and enable a rapid response to be made. In Australia the control of forests is less centralised, but the *Sirex* emergency of the 1950s and early 1960s led to the establishment of the National *Sirex* fund in 1962 funded by all States except the Northern Territory to finance research and surveys. This successful programme was the progenitor of a standing committee for Forest Pests and Diseases which can act swiftly following an outbreak and contributed personnel and funds for recommended actions.

In the particular instance of a western gall rust outbreak it seems probable that an eradication programme would be unsuccessful. In common with other rust fungi, sporulation is prolific. Also the pathogen exhibits a latent phase of indeterminate length between infection and gall maturation which makes the identification of infected trees extremely difficult. Although Peterson (1973) found that dispersal of spores was very inefficient, a few infections by wind-blown spores remote from the initial disease focus would be enough to thwart an eradication attempt.

Once the disease was established however, it seems likely that suitable management practices could minimise the impact on the crop. Nurseries could be located remote from heavily infected stands. Peterson (1968) found that reduction of infected trees in the vicinity of a nursery by wildfire and thinning was apparently related to reduction of infection of nursery stock from 4% to 0.5%. Appropriate management, especially pruning of lower branches and thinning infected trees, both desirable from other silvicultural standpoints could be expected to reduce the disease incidence in planatations.

In the long term the limitation of disease by growing trees with enhanced resistance to infection and disease expression is a desirable option. The approach has been used with significant success in the USA for control of fusiform rust of southern pines (Powers et al. 1981). Observations in native stands and plantations have suggested that at least three types of differences in susceptibility to western gall rust exist:

1. Mature tree resistance, as described above.
2. Differences between individual trees of the same provenance.
3. Differences in susceptibility to the rust between trees originating from the 5 native provenances of *P. radiata*. There are three mainland populations at Monterey, Arno Nuevo and Cambria and small populations on the two Mxican offshore islands of Guadalupe and Cedros. Examination of a collection of representatives of the native populations of *P. radiata* and the closely related *P. muricata* suggested that the other provenances (Old, 1981). A cooperative project set up by the University of California and supported by CSIRO, New Zealand private forestry companies and USA industry funds seeks to assess variation in resistance to western gall rust in all 5 native provenances of *P. radiata*. The possibility of resistance will be studied using selected between-provenance hybrids and seed families from individual trees located in seed orchards in California, Australia and New Zealand.

Identification of genetically controlled resistance will allow an evaluation of the tree breeding option for control to be made. Sources of resistance to the disease either at the population or individual tree level should be useful to growers of the species in California and possibly to the Australian industry in the event of the introduction of western gall rust.

REFERENCES

Brown, A.H.D. and Moran, G.F. (1980). In *Proc. Symp. 'Isozymes of North American Forest Trees and Forest Insects'*. Univ. Calif. Berkeley, 1979.

Byler, J.W., Cobb, F.W. Jr. and Parmeter, J.R. Jr. (1972). *Can. J. Bot*. 50, 1061.

Gibson, I.A.S. (1979). *Diseases of forest trees widely planted as exotics in the tropics and southern hemisphere. Part II. The genus Pinus*. C.M.I., Kew, CFI Oxford.

Gibson, I.A.S. and Jones, T. (1977). In *Origins of Pest Parasite Disease and Weed Problems*. (J.M. Cherrett and G.R. Sagar, eds.) p. 139 Blackwell Scientific Publications, Oxford.

Greaves, R. (1967). *Aust. For. Res*. 3, 36.

Hepting, G.H. (1964). In *F.A.O./I.U.F.R.O. Symp. on Internationally dangerous forest diseases and insects*. p 1. Oxford, 1964.

Nelson, D. (1973). *The ecology and pathology of pine stem rust in California*. PhD. Thesis. University of California, Berkeley.

Offord, H.R. (1964). *Diseases of Monterey pine in native stands of California and in plantations of western North America*. U.S.D.A. Forest Service. Report PSW-14.

Old, K.M. (1981). *Aust. For*. 44, 178.

Peterson, G.W. (1973). *Phytopath*. 63, 170.

Peterson, R.S. and Jewell, F.F. (1968). *Ann. Rev. Phytopath*. 6, 23.

Powers, H.R.,Schmidt, R.A. and Snow, G.A. (1981). *Ann. Rev. Phytopath*. 19, 353.

Zagory D. and Libby W. — 1985. *Phytopath*. 75, (in press).

Chapter Twenty-Six

THE PERSISTENCE OF INFECTIOUS DISEASES WITHIN WILDLIFE POPULATIONS: RABIES AND BOVINE TUBERCULOSIS

Roy M. Anderson

Various exotic infectious diseases of man and his domestic animals depend on reservoir host species within wildlife populations for their long term persistence and population stability. Rabies and bovine tuberculosis are important examples; both pathogens are able to infect a wide range of host species.

Rabies is a directly transmitted viral infection of the central nervous system to which all mammals are thought to be susceptible although not equally so. Bovine tuberculosis, a bacterial infection (*Mycobacterium bovis*), is less cosmopolitan in its utilisation of different host species, but the bacterium is able to infect a variety of domestic animals and wildlife mammalian and marsupial species. Both infections are of considerable economic significance throughout many regions of the world, both with respect to the risk they pose to human health and their impact on livestock production and management.

The control of these infections is based on two approaches, namely:

1. to minimize the degree of contact between man and livestock, and the reservoir host species; and
2. to reduce the reservoir host density aiming to eradicate the disease within wildlife populations in regions where the risk of human or domestic stock infection is high.

In Europe, for example, current rabies control programmes aim to substantially reduce the density of the principal reservoir host species, the red fox (*Vulpes vulpes*) (over 70% of the cases reported each year in Europe are fox infections (W.H.O., 1980). Similarly, in the United Kingdom the control of bovine tuberculosis in cattle is currently based on the removal of infected badger populations (*Meles meles*) in cattle farming regions. The badger is thought to be responsible for disease transmission to cattle (Zuckerman, 1980; Anderson and Trewhella, 1985).

The success of these attempts to control infection in wildlife host species will, in part, depend on the transmission dynamics and population stability of the host-pathogen association. Certain population attributes of the interaction will, for example determine the degree to which the incidence or prevalence of infection changes as overall host density declines. An understanding of the population dynamics of the pathogen and its wildlife host species can therefore provide important insights concerning the likely success of different approaches to disease control. In this short paper, the principal features of the transmission dynamics of rabies in fox populations and bovine tuberculosis in badger populations are examined and compared. Emphasis is placed on the stability of the association between host and pathogen populations and the likely impact on disease prevalence of reductions in reservoir host density.

FOX RABIES IN EUROPE

The intrinsic population growth rate, r, of the red fox in Europe is typically high, being of the order of 0.5 year^{-1}. In the absence of rabies, fox populations appear to increase to some characteristic density (the carrying capacity, K, of the habitat) which is determined by the mixture of habitat types in a defined area (K is typically within the range 0.1 -4 foxes km^{-2}, although much higher densities of up to 20km^{-2} have been recorded in suburban areas (Macdonald, 1980). The principal density-dependent constraint on population growth is thought to act on juvenile and cub mortality. The dynamics of growth of disease free populations is crudely captured by the logistic equation which describes the rate of change of fox abundance, N(t), with respect to time t:

$$\frac{dN}{dT} = \frac{rN(1 - N)}{K} \quad (1)$$

The net rate at which foxes acquire rabies infection is thought to be proportional to the number of encounters between susceptible foxes (of density X) and infectious or rabid foxes (of density Y); namely, βXY, whose β is a transmission coefficient with $1/\beta$ being proportional to the average time interval between fox contacts. The incubation period of the disease (of average duration $1/\sigma$ is variable in foxes although the average is of the order of 28–30 days. Infectious or rabid animals have the very short life expectancy of from 3 to 10 days, with an average of 5 days. In Europe few, if any, animals are thought to recover from the infection. On the basis of these simple observations a crude model can be formulated to describe the principal features of disease transmission within fox populations. The rates of change with respect to time of the numbers of susceptible (X(t)), incubating (I(t)) and rabid animals (Y(t)) can be expressed as

$$\frac{dX}{dt} = rX - \gamma XN - \beta XY \quad (2)$$

$$\frac{dI}{dt} = \beta XY - (\sigma + b + \gamma N)I \quad (3)$$

$$\frac{dY}{dT} = \sigma I - (\alpha + b + \gamma N)Y \quad (4)$$

Here α is the death rate of rabid animals ($1/\alpha = 5$ days), γ denotes the severity of density-dependent constraints on fox mortality ($\gamma = r/K$), b is the natural mortality rate of the fox in the absence of density-dependent factors, and N is total population size ($N = X + I + Y$). Note that infected animals (I and Y) are assumed not to contribute to the net rate of reproduction of the fox population (rX).

This simple model (the properties of which are well understood; (see Anderson, 1981; Mollison, 1984) is thought to capture the principal features of the interaction between host and pathogen and many of its properties are in broad agreement with observed epidemiological trends. These properties may be summarized as follows. Rabies will only be able to persist endemically within the fox population provided the density of the disease free population, K, exceeds a critical value K_T where

$$K_T = (\sigma + a)(\alpha + a)/\beta \sigma \quad (5)$$

and $a =$ per capita birth rate of the host. Epidemiological evidence suggests that K_T lies in the range of 0.25 to 1.0 foxes km^{-2} in Europe. These observations provide a target for control measures aimed at the reduction of overall fox density; the disease will be unable to persist if control measures maintain the density below K_T.

If the density of animals exceeds K_T, rabies will persist and regulate fox abundance below the disease free carrying capacity of the habitat (K). The regulated state with rabies endemic, however, is likely to be oscillatory particularly if the disease free carrying capacity of the habitat K is significantly larger than the threshold level K_T for disease persistence (Anderson, 1981). Under such circumstances, large amplitude oscillations will arise in both fox abundance and disease prevalence. The model suggests that the period of these cycles will lie in the range of 3 to 5 years. Such periods are precisely what are observed in Europe and North America where a striking feature of rabies epidemiology in areas of good fox habitat is the regular 3, 4 or 5 oscillations in fox abundance and rabies prevalence. In areas of poor fox habitat (small K) oscillatory behaviour will be less apparent and a stable endemic state may result.

As a consequence of its high pathogenicity, rabies persists within fox populations at very low levels of prevalence. This is an example of the commonly observed biological relationship between low 'standing crop' and high rate of turnover. Theory and observation reveal that the prevalence of infection will, in general, be in the region of 3–7%. Significant fluctuations arise as a consequence of the oscillatory behaviour of the interaction (and as a consequence of seasonal changes in fox contact and hence transmission) and during the trough of an epidemic cycle the infection is likely to become extinct in isolated fox populations. Spatial variability in fox density, the dispersal of juvenile foxes and movement between social groups of animals play an important role in the maintenance of the infection over large areas of fox habitat.

The epidemiology of fox rabies is, therefore, characterised by a degree of instability as reflected by the oscillatory behaviour of the population interaction. This is a direct consequence of the very high pathogenicity of the disease. Over large areas of good fox habitat (large K) local extinctions of the infection will be frequent with waves of disease moving back and forth over the area with roughly 3 to 5 years intervals between outbreaks in a given location. The oscillatory period of the epidemics is largely determined by the intrinsic growth rate of the fox (r) and the long incubation period of the infection relative to the life span of rabid animals. In areas of poor fox habitat (small K) the disease will be more stable but the standing crop of infections (rabid) animals will be low.

BOVINE TUBERCULOSIS IN BADGERS

Bovine tuberculosis is endemic within many badger populations throughout Britain but is particularly prevalent in areas of good badger habitat (typical densities in moderate habitat are in the range of 5–8 adults km^{-2}, while in good habitats a value of 20 adults km^{-2} is often observed (Cheeseman et al., 1981; Anderson and Trewhella, 1985). The badger (*Meles meles*) has a much lower intrinsic growth rate (r) than the fox, being of the order of 0.2 year^{-2}. This mammal species has a not insignificant maturation delay to first breeding (2–3 years), produces small litters of cubs which experience high rates of mortality in their first year of life and low thereafter, and exhibits limited powers of dispersal. The life expectancy of the badger from birth is roughly double that of the red fox, being of the order of 2–3 years. Population abundance is largely determined by habitat type and the often observed long term stability of badger populations appears to arise as a consequence of density-dependent constraints on fecundity. These constraints are thought only to come into play at densities close to the carrying capacity, K, of the habitat. Cyclic fluctuations in abundance in disease free populations may occur (due to the long maturation delay and density-dependence on fecundity) in areas of poor to moderate badger habitat. The oscillations are predicted to have a period of roughly 6–8 years but to be of small amplitude and hence difficult to detect by field survey (Anderson and Trewhella, 1984).

The principal features of the transmission dynamics of bovine tuberculosis in badger populations can be mirrored by simple mathematical models in a similar manner to that described for fox rabies (equations 2 to 4). The interaction is made somewhat more complex by the existence of a component of 'pseudo vertical' disease transmission (from parent to newborn offspring) in addition to horizontal transmission, the long duration of infectiousness of infected animals (low disease induced mortality rate) and the presence of 'carriers' and inactive cases of infection. All these components combine to create a high degree of population stability, in interactions between host and pathogen. 'Pseudovertical transmission' plus the presence of 'carrier' and inactive cases (infected animals who show no overt signs or symptoms of disease and who intermittently release bacteria into the environment) suggest that the critical density of badgers required to maintain the infection endemically is extremely low. This observation argues that the eradication of the disease within badger populations will be difficult to achieve by reductions in overall badger density. In unperturbed populations the prevalence of infection is high and is typically in the range of 10% to 20%. High densities of infectious animals result in significant levels of habitat contamination by bacilli released in badger urine, faeces and sputum. As such, environmental contamination presents a serious threat to cattle production in areas of good badger habitat.

The stability of the infection within badger populations is well illustrated by the observation that intensively applied control measures in the South West of England over the past 5 years, based on the removal of infected social groups of animals and the overall suppression of badger density, have only resulted in a 50% reduction in disease prevalence within populations of the reservoir host.

DISCUSSION

Rabies in foxes and bovine tuberculosis in badgers show very different epidemiological patterns within their respective host populations. Rabies is an epidemic disease which typically exhibits large fluctuations in prevalence from year to year with a low overall standing crop of infected animals. In contrast, bovine tuberculosis in badgers is characteristically stable, with the prevalence remaining relatively constant through time with a high standing crop of infected animals. These different patterns arise as a consequence of differences in the pathogenicities of the disease, their incubation periods, the degree of infectiousness of infected animals, and their modes of transmission.

At first sight, rabies seems to be a relatively easy infection to control by means of reduction in reservoir host density. It would appear that the instability of the infection with fox populations could be easily exploited to enhance the frequency with which local disease extinctions occur during the troughs in prevalence and fox abundance between epidemic cycles. Unfortunately, however, the population growth characteristics of the reservoir host complicate the issue. The fox has a high intrinsic growth rate for a mammal species, considerable powers of dispersal, and the ability to colonise very varied habitat types (from open moorland to gardens n suburban areas and cities). As such, red fox populations are able to recover rapidly to their precontrol levels of abundance once control measures cease (hunting, gassing, poisoning or trapping). The characteristic recovery time of the population following a perturbation is proportional to the inverse of the intrinsic growth rate r (i.e. a high value of r implies a raid recovery time and *vice versa*). Thus, despite the instability of the infection, intense and sustained effort is required in areas of good fox habitat to maintain fox populations at a sufficiently low level of abundance to eradicate the disease. In areas of poor fox habitat reductions in overall density are likely to be more successful. In both instances control measures should ideally be applied most intensively during inter-epidemic periods.

Control of bovine tuberculosis by reductions in badger density is equally problematic. The disease is highly stable within its host population and appears to be able to persist in low density badger populations. The component of 'pseudovertical transmission' and the long life span of infectious animals are of great significance in this respect. Eradication of the infection would therefore appear to require the virtual eradication of the reservoir host species in areas of high cattle farming activity. On the other hand, however, the badger has a low intrinsic growth rate and hence once populations are suppressed, recovery to precontrol levels will in general take many years. Bovine tuberculosis control in badger populations therefore require intense control effort to substantially reduce host abundance, applied at relatively infrequent intervals. In contrast, rabies control in fox populations requires less intensive effort at one point in time, but sustained effort at frequent intervals.

These conclusions (which are based upon an understanding of the population dynamics of the host and the pathogen) are largely confirmed by recent experiences with rabies control in Europe and bovine tuberculosis control in England. The principal message to emerge from these two examples is that the eradication or control of disease sources within reservoir host populations is always likely to be difficult unless control measures are applied very intensively and are sustained over long periods of time. The renewable nature of the reservoir host resource (by natural births) creates the greatest barrier to effective long term disease control. To overcome these problems attempts must be made to reduce the degree of contact between reservoir host populations and man and his livestock. This can be achieved in the case of rabies by the vaccination of domestic animals (particularly dogs and cats, who often provide the transmission link between wildlife hosts and man). A reduction in badger-cattle contact for the control of bovine tuberculosis, however, is more difficult to achieve in England in areas of good badger habitat and high cattle farming activity. Future progress in disease control is likely to depend on alterations in farm management practices and investigation of the feasibility of cattle vaccination against tuberculosis infection.

REFERENCES

Anderson, R.M. (1981). In: *The mathematical theory of the dynmics of biological populations*. (R.W. Hiorns and D. Cooke, eds) p. 47 Academic Press, London.
Anderson, R.M. and Trewhella, W. (1984). *Phil. trans. Roy. Soc. Series B*. (in press).
Cheesman, C.L. Jones, G.W., Gallagher, J. and Mallinson, P.J. (1981). *J. Appl. Ecol. 18*, 795.
MacDonald, D.W. (1980). *Rabies and Wildlife. A Biologist's Perspective*. Oxford University Press, Oxford.
Mollinson, D. (1984). *Nature. 310*, 224.
Neal, E.G. (1977). *Badgers*. Blandford Press, Poole, Dorset.
World Health Organisation (1980). *Rabies Bulletin for Europe*. Vols. 3, 4 and 5. W.H.O. Collaborating Centre for Rabies Surveillance and Research. Tubingen, West Germany.
Zuckerman, O.M. (1980). *Badgers cattle and tuberculosis*. Her Majesty's Stationery Office, London.

Chapter Twenty-seven

FROM EXOTIC DISEASE TO OCCUPATIONAL HAZARD: THE RECOMBINANT DNA SAFETY DEBATE

Ditta Bartels

The early 1970's can be seen as the dawn of genetic engineering, with methods being discovered for isolating individual genes and for transferring them from one organism to another. Concerns immediately arose about the safety of this research with much private and public discussion leading to major meetings like the conference at Asilomar in California in February 1975 (Watson & Tooze, 1982). In those days it was suspected that organisms containing manipulated genes might be a hazard. However, in the meantime, many tests have shown that manipulation of recombinant DNA is safe (Watson & Tooze, 1982), and the suspicions and fears of the early 1970s about recombinant DNA safety have now largely been allayed. But currently new types of experiments are being done, and the confidence that resulted from success in dealing with old problems appears to be preventing us from assessing potential future problems (Bartels, 1983).

In this chapter I would like to draw your attention to:

— what was suspected then;
— what we know now;
— what as yet we do NOT know.

EXOTIC DISEASE?

Back in the days of the Asilomar conference, the hazards of recombinant DNA were thought to be like those of an exotic disease. To capture the fears that prevailed at the time I cannot do better than quote from Michael Rogers, who, at that conference, gained fame from his irreverant 'pop' description of the problem (Rogers, 1982):

"The long imprisoned *E. coli*, laden with a brand-new bit of biological ability, suddenly finds itself liberated; floating in a minute droplet on a technician's finger, then onto a tuna-fish sandwich, and thence into a luckless human gut. Or, in a culture not quite completely killed, down some stainless steel laboratory sink and thus into a sewer system teeming with billions of close relatives.

And now what? Precisely what could our artificially mutated *E. coli* do with its sudden freedom?

An epidemic cancer that spreads through the sewer system? A once conquered disease like bubonic plague now, abruptly, again incurable? Or a brand new disease, sudden and mysterious, that has never before appeared in human beings?"

By now we do know how to answer Rogers question as to what the articifically mutated *E. coli* could do with its sudden freedom. The answer is nothing, absolutely nothing — no cancer epidemic, no plague, no brand new disease. And we can be pretty sure that this answer is correct, because in the last eight years, *E. coli* with brand new bits of biological ability have been handled in hundreds of laboratories, by thousands of technicans. It is most probable that over and over again, such *E. coli* have floated onto tuna-fish sandwiches, into human guts, down laboratory sinks, and into the sewerage system. And yet nothing untoward has happened. As scientific evidence goes this is as good as any we have got. We can therefore be confident that *E. coli*, and by extension other micro-organisms, which contain recombinant DNA do not present a biohazard.

OCCUPATIONAL HAZARD?

But to the scientists engaged in recombinant DNA work only handle recombinant *organisms*? No, they most frequently handle the DNA of a particular gene and they handle large amounts of such DNA. The technique of bacterial cloning has made it possible to purify, isolate and concentrate particular DNA molecules. Over the last few years many different genes have been handled in this way, and again nothing untoward has happened, however recently an exciting new set of genes have been cloned, namely those involved in cancer induction, the so-called oncogenes (Newmark, 1983).

So a new question has arisen: is *oncogene DNA*, isolated by means of the recombinant DNA technique, safe to handle? I believe it is likely that oncogene DNA is a hazardous material, and that experiments should be done to assess the hazard and ways of avoiding it. Especially so since it has been known for some time that tumour virus DNA will transform cells in culture, in certain cases, more effectively than intact virus. Bovine papilloma virus (BPV) DNA transforms foetal bovine skin cells (Boiron et al., 1965), while the polyomavirus BKV-DNA transformed hamster and rat cells with greater frequency and shorter time than intact virus (van der Noordaa, 1976).

Let me summarize the evidence which indicate that oncogene DNA could endanger those who work with it. First, when oncogene DNA is put into the fluid medium around a particular culture of mouse cells, known as NIH3T3, the cells become transformed (Shih & Weinberg, 1982).

Since oncogene DNA undoubtedly changes these cells growing in culture, is such a change also likely to occur in a living organism — for example in a laboratory worker? One reassuring answer that I have received from the Australian Recombinant DNA Committee is that the NIH3T3 cells are unlike the laboratory worker, as the 3T3 cells are already in a pre-cancerous state, having been 'immortalised' by their repeated passage in culture. However recently Weinberg and his colleagues at the Massachusets Institute of Technology have checked whether different known oncogenes can transform several types of cells into transformed cells (Weinberg et al., 1983). Their results are:

TABLE 1. *Effect of oncogenes on cultured cells*

Oncogene	Cells	Outcome
ras	NIH3T3	transformation
myc	NIH3T3	no transformation
ras	normal cells	no transformation
myc	normal cells	no transformation
ras + myc	normal cells	transformation

These results show tht normal, as well as precancerous cells can be transformed when oncogene DNA is added but, in these experiments, only by a mixture of *ras* and *myc* oncogene DNA. Thus the analogy can be upheld between the results obtained with cultured cells and intact organisms.

Work done by Fung and his colleagues (Fung et al., 1983) at the University of California, San Francisco goes a step further still. They injected oncogene DNA of the v-*src* type into chickens and small tumourous nodules formed at the injection site. The oncogene used by Fung, v-*src*, is similar to the oncogene used by Weinberg, c-*ras*, in that it transforms only the special NIH3T3 cells and not normal cells in culture. Nevertheless tumours formed in the animals when the oncogene DNA was injected.

A more suitable test

The question then arises, would the injection experiments be more convincing if a combination of oncogenes were to be used, say *ras* plus *myc*? Would the tumours be larger? Would they be of the spreading type and kill the animals? Let me tabulate the experimental comparisons that could be made:

TABLE 2. *Experiment to test oncogene combinations and intact animals*

Oncogene	Test system	Outcome
src	NIH3T3	transformation
src	normal cells	no transformation
src	intact animals	tumour nodules
ras + myc	normal cells	transformation
ras + myc	intact animals	?

I have approached Weinberg regarding the questioned outcome and he tells me that he does not know the answer because the appropriate experiments have not been done. But the point is that experiments such as the one suggested here *should have been done* as part of a rational recombinant DNA safety assessment programme. Due to the excitement of oncogene research, and the successes achieved in the last two years, laboratories around the world, including many in Australia, are gearing up to do oncogene work. In this work, large quantities of oncogene DNA are isolated, sequenced, analysed or added to cell cultures. In short, oncogene DNA is repeatedly transferred from one vessel to another. It is more than likely that in the course of these experimental manipulations, oncogene DNA will somehow enter the bodies of laboratory workers engaged in these activities.

Why is there a general lack of appreciation of the problem which I am outlining here? In part the answer can be summed up with the motto 'been there, done that'. In other words, the community of molecular biologists probably feels that the question of recombinant DNA safety has been aired at length, that extensive safety testing experiments have been done, and that no hazard has been demonstrated. To be sure, Martin and his colleagues at the National Institute of Health in the U.S.A. have done a detailed series of safety testing experiments with a virus which causes tumours and is called polyoma papovavirus. To quote from Martin et al. (1979):

> "Out task was to evaluate whether recombinants containing a viral DNA segement could transfer out of E. coli. The answer to that question is an emphatic. NO"!.

We must note that Martin's answer pertained only to recombinant organisms. The polyoma experiments shed no light at all on the question of oncogene DNA safety. First, the oncogenes manipulated by molecular biologists at this time, ras, myc and src, were not part of the polyoma test. Second, the experiments of Martin et al. tested the likelihood of recombinant DNA getting out of the E. coli cells, and did not concern the hazard of pure, concentrated oncogene DNA. Thus, regarding the latter, we have not as yet 'been there, done that'.

CONCLUSIONS

We need to KNOW the nature and extent of the potential occupational hazard, that I have outlined. After all, we may be in a situation like that of asbestos workers a decade ago, who worked in a hazardous environment that was not generally recognised as such. I believe that our earlier perception of recombinant DNA safety in terms of an exotic disease has prevented us from recognising or indeed examining, the safety of oncogene work from the perspective of occupational health. I do not think we should allow this situation to continue.

REFERENCES

Watson, J.D., and Tooze, J. (eds.) (1982). *The DNA Story: A documentary history of gene cloning* Freeman, San Francisco.
Bartels, D. (1983). *Search 14 (3-4)*: 88.
Boiron, M., Thomas, M. and Chenaille, P.H. (1965). *Virology 26*, 150.
Rogers, M. (1982). In *The DNA Story: A documentary history of gene cloning* (J.D. Watson, J. Tooze, eds) p. 28, Freeman, San Francisco.
Newmark, P. (1983). *Nature 305*: 470.
Shih, C. and Weinberg, R.A. (1982). *Cell 29*: 161.
Fung, Y-K, T., Critienden, L.B., Sadly, A.M. and Kung, H.J. (1983). *Proc. Natl. Acad. Sci. 80*: 353.
Martin, M.A., Chan, H.W., Israel, M.A., and Rowe, W.P. (1979). In *Recombinant DNA and Genetic Experimentation* (J. Morgan and W.J. Whelan, eds.) p. 209 Pergamon Press, Oxford.
Van der Noordaa (1976). *Journal of General Virology, 30*, 371.
Weinberg, R.A., Land, H. and Parada, L.F. (1983). *Science 222*: 771.

Chapter Twenty-Eight

BENEFICIAL USE OF AN EXOTIC PHYTOPATHOGEN, *PUCCINIA CHONDRILLINA*, AS A BIOLOGICAL CONTROL AGENT FOR SKELETON WEED, *CHONDRILLA JUNCEA*, IN AUSTRALIA

E.S. Delfosse, S. Hasan, J.M. Cullen, and A.J. Wapshere

Skeleton weed, *Chondrilla juncea* L. (Compositae), probably evolved in the Transcaspian area of central Europe (Iljin, 1930). It now also occurs naturally in Mediterranean Europe and north Africa and is the only *Chondrilla* spp. in the western Mediterranean (Cullen and Groves, 1977). It was first recorded in Australia, near Marrar, NSW, in 1917 (Maiden, 1918), and since then it has become a serious weed in cereal growing areas of south-eastern Australia. Heavy infestations of the weed reduce grain yields significantly, and also interfere with harvesting (Wells, 1971). It has never been a major weed in Mediterranean Europe (Wapshere *et al.*, 1974; 1976).

Skeleton weed is perennial, reproducing sexually from seed, and vegetatively from rosettes produced by the rootstock. Its life cycle hs been described by several authors (Wapshere *et al.*, 1974; 1976; Cullen and Groves, 1977; Groves and Cullen, 1981). The life cycle begins either with seed germination or with regeneration of new rosettes in autumn, which persist until spring. Aerial flowering stems are produced in late spring to early summer from apices and the centre of rosettes. They elongate, branch and produce buds from late spring to early summer. Many flowers, each producing about 11 seeds, are produced in summer. Seeds (actually achenes) have pappi, are dispersed by wind, are immediately viable with no secondary dormancy (Grant Lipp, 1966), and are effectively viable for about 6 months (Cullen and Groves, 1977). At the end of summer, stems die back, and new rosettes regenerate or seeds germinate in autumn. Stages may overlap, especially at the beginning and end of each season.

The weed was considered to be such a major threat to the Australian cereal industry that a *Chondrilla* Biological Control Unit (now the CSIRO Biological Control Unit) was established by CSIRO in Montpellier, France, in 1966 to search for natural enemies of the weed. At the same time, an increased ecological study of the weed began in Canberra. In this chapter we discuss the skeleton weed rust fungus, *Puccinia chondrillina* Bubak and Syd., one of the natural enemies found during the European surveys. The introduction of this rust as a classical biological control agent was a milestone in biological weed control in Australia, and remains the best example of the use of an exotic pathogen as a biological control agent for a weed.

BIOLOGY AND LIFE CYCLE OF *P. CHONDRILLINA*

P. chondrillina is found in Iran, Turkey, Greece, Yugoslavia, Italy and France, Spain and Portugal in Mediteranean Europe, wherever skeleton weed occurs. In southern Russia the fungus is found on other *Chondrilla* spp. as well as on *C. juncea* (Hasan, 1972). It is a macrocyclic, autoecious rust; asexual (uredial and telial) and sexual (pycnial and aecial) stages occur only on *C. juncea*.

The life cycle of *P. chondrillina* has been described by Hasan (1972), and is summarized in relation to the life cycle of *C. juncea* in Australia in Fig. 1. Seedlings and rosettes of skeleton weed are infected by uredospores in autumn. Brown uredosori are soon produced which, in turn, produce further uredospores, the process continuing over several generations on rosettes from autumn to spring. In late spring to early summer the flower shoot appears, and rosette leaves die back. Flower shoots are also infected by uredospores, producing uredosori, which give rise to further uredospores. These continue to infect stems, and even calyces of flowers, until stems die in late summer to early autumn. Black teleutosori bearing bicellular teleutospores appear at the end of the flowering season in Europe, but these do not play an important role in the life cycle of the rust in Mediterranean areas. In fact, basidiospores produced are not capable of infecting *C. juncea* under glasshouse conditions (Hasan and Wapshere, 1973).

From laboratory experiments, optimum levels of infection and highest number of sori were obtained at 20°C with 12–16 h of moisture, although some infection occurred from 10–30°C and 12–16 h of moisture (Hasan and Jenkins, 1972). No pustules appeared at an inoculation temperature of 5°C. In other tests on the effect of temperature on incubation period of *P. chondrillina*, Hasan and Jenkins (1972) found that pustules appeared after 28 d at 10°, 20 d at 15°, 8 d at 25° and 6 d at 30°C. Multiple pustules appeared at 5–30°C, but heaviest infection occurred at 20, 25 and 30°C, with poor infection developing at temperatures over 33°C. Uredospores germinate in water at temperatures from 0–36°C, with optimum germination at 18°C (Hasan and Wapshere, 1973). Each spore gives rise to 1–2 germ tubes, which appear after 2 h at room temperature. Growth of germ tubes is optimal at 10°C (667 μm over 48 h), but occurs from 0–36°C. Spore germination and penetration are normally complete 16 h post-inoculation. Germ tubes penetrate the plant only through stomata (Hasan and Jenkins, 1972).

SPECIFICITY OF *P. CHONDRILLINA*

C. juncea is a triploid, obligate apomict (Rosenberg, 1912; Battaglia, 1949), and occurs in three forms in Australia (A or narrow-leaf, B or intermediate-leaf, and C or broad-leaf) (Hull and Groves, 1973), which are distinct morphologically and isozymatically (Burdon et al., 1980). The plant occurs in many forms in Eurasia (Wapshere et al., 1974, 1976) each of which is attacked by a pathotype or pathotypes of *P. chondrillina*. Thus, the problem of determining if the rust was sufficiently host-specific to skeleton weed to be safely introduced into Australia as a biological control agent was initially affected by the need to find rust strains specific to Australian forms of the weed.

Thus, the study proceeded via two series of experiments (Hasan, 1972): the three Australian forms of skeleton weed were exposed in the laboratory to spores of *P. chondrillina* samples from a large number of European sites; and cultivated plants were exposed to the rust strains which attacked the Australian forms of the weed to determine its host range. As well, morphological matching of the three forms of skeleton weed with the many European morphotypes continued in Canberra. Plants of at lease one collection from Vieste, Italy, (IT32) morphologically resembled the Australian form A plants (R. Groves, pers. comm., 1984).

Pathogenicity of spore samples of *P. chondrillina* against Australian skeleton weed was assessed in two ways: by measuring the proportion of plants on which uredosori developed; and by grading the reaction type on a four point scale (Hasan, 1972). A strain of the rust collected from these morphologically identical plants from Vieste was found to be the most virulent against the Australian narrow-leaf form of skeleton weed (the most common and widespread of the three forms at the beginning of the biological control program). Strain IT32 of *P. chondrillina* was therefore used in host-specificity tests.

Fifty-eight cultivated plant species belonging to 30 families were tested with strain IT32 of *P. chondrillina* (Table 1), 25 of which were known to be attacked by other *Puccinia* spp. No attack was found on these plants by the rust. Thus, *P. chondrillina* exhibits a very high degree of host specificity.

In subsequent tests to determine if temperature could influence infectivity of *P. chondrillina*, additional experiments were conducted with the strain IT32. Plants closely related to *C. juncea* in the Cichorieae, plus *Chrysanthemum* sp., were tested (Table 1). While spores germinated on all test plants in the family Compositae, and in a few cases hyphae entered stomata, no disease was developed in any species except *C. juncea* (Hasan and Jenkins, 1972). *P. chondrillina* was therefore considered to be safe to introduce as a biological control agent (Hasan, 1972). Since its release in Australia in 1971 this has been confirmed: no plants other than the narrow-leaf form of skeleton weed have been attacked by IT32.

EFFECT OF *P. CHONDRILLINA* ON *C. JUNCEA*

Hasan and Wapshere (1973) found that death of *C. juncea* plants caused by *P. chondrillina* in natural European sites was frequently 50–70%, and that seedling death in experimental plots varied from 90–100% (2 to 3.6 times the mortality of uninfected controls). Similarly, mortality of mature rosettes infected by *P. chondrillina* was 2.1 to 4.6 times higher than uninfected controls. Rusted plants had significantly fewer flower shoots and basal rosettes, and regeneration from rootstocks was 13.3%, as opposed to 86.4% in controls (Hasan and Wapshere, 1973).

Thus, *P. chondrillina* seemed to have the potential to be an effective biological control agent (Wapshere, 1970; Hasan, 1972):

1. it germinates, infects plants and produces uredosori over a wide temperature range;
2. infection only require a short period of high humidity under dark conditions (i.e. overnight dews);
3. it reproduces year round via the asexual uredophase;
4. spores are powdery and wind-dispersed;
5 all aerial parts of its host are attacked;

TABLE 1. *Plant species used in host-specificity testing of form IT32 of Puccinia chondrillina Bubak & Syd.*

Weeds related to *C. juncea*

Cichorium intybus L. 1,2,3,
Crepis taraxacifolia Thuill 1,2,3,
C. foetida L. 1,3,
Hieracium pilosella L. 1,3
Hypochaeris radicata L. 1,3

Sonchus asper All. 1,2,3,
S. oleraceus L. 1
Taraxacum officinale Wiggers 1,3
Urospermum dalechmpii Desf. 1,2,3

Other composite weeds in Australia

Carthamus lanatus L. 1
Onopordum acaulon L. 1
Picris hieracioides L. 3

Scorzonera hispanica L. 3
Tragopogon sp. 3

Cultivated plants in the Compositae

Bellis spp. 3
Carthamus tinctorius L. 1,3
Cichorium endiva L. 1,2,3
Chrysanthemum spp. 1,2,3
Cynara scolymus L. 3
Dahlia sp. 3

Helianthus annuus L. 1,3
H. tuberosum L. 1,3
Lactuca sativa L. 1,2,3
L. scariola L. 3
Tagetes sp. 1,3
Zinnia sp. 3

Unrelated cultivated plants

Acacia dealbata Link
Allium cepa L. 1
Apium graveolens L.
Asparagus officinalis L.
Avena sativa L.
Beta vulgaris L.
Brassica oleracea L.
B. rapa L.
Capsicum annuum L.
Citrullus vulgaris L.
Cucumis melo L.
Cucurbita maxima Duch.
Dactylis glomerata L.
Gossypium spp. 1
Humulus lupulus L.
Ipomaea batatas Lam.
Linum usitatissimum L.
Lolium perenne L. 1
Lycopersicum esculentum Mill.
Medicago sativa L.

Nicotiana tabacum L.
Oryza sativa L.
Pastinaca sativa L.
Phaseolus vulgaris L.
Pisum sativum L.
Primula sp.
Saccharum officinarum L.
Solanum melongena L.
Sorghum vulgare Pers. 1
Trifolium repens L.
T. subterraneum L.
Triticum sp. 1
Vicia faba L.
Vitis vinifera L.
Zea mays L. 1

1. these species known to be attacked by other *Puccinia* spp;
2. these also tested for effects of temperature on infectivity;
3. also used in testing of TU21 and IT36.

---

6. growth, seeding, regeneration and possibly root reserves are affected; and
7. mortality of its host is high.

As these results were noted in the presence of parasites of *P. chondrillina*, such as the phytophagous rust fungus parasite, *Darluca filum* Cast. [= *Eudarluca caricis* (Fr.) O. Erikss.) (Hasan and Wapshere, 1973), which would be eliminated before being released in Australia, it appeared that even better results were possible in Australia. Although *D. filum* occurs in Australia (McAlpine, 1906), it was quite possible that there was no strain well adapted to this new species of *Puccinia*. In fact, *D. filum* was later observed to attach *P. chondrillina*, but only in dense, well-established infections. Thus, introduction of the strain IT32 of *P. chondrillina* was approved by the Department of Health, and was released in the field in June 1971, following three years of testing and discussion (Hasan, 1972; 1974; Cullen *et al.*, 1973). Nine laboratory-infected plants of *C. juncea*, each bearing 100–200 uredosori, were released near Wagga Wagga, N.S.W. (other release sites were also used from June to December 1981) (Cullen *et al.*, 1973). After one generation, about 50 uredosori were found on plants up to 1 m from the point of release. By the fifth generation (November) most plants at the release site were infected, some with 500 to 2000 uredosori/plant, with infection detectable up to 27 m from the point of release. Subsequent field examinations revealed that foci were formed at least 8, 24, 80, 160 and 320 km after *ca.* 4, 5, 8, 10 and 12 generations, respectively. The rapid initial spread was only slightly less than that for wheat stem rust epidemics (Cullen *et al.*, 1973) and was aided by a high host population, optimal field conditions and a virulent strain of *P. chondrillina* (Fig. 3). Within 1 year virtually all areas with narrow-leaf skeleton weed were infected by this strain of *P. chondrillina* (Cullen, 1978).

Very high levels of control of the narrow-leaf form of *C. juncea* have been achieved by this strain of *P. chondrillina* (Cullen and Groves, 1977; Cullen, 1978). The main limiting factors have been availability of spores and suitable conditions for germination (Cullen, 1973). The effect of the rust can cause death of plants directly, or by weakening the host, it is made more susceptible to competition from other plants, e.g., *Trifolium subterraneum* L. (Groves and Williams, 1975).

Fig. 1. Life stages of the rust fungus, *Puccinia chondrillina* Bubak & Syd., and its host, skeleton weed (*Chondrilla jundea* L.) in Australia.

BIOLOGICAL CONTROL OF THE INTERMEDIATE — AND BROAD-LEAF FORMS OF *C. JUNCEA*

As the narrow-leaf form of *C. juncea* diminished due to effects of biological control and subsequent plant competition, the other two forms have increased in density and distribution (Fig. 2) (Burdon *et al.*, 1981), as predicted (Hull and Groves, 1973). It was thus necessary to find strains of *P. chondrillina* that are virulent against these two weed forms (Hasan, 1981).

Spore samples of *P. chondrillina* were collected in Europe and western Asia and were tested against the Australian forms of *C. juncea*. Very few strains from Spain, France, Italy and Greece attacked these forms (Hasan, 1972). The survey region was therefore extended eastwards to the Adriatic coast of southern Italy, other parts of Greece, western Turkey and the Black Sea coastal areas of Bulgaria and Romania (Hasan, 1981).

Most of the spores collected during these later surveys were not pathogenic to Australian skeleton weed forms. Two isolates, TU21 from Manisa, Turkey, and IT36 from southern Italy, attacked the intermediate-leaf form. A reduced list of plants were tested against these isolates (Table 1); no plants other than *C. juncea* were

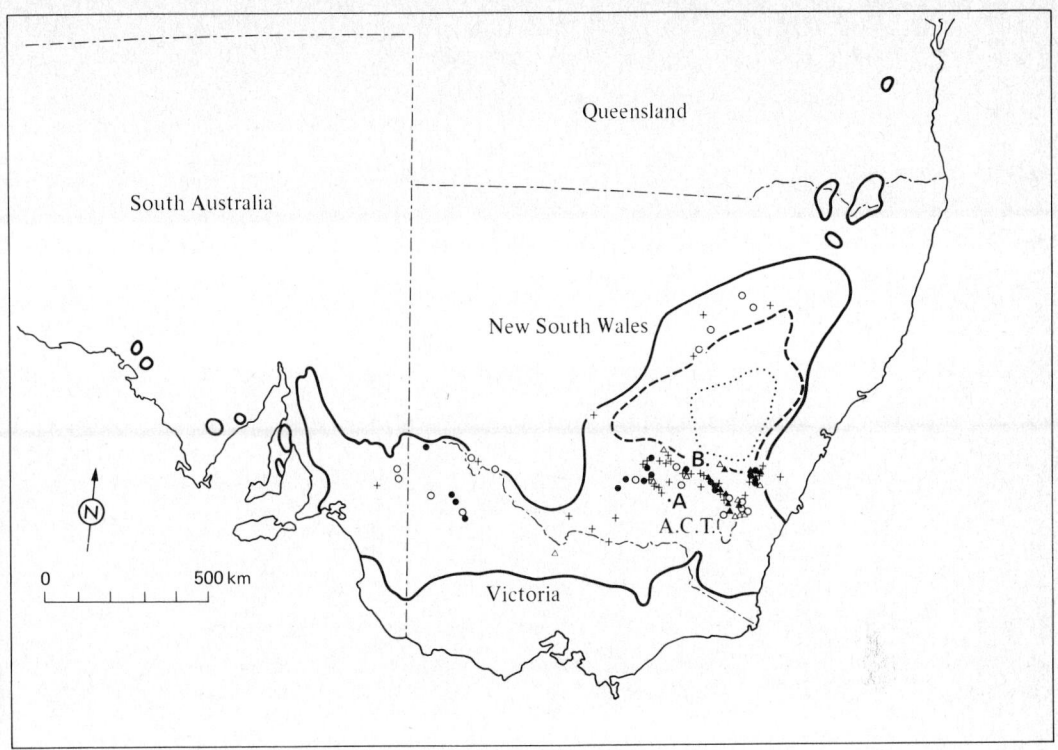

Fig. 2. Geographic distribution of the three forms of skeleton weed, *Chondrilla juncea* L., and the three strains of the rust fungus, *Puccinia chondrillina* Bubak & Syd., in southeastern Australia. Areas enclosed by solid, dotted and dashed lines represent the 1968 distributions of the narrow-, intermediate- and broad-leaf forms, respectively (Hull and Groves, 1973). Triangles and circles, respectively, represent sites at which the intermediate- and broad-leaf forms were found up to 1980. Unfilled symbols represent occurrences of a single plant or patch; filled symbols the close occurrence of many patches; and crosses the occurrence of plants of either form which could not be identified precisely. Strain IT32, released near Wagga Wagga (A) in June 1971, had spread throughout the region by December 1972. Strains IT36 and TU21, released near Young (B) from 1981–83, have not spread beyond the release plot, and have apparently died out. Modified, with permission, from Burdon et al. (1981).

attacked. TU21 was imported, and after normal quarantine procedures, released in the field near Young, N.S.W., in September 1980 (Burdon et al., 1981; Delfosse and Cullen, 1982) at a site of predominantly intermediate-leaf skeleton weed. The strain disappeared from the site during January 1981. As that was a very dry season, it was not known if it has died out due to poor conditions for germination or due to a low level of virulence. It was thus released a second time in June 1981. The level of inoculum increased initially, reached a peak in late spring, but again had disappeared by late summer. This result, combined with laboratory observations on its biology, forced the conclusion that TU21 was insufficiently virulent, particularly with regard to pustule size and spore production.

Isolate IT36 was imported in 1982 and released at the Young site in September. It performed well initially, but the season was again very dry, and plants dried out before IT36 became established. Uredospores persisted into February, but died out before the drought breaking rains of autumn 1983. IT36 was thus released a second time in April 1983, and initially spread well — up to 6 m from the point of release by early winter. However, an extremely dense cover of *T. subterraneum*, producing excessively moist conditions at rosette level (which caused spores to lose viability or to germinate *in situ* on old leaves rather than transferring to new leaves) caused the infection to die out. IT36 was released a third time in November 1983, initially spread up to *ca.* 10 m from the point of release, but again apparently died out in early 1985.

Several other promising strains of *P. chondrillina* which attack the intermediate and broad-leaf forms of *C. juncea* have been isolated from samples collected in Greece and Turkey, and these are being studied at Montpellier (Hasan, unpubl. data).

Fig. 3. Effects of the rust fungus, *Puccinia chondrillina* Bubak & Syd., on skeleton week, *Chondrilla juncea* L., at Wagga Wagga, N.S.W., spring 1972. A: skeleton weed rosettes; B: same plot, showing effect of the rust fungus on rosettes, 6-wks after A.

DISCUSSION

In some of the papers in these proceedings, the diseases are the problems, and ways to lessen their impact are discussed. In this case, the disease is beneficial, is exotic, was purposely imported as a biological control agent against a very damaging host weed, and is very restricted in its host range. The biological control of skeleton weed has been one of the most successful programs in recent times. *P. chondrillina* is one of four exotic agents to be established in Australia for management of the weed (Delfosse and Cullen, 1982). (The other agents are a midge, *Cystiphora schmidti* Rubs., a mite, *Aceria chondrillae* G. Can., and a moth, *Bradyrrhoa gilveolella* Tr. *P. chondrillina* seems to be the most damaging agent in this suite of natural enemies). The Industries Assistance Commission conducted a cost:benefit study of this program (Marsden *et al.*, 1980), and found that the total benefits attributable to this program (1960–2000) varied from $414.9 to $194.9 million (5 and 20% discount rate, respectively).

Despite some fears about the safety of *P. chondrillina* as a biological control agent, the rust has not attacked any other plant species since its introduction. In fact, as indicated above, it is so specific that it does not even attack the other forms of the same weed species. This feature has made it necessary to search for additional form-specific isolates of the rust.

Many other host-specific pathogens, particularly rusts, are known for other major weeds (Charudattan and Walker, 1982). After adequate testing of the host-specificity of each species of pathogen, many of these could be used as biological control agents in the future.

ACKNOWLEDGMENTS

Technical assistance of T. Hartley, E.E. Lewis, R.C. Lewis, A.D. Moore, N. Stahl, C.S. Smith and A. Walker is gratefully acknowledged. Mrs I. Pumpurs expertly typed this manuscript, which was appreciated. Permission to duplicate Fig. 1, with modification, from Dr J.J. Burdon is appreciated. Grant Funds from the Wheat Industry Research Council of Australia are gratefully acknowledged.

REFERENCES

Battaglia, E. (1949). *Caryologia* 2, 23.
Burdon, J.J., Groves, R.H., and Cullen, J.M. (1981). *J. appl. Ecol.* 18, 957.
Burdon, J.J., Marshall, D.R., and Groves, R.H. (1980). *Aust. J. Bot.* 28, 193.

Charudattan, R., and Walker, H.L. (1982). *Biological Control of Weeds with Plant pathogens*. John Wiley & Sons, New York.
Cullen, J.M. (1973). In *Proc. III Int. Symp. Biol. Control Weeds*. Commonw. Inst. Biol. Contr. Misc. Publ. No. 8, p. 111.
Cullen, J.M. (1978). In *Proc. IV Int. Symp. Biol. Control Weeds*. Inst. Food Agric. Sci., Florida, p. 117.
Cullen, J.M., and Groves, R.H. (1977). In *Exotic Species in Australia. Their establishment and Success*. (D. Anderson, ed) Chap 2 *Proc. Ecol. Soc. Aust. 10*, 121
Delfosse, E.S., and Cullen, J.M. (1982). *Aust. Weeds 1(3)*, 25.
Grant Lipp, A.E. (1966). *Fld Stn Rec. Div. Pl. Ind. CSIRO (Aust.) 5*, 17.
Groves, R.H., and Cullen, J.M. (1981). In *The Ecology of Pests. Some Australian Case Histories*. (R.L. Kitching and R.E. Jones, eds) Chap 2 CSIRO: Melbourne, 6.
Groves, R.H., and Williams, J.D. (1975). *Aust. J. Agric. Res. 26*, 475.
Hasan, S. (1972). *Ann. Appl. Biol. 72*, 257.
Hasan, S. (1974). *Phytopath. 64(2)*, 253.
Hasan, S. (1981). *Ann. Appl. Biol. 99*, 119.
Hasan, S., and Jenkins, P.T. (1972). *Pl. Dis. Reptr 56(10)*, 858.
Hasan, S., and Wapshere, A.J. (1973). *Ann. Appl. Biol. 74*, 325.
Hull, V.J., and Groves, R.H. (1973). *Aust. J. Bot. 21.* 113.
Iljin, M.M. (1930). *Trudy Prikl. Bot. Genet. Selek. 24*, 147.
McAlpine, D. (1906). *The Rusts of Australia*. Dept. of Agric., Vic. Brain: Melbourne.
Maiden, J.H. (1918). *Agric. Gaz. N.S.W. 29*, 330
Marsden, J.S., Martin, G.E., Parham, D.J., Ridsdill Smith, T.J., and Johnston, B.G. (1980). In *Returns on Australian Agricultural Research. Joint IAC-CSIRO Cost-Benefit Study*. Chap 16 CSIRO, Melbourne.
Rosenberg, O. (1912). *Svensk Bot. Tidskr. 6*, 915.
Wapshere, A.J. (1970). In *Proc. I Int. Symp. Biol. Contr. Weeds, Misc. Publ. CIBC, 1*, 81.
Wapshere A.J., Caresche, L., and Hasan, S. (1976). *J. Appl. Ecol. 13*, 545.
Wapshere, A.J., Hasan, S., Wahba, W.K., and Cresche, L. (1974). *J. Appl. Ecol. 11*, 783.
Wells, G.J. (1971). *J. Aust. Inst. Agric. Sci. 37*, 122.

Chapter Twenty-nine

POTATO SPINDLE TUBER VIROID AND AVOCADO SUNBLOTCH VIROID

J.L. Dale

Viroids are the smallest known infectious agents or pathogens and to date have only been found in plants. They have characteristics unlike any other group of organisms and while the diseases they cause have been, in many cases, known for some considerable time, their unique characteristics only began to emerge about seventeen years ago. Viroids were found to be closed circular, single-stranded ribonucleic acid (RNA) molecules; the smallest viroid being only 246 nucleotides long and the largest 371 nucleotides. These RNA molecules are highly based-paired and naked; they have no protective coat for their genome like viruses do. The base-pairing appears to 'substitute' for a protein coat by protecting viroids from the hostile, ribonuclease-filled, environments in which they exist. The RNA is not known to code for any proteins which implies a total dependence on the host for replication. The mechanism of viroid replication is not known but is generally accepted that viroids are replicated by host enzymes using RNA templates and that it is possible a rolling circle-type mechanism with the circular viroid serving as a template (Diener, 1984).

The relationships of viroids to each other and the groups' relationships to other pathogens or nucleotide sequences have been the subject of great interest and study. The complete nucleotide sequence of eight viroids is now know, and from this data, it can be seen that, except for avocado sunblotch viroid, all have a central region with a closely similar sequence. This may be because they are phylogenetically related or the result of convergent evolution because of a common functional requirement, neither of which can be excluded at this stage.

Nucleotide sequence data has also initiated speculations on the origins of viroids. The first hypothesis suggested that viroids were circularised spliced-out 'intervening sequences' or 'escaped introns' (Diener, 1979) but more recently (Kiefer et al., 1983) it has been proposed that viroids originated from transposable elements. This hypothesis is based on the sequence similarities between viroids and the ends of transposable elements and especially those of retroviral proviruses.

Though it is the biochemical and biophysical properties of viroids that cause the most interest, it is the biological effects that cause most concern; the ten best-described viroids (Table 1), with the exception of

TABLE 1. *A list of the best characterized viroids*

Name	Mode of Commercial Propagation of Host
Potato spindle tuber viroid	Vegetative
Citrus exocortis viroid	Vegetative
Avocado sunblotch viroid	Vegetative
Chrysanthemum stunt viroid	Vegetative
Chrysanthemum chlorotic mottle viroid	Vegetative
Hop stunt viroid	Vegetative
Cucumber pale fruit viroid	Seed
Coconut cadang-cadang viroid	Seed
Tomato planta macho viroid	Seed
Tomato apical stunt viroid	Seed

cucumber pale fruit viroid and chrysanthemum chlorotic mottle viroid, are all responsible for serious and sometimes devestating economic losses. However, the symptoms of the diseases are rarely severe, they are normally diffuse or poorly defined. Stunting, leaf distortion and fruit discoloration and distortion as well as yield reduction are the most commonly encountered symptoms of infection.

None of the viroids have been shown to have arthropod vectors in nature and viroids depend very heavily on man's activities for transmission. Most viroids primarily infect plants that are propagated from vegetative material such as cuttings, buds for grafting and tubers. The result of this is that if the plant from which the vegetative material is taken is infected then all the resulting plants will also usually be infected. The four exceptions to this (Table 1) are cucumber pale fruit viroid, cadang-cadang of coconut, tomato panta macho and tomato apical stunt viroid; however, cucumber pale fruit viroid seems to be a strain of hop stunt viroid.

The major mode of transmission is mechanical; on cutting and pruning tools and also farm implements such as ploughs. The third method of transmission, by seed and/or pollen, has only been shown for avocado sunblotch viroid (ASBV) and potato spindle tuber viroid (PSTV). As yet, the mechanism of spread of cadang-cadang viroid in coconuts in the Philippines has not been determined.

DETECTION OF VIROIDS

The task facing agriculturalists is the eradication and exclusion of viroids from crops. This seems feasible because, in general, viroids depend on man for their spread. If man's activities can be modified in such a way as to stop transmission, then eradication can be achieved. Of utmost importance in any program to eradicate or exclude a viroid is the ability to accurately detect infected plants. Infection cannot unequivocally be determined by symptoms alone because:

1. the symptoms are unclear and can easily be missed or confused with other disorders, even by the trained observer;
2. infections may be symptomless. These result from symptoms being masked because of unfavourable environmental conditions, or, because of the often long incubation period, the plant has yet to display symptoms.

Thus any system based on visual inspection alone for the detection of viroid infections is fraught with danger. However, many of the sensitive detection methods used for viruses cannot be applied to viroid detection. Serological methods such as gel diffusion and ELISA (Chapter 13) cannot be used because there are no viroid-specific proteins against which to make antibodies, and electron microscopy is of no use because viroids can only be seen when in a carefully purified state.

Nonetheless, there are methods available for viroid detection, and three are currently being used:

1. Biological indexing has been in use for many years. It requires the transfer of living tissue (by leaf, bark, bud or petiole grafting) from the plant to be tested, or indexed, to an indicator plant. This is normally a particular species or cultivar which is sensitive to and reacts to infection by a certain viroid in a characteristic and unmistakeable way. For instances, PSTV is inoculated onto tomato (cultivar Rutgers),CEV is inoculated onto citron (*Citrus medica* cv Arizone 861) and ASBV is inoculated onto avocado (cv Hass). This method has proved to be successful but has a number of limitations. For example, each plant to be indexed must be inoculated onto a number of indicator plants. These indicators must be observed for extended periods of time to ensure that all infected plants show symptoms. In the case of ASBV this may be up to three years, with most other viroid indexing systems being of somewhat shorter duration. This means that biological indexing is very expensive both in time and space and, therefore, is not conductive to large scale indexing.
2. Polyacrylamide gel electrophoresis (PAGE). This requires the partial purification of the viroid (normally a preparation containing low molecular weight nucleic acids) and the electrophoresis of these partially purified preparations through a polyacrylamide gel to separate the components on the basis of their molecular weight. The gel is then stained to reveal bands of nucleic acids. This technique has been used for the detection of PSTV, ABSV and CSV (Morris and Smith, 1977; da Graca and Mason, 1983), and is adaptable to large scale sampling and surveys. There have been few published comparisons of biological indexing and PAGE indexing but at least with PSTV, PAGE appears to be more sensitive than the tomato bioassay (Schumann *et al.*, 1978).
3. Nucleic acid hybridization. This is the most recently introduced method of nucleic acid hybridisation, and appears to be the most sensitive and the most adaptable to large scale indexing, and is already in use for the detection of PSTV (Owens and Diener, 1981) and ASBV (Palukaitis *et al.*, 1981; Allen and Dale, 1981). The technique involves the synthesis of a DNA molecule complementary cDNA 'probe' to the viroid RNA. This cDNA is labelled with a radioisotope (either ^{32}P or ^{3}H). The cDNA hybridizes to the viroid RNA under specified conditions (Chapter 12) and the amount of viroid present in a sample. As little as 40–125 pg (1pg = 10^{-12}g) of PSTV per gram of potato tissue may be detected (Owens and Diener, 1981). Initially, the cDNA probes were produced directly from purified viroid RNA and the method of analysis was liquid hybridization (Palukaitis *et al.*, 1981) but, increasingly,

viroid RNA is being synthesised into double stranded DNA and cloned into bacterial plasmids or bacteriophage (Cress and Owens, 1981). The viroid probe can then be synthesised directly from laboratory cultures of the bacteria or bacteriophage containing the cloned viroid. Also, liquid hybridization is being replaced by 'dot blot' hybridization which is more adaptable to large scale indexing. Palukaitis et al., (1981) found liquid hybridization using a cDNA probe synthesised directly from ASBV-RNA to be considerably more sensitive than PAGE.

The methods for detecting viroids have become increasingly more sensitive and more adaptable to large scale indexing but with this has come a change in the expertise required of the 'indexer'. Initially, biological indexing demanded a knowledge of plant propagation and symptom expressing and the availability of large amounts of glasshouse space. However, with the advent of nucleic acid hybridization, the services of a molecular biologist are required along with a well equipped laboratory and there is little need for glasshouse space.

VIROIDS IN AUSTRALIA

Only four viroids have been recorded in Australia, namely potato spindle tuber viroid, avocado sunblotch viroid, citrus exocortis viroid and chrysanthemum stunt viroid. Here, I shall only describe PSTV and ASBV in Australia.

Potato Spindle Tuber Viroid

PSTV and coconut cadang-cadang viroid are probably the two most economically important viroids because of the large scale losses they can cause. PSTV has a number of effects on its host and as with most viroids the effects are variable, and appear to depend on the strain of PSTV, the environment and the host. Normally the plant is stunted but may not be. The tubers become elongated with prominent bud scales and sometimes severe growth cracks. However, the foilage symptoms are usually obscure. Symptoms tend to become more severe with increasing temperature. The symptoms of PSTV infection probably belie its destructive potential. The viroid is transmitted in the field from infected plants by various means. From these initial foci of infection, the viroid can spread easily and rapidly from plant to plant merely by rubbing of foliage, as well as on contaminated tools and machinery. Economic losses can be quite drastic varying from 20–70% yield loss when there is 100% infection, and the poorly shaped tubers attract a lower price. Singh et al., (1971) estimated there was a 1% yield loss in the total American potato crop because of PSTV.

Fortunately, there have been only two recorded introductions of PSTV into Australia. The first was a quarantine intercept in Victoria and the infected plants were not released from quarantine (Sward, 1979). The second incident had far greater impact. In 1981, potato specimens were sent from potato breeding experimental plots at Glen Innes, NSW, for indexing for potato leafroll virus. The disease was tentatively diagnosed as PSTV by bioassay onto tomato and also by PAGE (Schwinghamer and Conroy, 1983) and confirmed by 'dot blot' hybridization and RNA sequencing. Subsequent surveys of breeding lines at the Glen Innes plots showed that 243 of the 347 lines of the 'variety collection' were infected, 134 out of 210 seedling derived breeding lines were infected as well as 84 out of 210 seedling derived breeding lines were infected as well as 84 out of 248 plants (representing 37 lines) of the advanced seedling derived lines were infected. However, a limited survey of commercial crops on properties where 'advanced' seedling-derived lines had previously been grown showed no evidence of infection (Schwinghamer and Scott, 1984).

Breeding lines from Glen Innes had been distributed to the Potato Research Station in Victoria in 1975, 1980 and 1981 and to South Australia in 1980. At the time that PSTV was recognised in the Glen Innes breeding lines, all lines in Victoria from Glen Innes had been discarded and destroyed, but from a survey of 550 other lines grown at the Potato Research Station only two lines were found to be infected with PSTV (Mason and Heath, 1984). The most likely explanation is that PSTV spread, though very slowly, from infected material imported from Glen Innes. Of course, there is also the extremely slight chance that the two lines were infected from another source of PSTV. The survey carried out in South Australia (Cartwright, 1984) revealed only one infected line, which had been imported from Glen Innes.

Evidence from infection rates and occurrence in various lines suggests that PSTV was introduced into the Glen Innes potato breeding program between 1972 and 1979, and remained undetected until 1981. The origin of the PSTV infection is uncertain but it most likely would have been introduced in imported material from overseas. In 1972, viroids were still poorly understood and reliable indexing techniques had not been developed. By 1975, however, both bioassay and PAGE detection techniques had been developed for PSTV (Morris and Wright, 1975) even though nucleic acid hybridization indexing was not developed until 1981 (Owens and Diener, 1981). All these techniques, however, require positive screening of imported planting material whilst in quarantine rather than passive screening by inspection because of the obscure or even absent symptoms induced by PSTV.

Avocado Sunblotch Viroid

Like PSTV, the symptoms of avocado sunblotch viroid infection in avocado are very variable. The most common symptoms are sunken yellow or red streaks on the fruit and sometimes yellow, orange or white streaks and spots on the stems and petioles, and occasional variegation and distortion of the leaves (Dale et al., 1982).

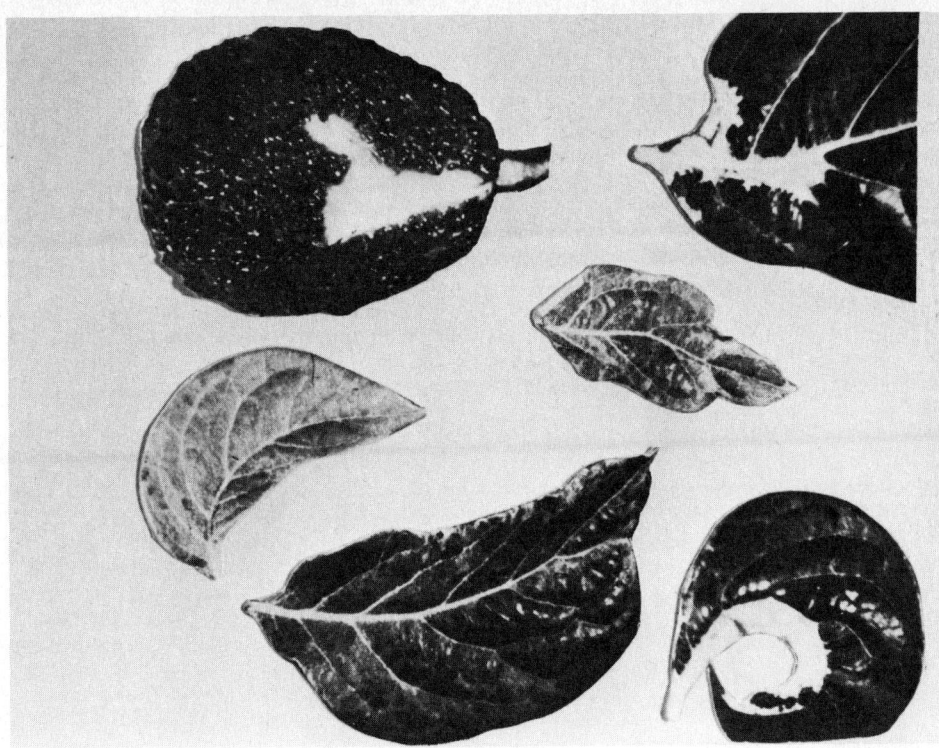

Fig. 1. Range of symptoms of avocado sunblotch viroid on avocado cv Hass including variegation and distortion of leaves and a typical fruit lesion.

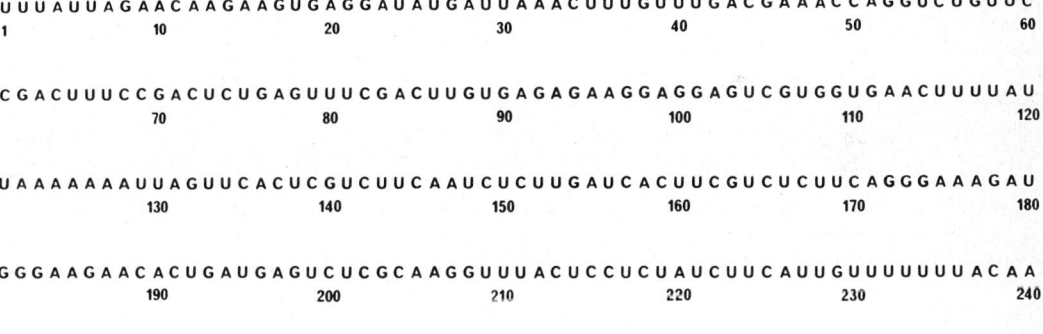

Fig. 2. The primary nucleotide sequence of avocado sunblotch viroid presented as a linear sequence (Symons, 1981). Symons, R.H. (1981). Nucleic Acids Res, 9, 6527.

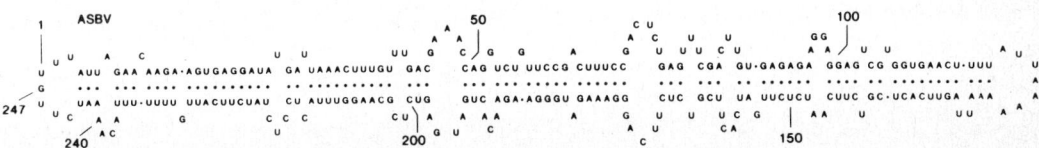

Fig. 3. The primary nucleotide sequence and proposed secondary structure of avocado sunblotch viroid (Symons, 1981). Symons, R.H. (1981). Nucleic Acids Res. 9, 6527.

All, none or combinations of these symptoms may be present in one infected tree but symptoms also vary between seasons. A further complicating factor is the occurrence to true symptomless carriers; trees that do not show symptoms during their lifetime.

Unlike PSTV, there is little or no transmission of ASBV in the field. Mechanical transmission has been demonstrated experimentally, and there was little pollen transmission (via bees) even under the most favourable experimental conditions. The two modes of transmission which appear to be epidemiologically important are by grafting and by seed. Both have become increasingly important because the method of propagation of avocados is by grafting onto seedling root stocks. Seed transmission rates vary greatly but generally the rate is greatest in seed from 'symptomless carriers' (80–100% and the resulting seedlings are also symptomlessly infected where as seed transmission rates from symptom-bearing trees are less (0–5%) and the resultant seedlings show symptoms (Wallace and Drake, 1962).

Transmission of ASBV through symptomlessly infected seed and cuttings would provide plenty of opportunities for introduction of the disease into new areas as has occurred in Australia. ASBV was first positively recorded in Australia in 1967 (Trochoulias and Allen, 1970) but was probably present for some considerable time before that. There are no records or evidence as to when and how often ASBV has been brought into Australia in infected tissue but the present distribution of ASBV suggests it occurred on more than one occasion. A number of isolated infected trees have been detected since sensitive indexing techinques have become available (Allen and Dale, 1981), however, more recently, large numbers of infected trees have been detected on the Atherton Tableland of Queensland (Peterson and Dale, unpublished results). This latest outbreak appears to have been derived from the dissemination of infected plants from one nursery.

The control of ASBV in Australia is considerably simpler than the control of PSTV because spread in the field is, at most, negligible and also because the avocado industry is much smaller than the potato industry. A pathogen tested scheme has been set up by the Australian Avocado Growers Federation, through which both seed and scion source trees are indexed for ASBV every five years by nucleic acid hybridization. While this scheme is voluntary and there is no legislation to destroy infected trees, economic considerations are inducing avocado growers to plant only indexed trees.

CONCLUSIONS

Viroids pose a real and considerable threat to agriculture in Australia. The diffuse or obscure symptoms of infection make visual diagnosis extremely difficult except in ideal circumstances and then only by trained personnel. Detection by other means is difficult because serological techniques are not possible, and because of the laborious and time consuming nature of graft indexing. The advent of nucleic acid hybridization indexing has vastly improved the accuracy and speed of detection but requires new expertise in the 'indexer'.

Viroids such as ASBV and also citrus exocortis viroid, which are already established in Australian crops, do not spread rapidly, if at all. Therefore they can be readily contained by the dissemination of viroid-free planting material. However, continued indexing of imported material and material in the pathogen-tested schemes is essential. The situation regarding PSTV is completely different. This viroid can spread rapidly in the field and has been shown to do so in Australia. The situation, at present, is that PSTV seems to have been eradicated in Australia and continued exclusion of this pathogen from Australia is of the utmost importance. This will require careful surveillance at quarantine ports of entry, and also continued surveying of pathogen-tested potato schemes using the most sensitive detection techniques.

Finally, it is important to note that there are a number of potentially destructive viroids not present in Australia, such as hop stunt viroid and tomato planta macho viroid. These must also be excluded. The possibility that PSTV and other viroids could be introduced in alternate hosts which are not subject to intense viroid indexing also poses a continuing threat.

REFERENCES

Allen, R.N. and Dale, J.L. (1981). *Ann. Appl. Biol.* 98, 451
Cartwright, D.N. (1984). *Australian Plant Pathology* 13,4.
Cress, D.E. and Owens, R.A. (1981). *Recombinant DNA 1*, 90.
Da Graca, J.V. and Mason, T.E. (1983). *Phytpath. Zeitschrift* 108, 262
Dale, J.L., Symons, R.H. and Allen, R.N. (1982). *CMI/AAB Descriptions of Plant Viruses No. 254*.
Diener,T.O. (1979). *Science* 205, 859.
Diener, T.O. (1984). *Intervirology* 22, 1.
Kiefer, M.C., Owens, R.A. and Diener, T.O. (1983). *Proc. Natn. Acad. Sci.* 80, 6234.
Mason, A. and Heath, R. (1984). *Aust. Plant Path.* 13, 20.
Morris, T.J. and Smith, E.M. (1977). *Phytopath.* 67, 145.
Morris, T.J. and Wright, N.S. (1975). *Amer. Potato J.* 52, 57.
Owens, R.A. and Diener, T.O. (1981). *Science* 213, 670.
Palukaitis, P., Rakowski, A.G., Alexander, D. McE and Symons, R.H. (1981). *Ann. Appl. Biol.* 98, 439.

Schumann, G.L., Thurston, H.D., Horst, R.K., Kawamoto, S.O. and Nemato, G.I. (1978). *Phytopath. 68*, 1256.
Schwinghamer, M.W. and Conroy, R.J. (1983). *Aust. Plant Path. 12*, 4.
Schwinghamer, M.W. and Scott, G.R. (1984). *Aust. Plant Path. 13*, 18.
Singh, R.P., Finnie, R.E. and Bagnall, R.H. (1971). *Amer. Potato J. 48*, 262
Sward, R.J. (1979). *Aust. Plant Path. 8*, 35.
Trochoulias, T. and Allen, R.N. (1970). *N.S.W. Agric. Gaz. 81*, 67.
Wallace, J.M. and Drake, R.J. (1962). *Phytopath. 52*, 237.

SUMMARY

REFLECTIONS ON THE SYMPOSIUM; AN AGRICULTURAL PRODUCER'S VIEW

Richard Warren

The summation of this collection of papers must, of necessity, be treated with a fairly broad brush.

The fine print of the technology and philosophy will be absorbed only by the closest reading of the papers themselves and further use of their references. The real importance is the interrelationship of each separate paper.

When faced with the 'first time' proposition of human, animal, avian and plant exotic diseases being discussed in one Congress Symposium, the first reaction may be to brand the idea as a good gimmick and an eye-catching theme for public relations. However the further the week's proceedings progressed, the more obvious it became that these were not separate disciplines. The problems of one were often paralleled by the problems of another.

More obviously the technology and research of human disease is entirely relevant to animal disease as man is an animal. On the other hand, there is also definite relevance between animal and plant exotic diseases and their place in modern science.

The papers in this Symposium should be read with the basic realisation that we in Australia, being an island continent, hav developed a 'Maginot Line' mentality in regard to exotic diseases.

In the early days of settlement we seemed somewhat safe, as distance from the settled world, the supply source of basic human, animal and plant resources, precluded major importations of exotic disease. Long sea voyages either led to the death of diseased material or the flux of time allowed recovery and cure.

Toward the latter end of the twentieth century we have a new phenomenon in world history. Advancement in physical communication has led to an extraordinary interchange of peoples and commodities on a scale undreamt of in quite recent times.

With Australia's huge area of land and small population, we see 2.4 million short-term human visits per annum already, as well as enormous increases in cargo tonnage volumes and speed of handling techniques.

We can add to this the new penchant for 'adventure tourism' and the difficulty of properly patrolling our vast coastlines, let alone supervision of the 1,600 unmanned air fields in our far north. It then becomes obvious that anyone, studying these proceedings, must be aware of the real world in which technology operates and the short-comings imposed by those demographic and geographical circumstances on the application of technology.

It should always be remembered that the crucial limitations of success in the technology of exotic diseases, may be the human frailties and the breakdown of relatively simple suport systems, both practical and political.

Dr D. Dowdle said, "Education and communication are an essential component of any program for the control of A.I.D.S.". This applies to any exotic disease program and is probably the single most important theme that occurred throughout this Congress.

Again, in review, the history of exotic disease control and research must never be ignored. This is classically highlighted by Professor Frank Fenner's paper on myxomatosis.

Although the use of this exotic disease was developed decades ago, it set the precedent for 'Off-shore' research and training including the testing of pathogens on fauna and domestic animals.

The paper on Patterson's Curse and the use of biological control is interesting as an example that no paper in this collection, however well presented, can be taken as the 'be-all and end-all' on any subject. It is an illustration of the ability of a small pressure group to interfere, rightly or wrongly, with the best laid plans of other men. Apart from being a plausible, but very one-sided case, it should provide the reader with food for thought. It should lead us all to consider more carefully man's responsibility to native flora and fauna with the introduction of exotica. This paper raised to some the question of how the (exotic!) imported honey bees must confound the purity of pollen in our unique floral society! There is often the need to remind ourselves that for every plus there may be a minus.

The paper of Professor Douglas Whalan emphasises that nothing can be achieved in the exotic disease field without the sanction and co-operation of the law of the land. In these days of over-government, continued pressure must be brought to bear not only for the unification of legal approaches amongst the multifarious bodies with legal inputs, but also the simplification of the law.

One of the most important papers of the Symposium was Professor Ron Johnston's. It highlights the inadequacies of major science policy decisions taken on an *ad hoc* basis, in a climate of special pleading and in the absence of a general policy guidance.

As the horizons of science are spread ever wider, competition for the elusive but essential dollar for research and development of technology becomes increasingly intense. The responsibility of the administrators of finance is no less than that of the recipient and user of that finance. There will indeed be more and more pressure for decisions to be made on an objective basis. In many cases, one can discern an urgent need for those seeking financial support to extend their skills beyond the test tube or the farm. They must develop the hither-to unsung skills of objective analysis, qualification and quantification of their special needs.

The economic implications for Australia of the introduction of an exotic disease are going to vary enormously. They may vary from a beneficial cost situation, through the range of non-cost to the possibility of a disease, such as foot-and-mouth disease, which could conceivably cost up to $3,000 million in the first year, and have a major effect on the national economy. Thus the importance of the papers that raise these matters should be recognised by those who may consider it outside their purer fields.

The economic and social welfare of the community will ultimately decide their priorities of support, whether it be in the field of human, animal or plant disease.

The major Symposium interest in the ecology of exotic pathogens forms the basis for the greatest excitement on the horizons of science. It is here that science fiction of yesterday is today's technology and will be tomorrow's accepted norm.

Rather than review the specifics of the appropriate papers it is important that each be read with a query paramount in one's mind:

"What other applications, what other research, what other field, can this particular matter be related to or influenced by?"

It will be well recognised that traditional techniques are changing and being superseded rapidly. The historical concepts of vaccines had not changed for nearly 200 years until, suddenly, genetic engineering is changing our thoughts radically. The great advances that led to electron microscopy, immunofluorescence, complement fixation and immunodiffusion are being replaced by contemporary procedures. These procedures will have advantages such as greater sensitivity, specifity, speed or portability.

This collection of papers that formed the 'Exotic Diseases Symposium' of the 1984 ANZAAS, under the title 'Horizons of Science', serves well to remind us that there is always more ahead of us than we can see at any given time and that there is an inter-dependence of all disciplines of research and technology of disease.

It is clear that all with an interest in exotic diseases and their influence on the survival and production of the human, animal, avian and plant life, belong to a unique and important single discipline no single section of which can ever be complete on its own.

LIST OF CONTRIBUTORS AND ADDRESSES

(Second and subsequent authors with the same address as the first are referenced to the first)

Anderson, R.M.
Department of Pure and Applied Biology,
Imperial College,
South Kensington, SW7 2BB, U.K.

Bartels, D.
History and Philosophy of Science,
University of N.S.W.,
Kensington, N.S.W.

Bath, S. (see Stear, M.J.)

Bean, W.J. (see Webster, R.G.)

Brandon, M.R.
Faculty of Veterinary Science,
Melbourne University,
Melbourne, VIC. 3052

Briggs, L.
Federal Council of Australian Apiarists'
Association,
R.M.B. 1030,
Glenrowan, VIC. 3675

Brown, F.
Wellcome Foot and Mouth Disease Research
Laboratory,
Pirbright, Working, Surrey GU24 ONQ, U.K.

Brown, S.C. (see Stear, M.J.)

Bults, H. (see Brandon, M.R.)

Burdon, J.J.
CSIRO Division of Plant Industry,
P.O. Box 1600,
Canberra City, A.C.T. 2601

Burgess, G.W.
Graduate School of Tropical Veterinary Science,
James Cook University,
Townsville, QLD 4811

Campbell, R.W.
Veterinary Field Services,
Victorian Department of Agriculture,
Melbourne, VIC. 3001

Cullen, J.M. (see Delfosse, J.M.)

Dale, J.L.
Department of Primary Industries,
Plant Pathology Branch,
Meiers Road,
Indooroopilly, QLD 4068

Delfosse, E.S.
CSIRO Division of Entomology,
G.P.O. Box 1700,
Canberra City, A.C.T. 2601

Della Porta, A.J.
CSIRO Australian National Animal Health
Laboratory,
P.O. Bag 24,
Geelong, VIC 3220

Dimmock, C. (See Stear, M.J.)

Dowdle, W.R.

Doyle, K.A.
Aust. Agricultural Health and
Quarantine Service,
Department of Primary Industry,
Canberra, A.C.T. 2600

Fenner, F.
John Curtin School of Medical Research,
Australian National University,
Canberra, A.C.T. 2601

Francki, R.I.B.
Waite Institute,
Glen Osmond, S.A. 5064

Geering, W.A. (see Meischke, H.R.C.)

Gibbs, A.J.
Virus Ecology Research Group,
Research School of Biological Sciences,
Australian National University,
Canberra, A.C.T. 2601

Gorman, B.M.
Queensland Institute for Medical Research,
Bramston Terrace,
Herston, QLD 4006

Hasan, S. (see Delfosse, E.S.)

Johnston, J.
Bureau of Agricultural Economics,
Department of Primary Industry,
Canberra, A.C.T. 2600

Johnston, R.D.
Department of History & Philosphy of Science
P.O. Box 1144,
University of Wollongong,
Wollongong, N.S.W. 2500

Kawaoka, Y. (see Webster, R.G.)

Kerin, J.C.
Minister for Primary Industry,
Parliament House,
Canberra, A.C.T. 2600

Kihm, U.
Swiss Vaccine Institute,
Socinstrasse 57,
4000 Basel, Switzerland

Laver, W.G.
John Curtin School of Medical Research,
Australian National University,
Canberra, A.C.T. 2601

Mackie, J. (see Stear, M.J.)

Mahon, R.
CSIRO Division of Entomology,
G.P.O. Box 1700,
Canberra, A.C.T. 2601

Marshall, D.R.
I.A. Watson Wheat Research Centre,
P.O. Box 219,
Narrabri, N.S.W. 2390

Meischke, H.R.C.
Australian Agricultural Health and
Quarantine Service,
Department of Primary Industry,
Canberra, A.C.T. 2600

Milner, R.J.
CSIRO Division of Entomology,
G.P.O Box 1700,
Canberra, A.C.T. 2601

Morris, B.
Department of Immunology,
John Curtin School of Medical Research,
Australian National University,
Canberra, A.C.T. 2601

Naeve, C.W. (see Webster, R.G.)

Newton-John, H.
Fairfield Infectious Diseases Hospital,
Melbourne, VIC 3078

Nicholas, F.W. (see Stear, M.J.)

Old, K.M.
CSIRO Division of Forest Research,
Banks Street,
Yarralumla, A.C.T. 2600

Randles, J.W.
Waite Institute,
Glen Osmond, 5064, S.A.

Scott, P. (see Johnston, R.D.)

Stear, M.J.
Department of Immunology,
John Curtin School of Medical Research,
Australian National University,
Canberra, A.C.T. 2601

Symons, R.H.
Department of Biochemistry,
Adelaide University,
Adelaide, S.A. 5000

Wace, N.M.
Department of Biogeography and Geomorphology,
Research School of Pacific Studies,
Australian National University,
Canberra, A.C.T. 2601

Wapshere, A.J. (see Delfosse, E.S.)

Warren, R.G.
Cattle Council of Australia,
G.P.O.Box 10,
Canberra, A.C.T. 2600

Webster, R.G.
St. Jude Children's Research Hospital,
Memphis, Tennessee, 38101, U.S.A.

Whalan, D.J.
Faculty of Law,
Australian National University,
Canberra, A.C.T. 2601

Wood, J.M. (see Webster, R.G)

INDEX

aboriginal people 7, 12
ADAPTATION 4–20, 40, 98–9, 104–7, 147–50, 152
Aedes aegypti (see mosquitoes)
African horse sickness (orbivirus) 28
African swine fever (iridovirus) 28, 54
aircraft 10, 12–15, 18, 20, 24, 38, 46–8, 66–7, 70
Akabane (bunyavirus) 48
alfalfa mosaic virus 42
Animal Health Emergency Information System 51
ANIMAL PRODUCTS 28–39, 49, 135–7
Animal Virus Research Institute, Pirbright UK 70–1, 76
Antarctica 3–4
antelopes 29, 32
ANTIBODIES; MONO and POLYCLONAL (see SEROLOGY)
aphids 15, 42, 45, 115–9
apples, 8, 40
Aquired Immune Deficiency Syndrome (AIDS) 24, 26, 125–7
ASSAY METHODS 8, 83, 95, 179
ATTENUATION OF VIRULENCE (see HOST PARASITE RELATIONS
Aujeszky's disease (herpesvirus) 28, 37
Australian Agricultural Council 64
Australian Agricultural Health and Quarantine Service 45
Australian Apiarists Association 154
Australian Commonwealth Bureau of Agricultural Economics 64, 67–72
Australian Commonwealth Department of Health 64–7
Australian Counter Disaster College, Macedon 50
Australian National Animal Health Laboratory 20, 43–9, 51, 59, 61, 63–73, 75, 77–78
Australian National Cattle Council 64
Australian National High Security Quarantine Facility (Fairfield) 23, 48, 77
Australian Veterinary Association 64
Avena fatua (wild oats) 106–7
avian influenza (fowl plague, see influenza)
avocado sunblotch viroid 87, 178–82

Bacillus thuringiensis 115–7
bacteria 40, 115–6
bacteriophage 86, 89
baculoviruses 118–9
badger 164–7
banana Sigatoka disease 41
bats 4
Beauveria (see fungi)
beekeeping 152–7
beetles 119

BIOGEOGRAPHY 3–8, 23–26, 104–7, 128–34
BIOLOGICAL CONTROL 32, 55, 75–9, 115–9, 152–9, 171–5
Biological Control Bill 1984, 158
birds, import and export 46
blackberry rust
bluetongue (orbivirus) 28, 30, 32–4, 38, 50, 54, 64, 67, 76, 79, 128–138
boats (see ships)
Boophilus spp. (see ticks)
Borna disease virus 28
bovine brucellosis 39
bovine ephemeral fever (rhabdovirus) 76
bovine leucosis (oncovirus) 110
bovine papilloma (papovavirus) 169
bovine pleuropneumonia 39
bovine tuberculosis 39, 164–7
bovine viral diarrhoea (pestivirus) 138
BREEDING 104–7, 108–110
broad bean stain (comovirus) 76
Brucella melitensis infection 28, 34, 38
bubonic plague 24, 45, 53
buffaloes 29, 30
bushfires 12, 20
Butler, Harry 48

cacao swollen shoot virus 40
camels 24
cancer 110, 169–70
canine brucellosis 28
canine distemper (morbillivirus) 30
CARGO (see TRADE)
cars 12–13, 18–20
Castanea spp. 5
cats 9, 37, 51, 75
cattle 8, 29, 31–3, 36–8, 109–110, 128–9, 135–9, 164–7
caulimoviruses 77
Centers for Disease Control, Atlanta 125–7
Cerastocystis ulmi 5, 41
Cercopithecus (see monkeys)
cereals 8, 40
cetaceans 8
Chagas disease 28
cheese 46, 137
cholera 24, 45, 53
Chondrilla juncea (Skeleton weed) 171–6
chromosomes 97
citrus canker 41
citrus dieback 41
citrus greening 41
CLIMATE 3–6, 8–9
COASTAL SURVEILLANCE (see EMERGENCY PLANNING)
cocoa 40
Cocos-Keeling Quarantine Station 48
coffee 40
coffee rust fungus 40

Commonwealth Scientific and Industrial
    Research Organization (CSIRO) 64–68,
    70–2, 128, 147, 153–7, 163, 171
Commonwealth Serum Laboratory 75, 128
Commonwealth/State interactions 50, 53–5, 59
COMPENSATION 50–2, 54–5, 58–62,
    112–4, 140, 147
computers 47, 51
CONCLUSIONS (see
    RECOMMENDATIONS)
contagious bovine pleuropneumonia 66
CONTINENTAL DRIFT (see
    BIOGEOGRAPHY)
CONTROL OF DISEASE 26, 30–38, 40–2,
    45–9, 50–2, 53–5, 58–62, 69–71, 75, 115–9,
    138–9, 140, 162–3
Cooperia (see worms)
Corynebacterium spp. 110
COSTS (see ECONOMIC EFFECTS)
cowpea (*Vigna* spp.) 102
cowpea mosaic (comovirus) 102
Creutzefeld-Jakob disease 34
Crimea-Congo haemorrhagic fever 23, 25
'CSIRO 368' (rhabdovirus) 76
Culicoides (biting midges) 33, 38, 130–3

deer 29, 32, 38, 129
dengue fever (flavivirus) 24, 46, 48
Deoxyribose nucleic acid (DNA) (see
    NUCLEIC ACID)
DIAGNOSIS 23–6, 29–38, 50–2, 63–73,
    75–9, 83–95, 96–102, 125–7, 137
dingo 7–8
DISEASE CONTROL (see CONTROL
    OF DISEASE)
DISEASE RESISTANCE (see HOST/
    PARASITE RELATIONS)
DISPERSAL (see ECOLOGY)
dogs 37, 51, 75
dot-blot assays 87–90, 94–5, 180
dourine 28
Drosophila 100
duck viral enterities 28
duck viral hepatitis (hepadnavirus) 28
Dutch Elm disease (see *Ceratocystis*)

East Coast fever 28
Ebola (filovirus) 23–6, 53
Echium spp. (Paterson's Curse, Salvation
    Jane) 9, 55, 152–8
ECOLOGY 4, 12–20, 23–7, 29–38, 40–2,
    77, 101, 137, 140, 144–5, 147–51, 157–8,
    164–7, 170–6, 179–82
ECONOMIC EFFECT (COSTS) 29, 40–2,
    46–9, 51, 54, 58–62, 66–9, 140, 149
EDUCATION (see EMERGENCY
    PLANNING)
electron microscope 84
electrophoretic analysis 101.

elms (see *Ulmus*)
embryos 38
EMERGENCY PLANNING 25–7, 30, 37–9,
    40–1, 45–9, 50–2, 58–62, 63–73
Endocronartium spp. (gall rust) 160–3
Endothia parasitica 5, 40
Entomophaga (see fungi)
Enzyme Linked Immuno-Sorbent Assay
    (ELISA) 91
epidemic typhus 24
EPIDEMICS 23–27, 28–38, 40–2, 104–7,
    115–9, 125–7, 132–3
epizootic lymphangitis 28
equine arteritis (togavirus) 28
equine babesiosis 28
equine encephalosis (orbivirus) 28
equine influenza (orthomyxovirus) 28
ERADICATION (see CONTROL)
Erwinia spp. 40–2, 46
Escherichia coli 109, 168–70
EVOLUTION 3–5, 98–102, 104–7
EXPORTS (see TRADE)

FERAL ANIMALS 11–13, 37–8, 50, 75, 153
field beans (*Vicia*) 42
figs 8, 40
fire blight (see *Erwinia*)
FISHING 48
flax (*Linum*) 104
flies 38, 100, 119, 145
flour beetles 9
Food and Agriculture Organization of the
    United Nations 63, 65, 70
foot-and-mouth disease (aphthovirus) 9, 13,
    28–30, 37–8, 46, 49, 51, 54, 59, 60–1,
    63–71, 92–3, 112–4, 128, 135–9
forestry 160–3
fowl plague (see influenza)
foxes 24, 37, 164–7
fungi 5, 40–2, 50, 67, 104–7, 115–9, 158,
    160–3, 171–6

garbage 37, 49, 138
geminivines 77
gene-for-gene hypothesis 104–5
General Agreement on Tariffs and Trade 46
GENETIC ENGINEERING 76–7, 85–95,
    97–9, 113–4, 168–170
GENETICS 5, 26, 96–102, 104–111, 113–4,
    119, 131–3, 144–5, 149–50, 160–3, 168–70
glanders 28
goat (capripoxvirus) 36
goats 8, 29, 30, 32–8, 112, 128, 135
gold 9
Gondwanaland 3–4
GOVERNMENT 15, 37–8, 45–52,
    58–62, 63–7
grapes (vines) 8, 45
grasshoppers 115, 119

Haemonchus (se worms)
haemorrhagic septicaemia 28
hair (see ANIMAL PRODUCTS)
Hantaan (bunyavirus) 25
heartwater disease 28
Heliothis spp. 115
hepatitis B (hepadnavirus) 112–114
herpesviruses 37
hides (see ANIMAL PRODUCTS)
HISTOCOMPATIBILITY COMPLEX, MAJOR BOVINE (see SEROLOGY)
HLTV-III (lentivirus) 126–7
horses 8, 67
HOST/PARASITE RELATIONS 5, 9–11, 13, 23, 27, 28–37, 40–2, 96–102, 104–7, 108–11, 118–9, 140–5, 149–50, 163–7, 171–6
HUMAN DISEASES 23–27, 31, 34, 37–8, 125–7, 164
HYPERSENSITIVITY (see HOST/PARASITE RELATIONS)

IDENTIFICATION (see DIAGNOSIS)
IMMUNOLOGY (see SEROLOGY)
IMPORTATION (DELIBERATE) OF PESTS, PATHOGENS and BIOLOGICAL CONTROL AGENTS 38, 40, 75–9, 115–9, 147–51, 152–8, 171
IMPORTS (see TRADE)
Indo-Malaysia 4–7, 9
infections bulbar paralysis (see Aujeszky's disease)
influenza (orthomyxovirus) 26, 28, 45, 54, 75, 79, 98–9, 114, 140–5
INSECT CONTROL (see PEST CONTROL)
INTERFERENCE (see VIRAL INTERFERENCE)
ISOLATION UNITS 23–7, 38, 45–8, 50–2, 63–73, 77–9

Japanese B encephalitis (flavivirus) 28
Jarrah dieback 42

Karnal bunt (see fungi)
Kennedya yellow mosaic (tymovirus) 101
Kimberley (rhabdovirus) 76
Korean haemorrhagic fever (bunyavirus) 24, 48
Kununurra (rhabdovirus) 76

Lagos bat (rhabdovirus) 76
Lantana spp. 9
Lassa fever (arenavirus) 23–26, 53
LEGAL LIABILITY (see COMPENSATION)
LEGISLATION 37–9, 45–52, 53–5, 58–62, 157–9
leprosy (see *Mycobacterium*)
leptospirs 8

LITIGATION 55, 155–6
LOSSES (see ECONOMIC EFFECTS)
louping ill (flavivirus) 28
lucerne (alfalfa) 117–8
lucerne aphids 42, 117–9
lumpy skin disease (capripoxvirus) 28

Maedi-Visna (lentivirus) 28
malaria 23–4, 46–8
man-made molecules 20, 55
Marburg disease (filovirus) 23, 26, 53
Marek's disease (gammaherpesvirus) 109, 112
marsupials 7, 33
Mastomys (see rodents)
measles (morbillivirus) 30, 112
meat (see ANIMAL PRODUCTS)
Mendel, Gregor 96, 104
Metarhizium spp. (see fungi)
mice (see rodents)
MICROBIAL INSECTICIDES 115–9
MINING 12–20
mink 37
MOLECULAR HYBRIDIZATION TESTS (see NUCLEIC ACID HYBRIDIZATION TESTS)
monkeys 25, 34
monoclonal antibodies (see SEROLOGY)
mosquitoes 24, 38, 48, 97, 100, 117, 147
Mucor spp. (see fungi)
mucosal disease (pestivirus) 30
Mycobacterium spp. 24, 53
mycoplasms 40
myxomatosis (poxvirus) 5, 77, 93–4, 147–151

nematodes 28, 36–7, 40, 115–6
Neoaplectana spp. (see nematodes)
Newcastle disease (paramyxovirus) 28, 46, 54, 67, 76, 79, 112
NEWSMEDIA (see EMERGENCY PLANNING)
Niarobi sheep disease (nairovirus) 28, 38
Nicotiana glauca 99–100
NUCLEIC ACID HYBRIDIZATION TESTS 85–90, 98–102, 131–2, 145, 179–82
NUCLEIC ACIDS 85–90, 96–102, 113–4, 131–2, 141–5, 168–70, 178–82

ocular squamous cell carcinoma 110
oranges 8
orbiviruses 76, 128–33
Oriental fruit fly 48
Oryctolagus spp. (see rabbits)

parasitoid wasps 116
Paterson's Curse (see *Echium* spp.)
pathogen characterization 76, 84
pears 8, 40

PEPTIDE IMMUNOGENS (see SEROLOGY)
PEST (and WEED) CONTROL 37, 45–9, 50–2, 75–9, 115–9
pests-des-petits-ruminants (morbillivirus) 28, 30
PESTS 116–9
photobiotin 88–9
Phylloxera (see aphids)
Phytophthora spp. 5, 40, 42
Pierce's disease of grapevine 41
pigs 8, 28, 30–2, 36–8, 60, 64, 71, 109, 135
pine needle rust (see fungi)
Pinus radiata (Monterey pine) 160–3
plague (see bubonic plague)
plasmids 86
Plum Island Quarantine Disease Center, USA 70, 75
plumpox (potyvirus) 41–2
polio (poliovirus) 112
polyoma (papovavirus) 169
poplar rust (see fungi) 42
POPULATION DYNAMICS (see ECOLOGY)
POPULATION GENETICS (see GENETICS)
potato 40
potato black wart (see fungi)
potato blight (see *Phytophthora* spp.)
potato cyst nematode (see nematodes)
potato spindle tuber (viroid) 178–82
poultry 60, 76, 109
protozoa 116, 119
pseudorabies (see Aujeszky's disease)
psittacosis 24
PUBLIC RELATIONS (see EMERGENCY PLANNING)
Puccinia spp. (see fungi)
pulmonary adenomatosis (Jaagseikte) 28

QUARANTINE 9–15, 17, 20, 40–1, 23–6, 29–39, 45–9, 53–5, 66–72, 77, 138, 145, 147, 180

rabbits 5, 8, 9, 135, 147–151
rabies (rhabdovirus) 8–9, 24, 28, 46, 51, 53–4, 66, 75, 77, 112, 164–7
racoon 24, 37
radioisotopes 132
rats (see rodents)
recombinant DNA (see NUCLEIC ACIDS)
RECOMMENDATIONS AND CONCLUSIONS 20, 27, 38–9, 42, 55, 72–3, 79, 119, 139, 157–8, 116–7, 170, 182
REFUGEES (see TRAVEL AND TRANSPORT)
reoviruses 128
reptiles 118
RESERVOIRS (OF PATHOGENS) (see ECOLOGY)
RESISTANCE OF HOST ORGANISMS (see HOST/PARASITE RELATIONS)
respiratory syncytial (pneumovirus) 112
restriction endonucleases (see NUCLEIC ACIDS)
retroviruses 125–7
ribose nucleic acid (RNA) (see NUCLEIC ACIDS)
Rift Valley Fever (phlebovirus) 23–4, 28, 38, 48, 59
rinderpest (morbillivirus) 28, 30–1, 38, 45, 54, 66
RISK ASSESSMENT (see EMERGENCY PLANNING)
rodents 4, 9, 24–5, 34, 37–8, 45, 48, 83, 135
rubella (rubivirus) 112

SAFETY TESTING 118
Salmonella spp. 8
Salvation Jane (see *Echium*)
Scolytus spp. 5
scrapie (agent) 28, 31–2, 38, 48, 58, 60
SEA LINKS (see TRADE also TRAVEL AND TRANSPORT)
seals 4, 7–8
SELECTION r- and K- (see ADAPTATION also BREEDING)
SEROLOGY 84–5, 91–5, 101, 108–114, 130–3, 137, 141–4
sheep (capripoxvirus) 24, 48, 53
sheep 8, 24, 29, 30, 32–8, 112, 128–9, 135
sheep blowfly 119
sheep scab 28
ships (see TRAVEL AND TRANSPORT)
sickle cell anaemia 101–2
silkworm 115
Sirex wood wasp 116, 162
Skeleton weed (see *Chondrilla*)
smallpox (orthopoxvirus) 112
SMUGGLING (see TRAVEL AND TRANSPORT)
soybean dwarf (luteovirus) 76
sparrows 9
spiders 13
Standing Committee on Agriculture 153
STATE POWERS (see LEGISLATION)
subterranean clover red leaf (luteovirus) 76
sugarcane smut (see fungi)
Surra disease 28
swine fever 28, 54
Sylvilagus spp. (see rabbits)

Tasmania 6
taxonomy 129
tea 40
thistles 11
ticks 38, 108–9
Tilletia spp. (see fungi)
tobacco blue mould (see fungi)
tobacco mosaic (tobamovirus) 114

tobacco-ringspot (nepovirus) 76
Torres Strait 48
TOURISTS (see TRAVEL AND TRANSPORT
toxins 117
TRADE 8–20, 45–7, 50–2, 58–62, 67–9
TRAINING (see EMERGENCY PLANNING)
transmissible gastro-enteritis 28
transmissible mink encephalopathy 34–5
TRANSMISSION OF PATHOGENS (see ECOLOGY)
TRAVEL AND TRANSPORT 7–20, 23–7, 37–8, 40–2, 45–9, 50–2, 67, 75
trichnosis (see nematodes)
true broad bean mosaic (comovirus) 76
typhus fever 53

Ulmus spp. (elms) 5, 40

Vaccination 28–38, 46, 64, 66, 69–71, 75, 109, 112–4, 138–9
vaccinia (orthopoxvirus) 112, 114
vesicular diseases 50–1, 64, 71, 135–9
vesicular exanthema (calicivirus) 28, 30, 138
vesicular-stomatitis (vesiculovirus) 28, 30, 76, 114, 138
vines (see grapevines)

viral haemorhagic fever (arenaviruses) 23
VIRAL INTERFERENCE 142–5
viroids 40, 178–82
VIRULENCE (see HOST/PARASITE RELATIONS

warble fly 28
WEED CONTROL (see PEST AND WEED CONTROL)
WEEDS 4–6, 8–11, 17–20, 152–7, 171–6
weevils 9
Wesselsbron disease (flavivirus) 28, 38
Western blot assays 94–5
wheat stripe rust (see fungi)
WILD LIFE RESERVOIRS (see ECOLOGY)
wind 38
wood borers 48
World Health Organization 47, 75, 117
worm parasites 108–9

yellow fever (flavivirus) 24, 45, 48, 53, 112
YIELD LOSS (see ECONOMIC EFFECTS)

ZOONOSES (see ECOLOGY)
Zoophthora radicans 117–8